PROJECT
PROGRAM
CHANGE

PROJECT PROGRAM CHANGE

ROLAND GAREIS
LORENZ GAREIS

AN AUERBACH BOOK

CRC Press
Taylor & Francis Group
6000 Broken Sound Parkway NW, Suite 300
Boca Raton, FL 33487-2742

Printed on acid-free paper

International Standard Book Number-13: 978-1-138-50314-4 (Hardback)

Visit the Taylor & Francis Web site at
http://www.taylorandfrancis.com

and the CRC Press Web site at
http://www.crcpress.com

CRC Press
Taylor & Francis Group
6000 Broken Sound Parkway NW, Suite 300
Boca Raton, FL 33487-2742

Printed on acid-free paper

International Standard Book Number-13: 978-1-138-50314-4 (Hardback)

Visit the Taylor & Francis Web site at
http://www.taylorandfrancis.com

and the CRC Press Web site at
http://www.crcpress.com

Contents

Foreword

The aim was to create a project and program management initiative. It was late 1990, and I was on the phone with Dave Hornestay. He was a senior executive at NASA Headquarters and in the process of explaining to me their strong interest in having me come downtown to start creating a learning initiative focused on increasing the likelihood of project success. This was a few years after Challenger, and ideas were being explored about regaining capability and excellence.

It seemed like an absurd request. My organization psychology background was squarely about learning, team development, change management, and leadership. Project management at the time was about methods, formulas, and disciplined execution. I picked up several books about project management and was astounded and turned off to note the lack of writing about people, leadership, and strategy. I vowed as a first action to start a curriculum without project books. No need to reinforce the mistaken.

In reading *PROJECT.PROGRAM.CHANGE,* it has become clear how significantly things have changed! This is not a surprise, as I have known the authors for many years. They have been leading proponents of a practice of projects that squarely focuses on the strategic, human, adaptive, and integrated.

The book you hold in your hand represents an integrated, systemic, and balanced view of programs and projects that incorporates strategic, social, and execution factors necessary for success. A simple review of the table of contents illustrates a holistic understanding of modern project work. The challenge is about addressing the dilemmas of management and leadership. There is emphasis on planning, phases, roles, coordination, as well as team performance, culture, social competence, and developing necessary capabilities. There is recognition that project risk is within strategic, economic, technical, social platforms.

It is the right approach and one to be expected from experienced practitioners like Roland and Lorenz Gareis. My association with Team Gareis goes back to their successful and even joyful *Happy Projects!* series of conferences. It is a stretch to call projects "happy," and yet through carefully orchestrated engagements that weave the thoughtful, with the disciplined, with the social, they make it work and have done so for a long time. These are events that I eagerly anticipated as the former NASA Director of NASA Academy, as they annually would bring together a thoughtful community of project professionals and explore a diverse, but relevant, array of topics in a setting of active conversation. The chance to participate in a stimulating Viennese environment only made the discussions better, covering topics that would span the breadth of issues that are now covered in this book. For me this book represents the captured and shared knowledge of those conversations. The book represents wisdom from authors who have experience, practice, and dedication to excellence in program and project management.

This is a book that covers the necessities for program and project success. For any reader it is an excellent starting point. It also provides a foundation for asking difficult and provocative questions about leading a modern project. Such provocation is to be expected from natural rebels such as Roland and Lorenz. They are smart, deliberate, and smooth in their presentation, but any reading of the material will lead to the realization that leadership in this domain will require innovation, agility, and disruption.

Project Program Change and Success

This is a book intended to offer insights and learning on organizations, projects, and team success. The tools are focused around projects, programs, and change, yet the ultimate destination is organization success through business value. Many of the readers will be looking for answers. Some will come looking to learn as a starting point with hopes that the book will provide a strong foundation. What does the book say about success? Let me offer a few of the things that reinforced my personal experience.

1. Projects, Programs, Changes Require a Variety of Approaches

PROJECT.PROGRAM.CHANGE covers a lot of territory. There is a wide spectrum of disciplined methods that one would expect from a project management book. There is also a heavy emphasis on people, team, and social collaboration so that the human emphasis is strong. A third vital strand is about strategy, systems, culture, and process. These are often considered competing efforts, yet the strength of this work is that it demands the reader to understand that modern work is about balancing apparently different approaches. You can be innovative, adaptive, concerned about people, and still use disciplined tools that support structure and method.

2. Strategic Management is Essential

Let's look at the message on the importance of strategic managing. The beginning of any successful execution is accuracy of strategic insight. It is crazy to think that separating strategy from execution can promote a successful outcome. Yet for decades the notion has been spread that they are two completely separate things, done by different levels of an organization.

As the authors discuss, "The objective of the strategic managing of an organization is to ensure the sustainable development of this organization within this context." This opening sentence leads to a full discussion about the strategic implications of a project and the methods that must be considered. It accurately leads to consideration of project, program, and portfolio factors. Ultimately, it portrays the challenge and dynamics of the interaction between strategic organization business drivers with the reality of project implementation and real-time data. Projects are therefore best considered investments that are measured through value to an organization.

3. Culture, Change, and the Importance of Social Collaboration

The movement of a project is depicted in social, team, collaborative components. This is vital, in that we know success comes from the leadership and collaborative elements. Project management is not a logical, process-driven machine. It is ultimately about people, and this makes it dynamic, interpersonal, and messy. This is the most emblematic distinction of our current age of work. It is done across immense communities of discipline excellence, and the challenge is not finding the talent (finding talent is no longer hard, hiring it and maintaining it on a team may be a different story), it is managing how people collaborate. Success is the residue of effective design emphasizing collaborative cultures.

4. Performance Happens at the Team Level

One of my biggest mistakes in the early years of building the NASA Program and Project Leadership Academy was an exclusive focus on individual training and development. This was a common bias in the past, and it required painful mission failures to Mars in the late 1990s to recognize that having capable, competent, and confident individuals was the key ingredient for project success. This should have been obvious, since I had witnessed teams of exceptional talent stunningly fail due to dysfunctional interpersonal exchange. I had conversely noticed that some of our most effective teams were composed of solid (not great) individual talent that worked together extremely well. This led to my conclusion that performance happens at the team level. The sheer importance of project team design and development is perhaps the most common property of project outcome, and it is covered with thought.

Diversity of method, strategic connection to execution, social collaboration, and project team design are the broad factors that most resonate with my thinking about what makes projects, programs, change proceed toward success. The work of Gareis and Gareis in truth goes much further. It addresses the full complement of essentials for a leader and practitioner who endeavors the challenges of this work. You will undoubtedly recognize and value other critical ingredients of this book. I can say that if this were the book that I first read back in 1990, I would have easily embraced the field with excitement and passion. It is a challenging, complex, and diverse field of work, and *PROJECT.PROGRAM.CHANGE* ably shares the knowledge of wise teachers. Read on!

— Edward J Hoffman, PhD
CEO, Knowledge Strategies
Academic Director, Information and Knowledge Strategy
Columbia University School of Professional Studies

Preface

How the Book *PROJECT.PROGRAM.CHANGE* Came About

In recent years, projects as temporary organizations and the use of project management in industry and public administration have become increasingly important.

At the same time, criticisms of project management are also being voiced: The use of project management methods is too bureaucratic; there is too little customer and stakeholder orientation; the processes lasted too long; there was not enough response to changes in the business environment; etc. Agile approaches that promote flexibility, empowerment, and customer orientation are offered as alternatives. Frequent and rapid communication with digital media is required.

Over the last few years, we have continued to develop our approaches to project, program, and change managing: The concepts of sustainable development, empowerment, management for stakeholders, and agility, as well as requirements management, business analysis, and benefits realization management have also been included. The values underlying the management approaches have been defined and interpreted for the different management approaches. However, we have not yet adequately communicated these further developments!

We decided to publish *PROJEKT.PROGRAMM.CHANGE* to present observations from our management and consulting practice, to illustrate new relations between management approaches, and to clarify these approaches. This book represents a further development of the book *Happy Projects!,* which was first published in 2003. We want to offer readers a journey from *Happy Projects!* to "Values for Business Value". So in April 2017, the German version of the book was published. Now, about one year later, the English version is available for interested readers.

Readers of *PROJECT.PROGRAM.CHANGE*

This book is intended as a handbook for "intrapreneurs" of project-oriented organizations—namely, for project managers, program managers, change managers, and the owners of projects, programs, and changes. In addition, it is a textbook for management trainers and consultants, for researchers and teachers in universities and colleges, as well as for students.

Information as a Difference That Makes a Difference (Gregory Bateson)

The book *PROJECT.PROGRAM.CHANGE* provides information for the management community. Dealing appropriately with the dynamics and complexity in project-oriented organizations, securing quick wins in changes, using synergies in programs, involving stakeholders in management at an early state, applying methods consistently, and supplying contextual information to provide sense for members of organizations are examples of how readers might behave "differently" after processing the information provided in this book.

Chapter 1 presents possible perceptions of projects and programs, distinguishes small projects, projects, and programs from non-projects, and analyzes the contexts and benefits of projects and programs. This forms a basis for the distinction between mechanistic and systemic project management approaches provided in Chapter 2. Chapters 3 and 4 describe prerequisites for managing projects—namely, strategic managing and investing as well as managing requirements in sequential and iterative approaches.

Projects must be initiated professionally to provide a basis for efficient and effective project managing. The objectives, process, roles, and methods of the business process "Project initiating" are described in Chapter 5. The objectives, process, contexts, values, and benefits of the business process "Project managing" are presented in Chapter 6. The sub-processes of project managing—namely, project starting, project coordinating, project controlling, project transforming or repositioning, and project closing, and the methods to be used for these sub-processes, are described operationally in Chapters 9 to 12. Prior to this, Chapters 7 and 8 deal with models for designing project organizations, developing project-specific cultures, teamwork, and leadership in projects. Practical examples of the application of project management methods are provided by a case study "Values4Business Value".

The business processes "Program initiating", "Program managing", "Change initiating", and "Change managing" are dealt with in Chapters 13 and 14. The objectives, processes, roles, and methods to be used to perform these processes are described. A case study of an energy company is used to illustrate program managing. The integrative performance of project, program, and change managing is important. The relationships between these approaches are analyzed.

The subject of Chapters 15 and 16 is managing a project-oriented organization. Strategies, structures, and cultures of a project-oriented organization are presented. Specific business processes of a project-oriented organization—namely, "Project portfolio managing" and "Project (or Program) consulting"—as well as different processes for managing project personnel, are considered in detail. A vision for a project-oriented society is offered in Chapter 17.

The accompanying case studies of RGC's "Values4Business Value" shows that the management approaches presented are also relevant for small and medium-sized companies. Case studies of large telecom or utility companies are also presented.

Reading the Book

PROJECT.PROGRAM.CHANGE is not necessarily intended to be read in order from the first page to the last. Thus we would like to advise readers on how to read it efficiently.

Advice for Project Management Beginners

> Beginners in the topic—for example, students of project management or people who are preparing for a basic project management certification, can concentrate on Chapters 1 and 5–12. These cover basic concepts, processes, and methods as well as roles, organizational forms, communication formats, and leadership styles for projects.
> The other chapters can be skipped in the first instance and used to deepen understanding at a later stage.

Advice for Managers of Project-Oriented Organizations

> Managers who work as project, program, and change managers or owners and managers of management offices or expert pools should read everything.
> Sorry, but it pays off . . .

Advice for Those Interested in the New Application of Theoretical Models

> People who are interested in theory, such as well-informed managers of project-oriented companies, researchers, teachers, consultants, and trainers, can focus on new developments and newly established relations.
  - For example, Chapter 2 covers project management approaches and new values, Chapter 3 covers strategic management and investing, and Chapter 4 covers various methods for managing requirements.
  - In Chapter 6, the interpretation of values for project initiating and project managing, in Chapter 7 the use of Scrum sub-teams in projects, and in Chapters 13 and 14 the relation between project, program, and change managing should be of particular interest.
  - The interpretation of values for project portfolio management in Chapter 16 should also satisfy the curious . . .

. . . And for Our Fans

> Our fans will find new developments in each chapter, and in particular, attempts to implement the values of a systemic management paradigm.
> We would be very grateful for further ideas and feedback from you!

The Unique Selling Proposition of the Book

The book *PROJECT.PROGRAM.CHANGE* provides information about:

- social system theory and radical constructivism as the epistemological context of project, program, and change managing,
- using projects, programs, and changes to implement organizational strategies and investments,
- the importance of project, program, and change initiating for the successful performance of projects,
- the managing of requirements in sequential and iterative approaches,
- fulfilling solution requirements "by projects",
- the perception of project, program, and change managing as business processes of the project-oriented organization,
- the objectives, methods, and roles of the business processes "Project managing", "Program managing", and "Change managing",
- the values underlying the systemic management approach—for example, holistic boundaries, sustainable development, agility, empowerment, and resilience,
- the difference between processing change requests and change managing,
- the strategies, structures, and cultures of a project-oriented organization, and
- the redefinition of project, program, and change managers as "intrapreneurs" of project-oriented organizations.

The book *PROJECT.PROGRAM.CHANGE* presents selected concepts and models from the literature, but above all from the observations and experiences of the authors as managers and consultants. Its strong practical orientation is also the result of reflections and discussions with the peer review group established for reflecting on the topics covered in the book.

Production Process of the Book

The book publication was a major objective of the change "Values4Business Value", whose objective was the further development and communication of the RGC management approaches. The approaches were further developed by study of the literature, analysis of documents, interviews, brainstorming workshops, prototyping, presentations, reflection workshops, and self-observation.

As always, the production of this book was tedious and exhausting, but also pleasurable and fulfilling. It was tedious and exhausting because . . .

- The writing required a lot of discipline. Starting each new chapter required an act of overcoming.
- Of course, feedback from the peer review group had to be taken seriously and therefore necessitated significant changes.

- Contracts with RGC customers always had priority over internal innovation projects. Thus there were many “disturbances” which interrupted the rhythm of writing.
- Learning was necessary. We always apply new methods ourselves in RGC so that we can treat a topic or a method authentically in training and consulting situations, but also so that we can publish it. The opportunities for reflection provided, for example, by using an iterative approach in the change “Values4Business Value”, provided valuable experience. This learning led to changes in the objectives and a substantial lengthening of the project duration.

At the same time, producing the book was pleasurable and fulfilling, because . . .

- The further development of management approaches was a great creative process.
- The further development of the content provided unique opportunities for the authors to communicate with each other and with RGC colleagues, members of the peer review group, and customers.
- The quick wins achieved provided immediate “business value” for our customers and ourselves.
- The work had a very satisfying, tangible result.
- The finished book expresses essential parts of RGC’s identity.
- And because we expect that *PROJECT.PROGRAM.CHANGE* will be a classic for the next 10 to 15 years.

The book’s development process is illustrated in detail in the accompanying case study with examples and interpretations of the sub-processes of project and change managing.

Once the production of the German book was completed, we were happy that Taylor & Francis developed this English version of the book with us.

Acknowledgments

We have cooperated with MANZ Verlag in Austria for decades. Heartfelt thanks for having the confidence to travel with us from "Happy Projects!" to "Value for Business Value". We hope that *PROJECT.PROGRAM.CHANGE* will be a new classic for the management community.

To ensure that *PROJECT.PROGRAM.CHANGE* would have a practical orientation, we invited managers of project-oriented organizations to reflect on and discuss the contents of the book in a peer review group during the development process. We would sincerely like to thank the following members of the peer review group for this important contribution to quality assurance:

- Mag. (FH) Ulrike Danzmayr, Central Administration for Personnel, Organization, and Protocol at the Federal Ministry of Finance
- Bernhard Engl, Responsible for organizational development and cultural and system development at Rubner Holding AG
- Prof. (FH) Dr. Gerhard Ortner, Professor of Project Management, IT, and Business Management at the technical college BFI in Vienna
- Marcus Paulus, MBA, Head of the Project Management Office of Wien Energie GmbH
- Dipl.-Ing. Dr. Robert Schanzer, Head of Project and Program Management for IT Services at the Social Security GmbH and Board of Management at Projekt Management Austria
- Min. Council Dr. Hannes Schuh, MBA, Head of the Internal Audit Department of the Federal Ministry of Finance, Board of Controllers of the European Patent Organization
- Mag. David Spreitzer, MBA, Manager in the Program and Project Management Office at Borealis AG
- SR Dipl.-Ing. Helmut Wanivenhaus, Unit Head of Management Systems at the City Council of Vienna—Buildings and Technology

We are grateful to our RGC colleagues for having never given up the hope of finishing *PROJECT.PROGRAM.CHANGE.* Thank you, Michael Stummer, for many valuable suggestions about the content. Special thanks goes to Susanne Füreder, Patricia Ganster, and Lukas Weinwurm for editorial support. Without you the process would not have been so harmonious and efficient, despite the many delays to the production.

Also many thanks for very constructive and timely cooperation to the translator of the English version, David Nolland, and to the Senior Acquisition Editor from Taylor & Francis, John Wyzalek.

Happy Projects!
Roland and Lorenz Gareis
Vienna, March 2018

About the Authors

Lorenz Gareis **Roland Gareis**

Roland Gareis

Roland Gareis was born in 1948 in Vienna. He and his wife, Haldis, have two children, Luisa and Lorenz, and three grandchildren, Ella, Polly, and Emil. He used to be a professional soccer player for Rapid and the Wiener Sportklub; nowadays he plays tennis and skis. He likes to spend his free time in Reichenau on the Rax.

Roland studied at the Hochschule für Welthandel in Vienna, completed a doctorate at the WU Vienna University of Business Administration, and habilitated at the Technical University of Vienna. His professional experience is summarized below:

> Since 1994 Managing Director of RGC Roland Gareis Consulting GmbH Vienna
> Since 2017 lecturer at the University of Vienna
> Since 2015 Scientific Advisor for the Master's Program "Program and Investment Management" at the University of Political Studies and Public Administration, Romania
> 2014 awarded the IPMA Research Achievement Award
> 2005–2013 Managing Director of RGC Roland Gareis Consulting srl in Bucharest

- 1994–2013 Professor of Project Management at Vienna University of Economics and Business, PROJECT-MANAGEMENT GROUP
- 2007–2014 Academic Director, Professional MBA Project & Process Management (Vienna University of Economics and Business, Executive Academy)
- Guest professor at the Georgia Institute of Technology (Atlanta, Georgia 1979–1982), at the Swiss Federal Institute of Technology (Zurich 1982), at Georgia State University (Georgia 1987), and at the University of Quebec (Montreal, Canada 1991)
- 1998–2001 Director of Research of the IPMA-International Project Management Association
- 1988–1990 Project Manager of the IPMA World Congress on "Management by Projects" in Vienna
- 1986–2002 Chairman of PROJECT MANAGEMENT AUSTRIA, the Austrian project management association
- 1978–1993 lecturer at the Vienna University of Economics and Business and at the Vienna University of Technology, Director of the inter-university course "Project Management in Export"

Lorenz Gareis

Lorenz Gareis was born in 1985 in Vienna. He and his wife, Katharina, have a daughter, Ella. He has been passionate about hockey since he was six years old, playing formerly for the WEV, and now for the Wiener Wölfe. As an enthusiastic cook, his spare time is dominated by cooking, in addition to his family and sport. Reichenau at the Rax provides a place of balance and recovery.

Lorenz studied at the Vienna University of Economics and Business and the Universidade Nova de Lisboa. His professional experience is summarized below:

- Since 2017 Managing Director of Roland Gareis Consulting GmbH
- Since 2013 Lecturer in Project Management at the Technical College of bfi Wien
- 2013–2014 Lecturer in Project Management at the Technical College of Burgenland
- Since 2012 Principal Consultant and Trainer at RGC Roland Gareis Consulting GmbH in project & program management, process management, change management, and business analysis
- 2010–2012 Human Capital Consultant at Mercer in Frankfurt and Vienna with advisory focus on Talent Management, Rewards Management, and HR Strategy Management in Germany, Austria, and other European countries
- Extensive consulting experience in projects in the automotive, IT, consumer goods, pharmaceutical and finance sectors, as well as industry and public administration

Members of the Peer Review Group

> Mag. (FH) Ulrike Danzmayr, Central Administration for Personnel, Organization and Protocol at the Federal Ministry of Finance
> Bernhard Engl, Responsible for organizational development and cultural and system development at Rubner Holding AG
> Prof. (FH) Dr. Gerhard Ortner, Professor of Project Management, IT, and Business Management at the technical college BFI in Vienna
> Marcus Paulus, MBA, Head of the Project Management Office of Wien Energie GmbH
> Dipl.-Ing. Dr. Robert Schanzer, Head of Project and Program Management for IT Services at the Social Security GmbH and Board of Management at Projekt Management Austria
> Min.Rat Dr. Hannes Schuh, MBA, Head of the Internal Audit Department of the Federal Ministry of Finance, Board of Controllers of the European Patent Organization
> Mag. David Spreitzer, MBA, Manager in the Program and Project Management Office at Borealis AG
> SR Dipl.-Ing. Helmut Wanivenhaus, Unit Head of Management Systems at the City Council of Vienna—Buildings and Technology

1 Projects and Programs

The project management literature provides different definitions for projects and programs. Different perceptions of projects, such as perceiving projects as tasks, as temporary organizations, and as social systems, lead to different expectations regarding the way projects are managed and further on to different project management approaches.

A general clarification of the term "project" and the operationalization of project definitions in different organizational contexts are required. Small projects, projects, and programs are to be differentiated, so the adequate organization for performing different business processes can be selected.

Projects and programs are performed in contexts, which influence their success. Contexts are, for example, the strategies, structures, and cultures of the project-oriented organization, the investment implemented by a project or program, and the change delivered by projects.

The benefits provided by projects and programs are to be differentiated from the benefits provided by project managing and program managing.

1 Projects and Programs

1.1 Perceptions of Projects and Programs

1.2 Definitions of Projects and Programs

1.3 Categorization: Small Project, Project and Program, Non-Project

1.4 Projects and Programs: Contexts

1.5 Projects and Programs: Benefits

1.1 Perception of Projects and Programs

The project management literature and international project management standards provide different definitions for projects and programs.[1,2,3,4] This is relevant insofar as different project perceptions and definitions result in different project management approaches.

Perceiving projects as tasks with particular characteristics leads to a specific project management understanding—that is, of the objectives of project management to be achieved, the project management tasks to be fulfilled, the project dimensions to be managed, and the project management methods to be used. Perceiving projects as tasks leads to a different understanding of project management than perceiving projects as temporary organizations and as social systems.

Perceiving Projects as Tasks with Particular Characteristics

Traditionally, projects are seen as tasks with particular characteristics. These particular characteristics are the medium scope of the tasks to be fulfilled, the relative uniqueness of these tasks, their short-term to medium-term duration, as well as their associated risks and strategic importance. Projects are understood as goal-determined tasks, because goals relating to the scope, schedule, and costs can be planned and controlled.

Perceiving Projects as Temporary Organizations

Projects can be seen as temporary organizations established in order to perform relatively unique, short-term to medium-term, risky, and strategically important business processes of medium scope. Here projects are understood as temporary organizations for performing business processes with particular characteristics.

As with other organizations, projects have specific identities, expressed through specific project structures and project contexts. Due to the temporary nature of projects, establishing a project during project starting and dissolving it during project closing acquire particular significance.

Perceiving Projects as Social Systems

Perceiving projects as temporary organizations allows them to be also perceived as social systems. According to the social systems theory, organizations, and therefore

[1] Project Management Institute, 2013.
[2] International Project Management Association, 2006.
[3] Projekt Management Austria, 2008.
[4] DIN 69901-5: 2009-1, 2009.

also projects, are social systems which are clearly separate from, yet at the same time related to, their contexts. The specific characteristics of social systems, such as their social complexity, dynamics, and self-reference, are also relevant for projects. In the RGC project management approach presented here, projects are understood as both temporary organizations and social systems.

Projects can either be perceived as tasks with particular characteristics or as temporary organizations and social systems. These different project perceptions lead to different project management approaches.

As stated by Luhmann,[5] "A system is [. . .] anything in which it is possible to distinguish between inside and outside. The inside-outside distinction implies that an order is determined, one which cannot be arbitrarily expanded, but whose internal structure and particular form of its relationships creates boundaries".

Luhmann principally divides social systems into interactions, organizations, and societies. Here a further distinction is made between permanent organizations such as companies, divisions, and departments, and temporary organizations such as projects and programs, which can also be perceived as social systems (see Fig. 1.1).

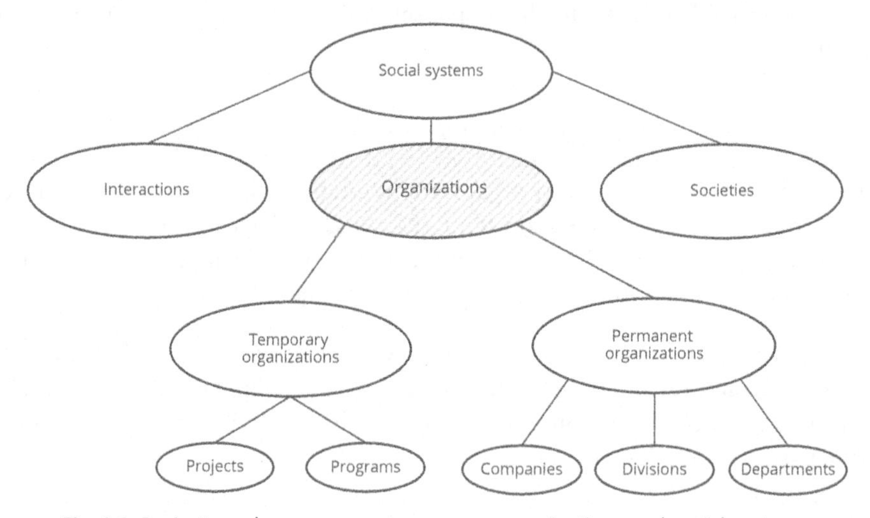

Fig. 1.1: Projects and programs as temporary organizations and social systems.

The reason for differentiating between different social systems is that drawing boundaries creates systems which are less complex than their particular environment.[6] Projects as subsystems of companies are less complex than the company as a whole. By reducing complexity each social system can be successfully managed.

[5] Luhmann, N., 1964, p. 24
[6] Cf. Kasper, H. 1990, p.156

Excursus: Projects and Programs as Social Systems

A social system is characterized as having specific structures which differentiate it from its environment, while simultaneously placing it in contexts which create dependencies. Context dimensions are the stakeholder relations (see Fig. 1.2), the superordinate social system, the history of the social system, and expectations about its future.

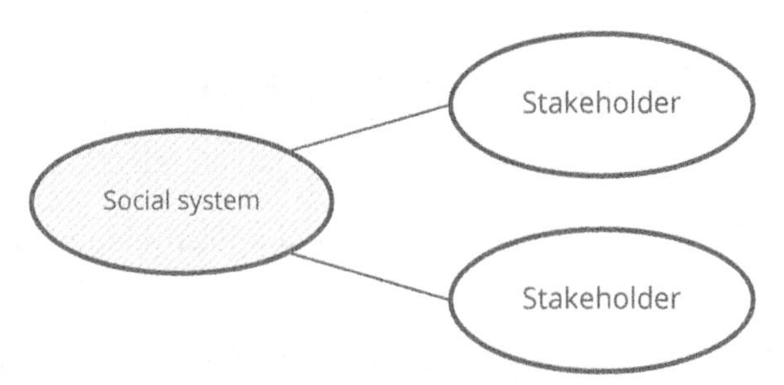

Fig. 1.2: Social system and its stakeholders.

An organization has both external and internal stakeholders. Project stakeholders external to a company include customers, suppliers, competitors, and media, while stakeholders within a company may include the management and individual departments. Every social system is shaped by its history. Many of a system's "peculiarities" can only be explained and understood with reference to past events. On the other hand, it is also future expectations of the social system which determine current actions. Consequently, the results of analyses of the pre-project and post-project phases provide the orientation for action in a project. A project's "superordinate" social system is the organization undertaking that particular project. A project contributes to fulfilling this organization's objectives and strategies.

Social systems are complex, self-referencing, and dynamic, and so these characteristics apply to both projects and programs. Luhmann understands communications as elements of social systems. He defines the following factors for evaluating a social system's degree of complexity:

- Number of system elements
- Number of potential relationships between these elements
- Diverseness of these relationships
- Development of these three factors over time

The formation of social systems both reduces and builds up complexity. The ability of a social system to survive is largely determined by its ability to develop an appropriate degree of complexity as a means of adequately dealing with the complexity of its environment. In project management practice, it is observed that the willingness to develop the appropriate level of project complexity is often small. For example, the involvement of project stakeholders in project management, the analysis of project risks, and the consideration of ecological and social consequences of project management are often not practiced.

"A system can be described as self-referencing when the elements of which it is composed regenerate themselves". Projects and programs are able to reflect. Thus project communications lead to new communications. Aggregations of communications lead to project roles, project rules, etc. "The dynamics of system processes depend to a large part on the dynamics of the environment, as well as the degree to which the system is open to this environment". Due to their relative uniqueness, projects and programs are generally very dynamic.

The boundaries of social systems, their structures, and their contexts are social constructs. Constructivism deals with the creation of realities by people or social systems. Watzlawick[7] states that there is no objective reality, but rather only subjective constructions of reality. Social systems theory and constructivism are the two theoretical models which offer a basis for systemic thought. Social systems theory deals with the "world of objects", and constructivism with human recognition, thought, and judgment.

1.2 Definitions of Projects and Programs

Project Definition

A project is a temporary organization to perform a relatively unique, short-term to medium-term, strategically important business process of medium scope.

Projects are used to perform relatively unique business processes. The more unique the objectives and tasks of the process to be performed, the greater the associated risks. There is often little opportunity to make judgments based on experience. Projects are of short to medium duration. They should be performed as quickly as possible—that is, within a period of several months up to one year. The duration of a project depends basically on the project type. For example, a conception project should not last longer than three to five months. Infrastructure projects, such as construction or engineering projects, are an exception to this general "rule" and usually have a duration of over a year. Longer project durations can be avoided by "constructing" a chain of projects or even programs.

Program Definition

A program is a temporary organization to perform a unique, medium-term business process which is large in scope and of significant strategic importance. A project is no longer sufficient for performing processes with these characteristics. Programs include several projects linked by overall program objectives. Objectives of programs may be to fulfill services (a contracting program), to establish a new infrastructure (a construction or an IT program), or to reorganize (reorganization program). The term "program" is used in an organizational sense, not to be confused, for example, with a software or TV program.

Definition: Project and Program

A project is a temporary organization to perform a relatively unique, short-term to medium-term, strategically important business process of medium scope.

A program is a temporary organization to perform a unique, medium-term business process which is large in scope and of significant strategic importance. Programs include several projects linked by overall program objectives.

[7] Watzlawick, P. 1976

Business Processes, Projects, and Programs

The relationship between business processes and projects and programs is, that in order to ensure the success of the fulfillment of selected business processes, these shall not be performed by the permanent line organization, but by temporary organizations. This assures the appropriate management attention for their successful fulfillment. A differentiation between those business processes to be performed by the permanent organization and those performed by temporary organizations is made.

The process map of an organization provides the basis for identifying those business processes which require projects or programs for their fulfillment. It is possible to distinguish between primary, secondary, and tertiary processes. Projects for performing primary processes are tendering projects and contracting projects. Projects for performing secondary processes are, for example, product developing projects or reorganizing projects.

In the past, many industries, such as construction, engineering, and IT, used projects in order to perform primary processes. It is only in recent years that a broader project orientation, and consequently the use of projects to perform secondary and tertiary processes, has become perceptible.

Another relationship between business processes and projects is, that both project initiating and project managing are business processes. Their process quality can be defined and controlled. Further similar methods can be applied in process managing and project managing, including defining boundaries, analyzing stakeholders, breakdown structuring, and responsibility charts.

Excursus: Business Process Management

A business process can be understood as a logical flow of tasks with defined objectives and with defined start and end events. A process requires cooperation between several roles of one or more organizations. Elements of processes are tasks and decisions, as well as their interdependencies. A process extends across departments or divisions and, therefore, proceeds horizontally through one or more organizations (see Fig. 1.3).

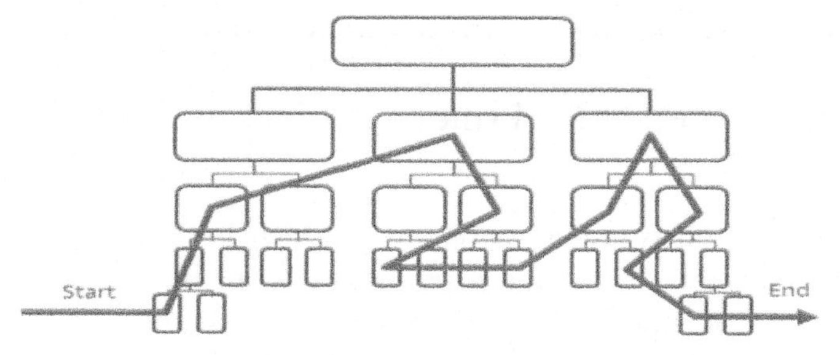

Fig. 1.3: Business process as logical flow of tasks to be performed by one or more organizational units.

Business processes can be distinguished into core, support, and management processes, or into primary, secondary, and tertiary processes. Differentiation into primary, secondary, and tertiary processes is made depending upon the relevance of the processes for clients of the organization. Primary processes are business processes associated with services for clients. Secondary processes directly support the primary processes. Tertiary processes support the primary processes only indirectly. The typical primary processes in an IT company, for example, are tendering and contracting. A typical secondary process is product developing. A typical tertiary process is strategic planning.

Business process management includes modelling, controlling, and optimizing business process portfolios and individual business processes. Business process management ensures an integrative process view, team orientation, concentration on core competences, elimination of activities which do not create value, and minimization of the process costs.

Process-oriented management approaches are lean management, total quality management and business process re-engineering. The paradigm shift in management characterized by client orientation, teamwork and networking with suppliers and partners promotes a business process approach.

Macro-business process management focuses on an organization's process portfolio, differentiates between process types, and creates links between processes. The tasks of macro-process management include preparing process management standards, identifying processes, structuring the process portfolio, the definition of process managers, and qualification of the process management personnel. Business processes are derived from an organization's objectives and strategies.

Micro-business process management considers the single processes in an organization. Micro-process management methods include process descriptions, process breakdown plans, flow charts and responsibility charts, process ratios and process reports. Process managers are needed to perform micro-process management. Process managers can be supported by process management teams. A process management office can be responsible for macro-process management. A distinction must be made between the process management roles outlined above and the roles for performing single processes.

With a dynamic, team-oriented approach, business process management leads to a new understanding of management within organizations. Business process management not only supports decision-making for process optimizations, it also strives to influence the behavior of those performing the process. Consequently, process management promotes organizational learning within a company.

1.3 Categorization: Small Project, Project and Program, Non-Project

By differentiating business processes according to their strategic importance, scope, organizational units involved, etc., it is possible to define the appropriate organizations for their performance. Possible organizations for performing business processes are the permanent line organization and temporary organizations, namely small projects, projects and programs.

The line organization is most suitable for the performance of routine processes. Business processes which are small to medium in scope and of little strategic importance, such as performing a small event, preparing a brochure, or processing a smaller contract, can be performed as small projects. Small projects use fewer project management methods and the project plans are less detailed than those required for projects. For example, it is usually sufficient to develop the work breakdown structure down to the third level. Small projects require a less differentiated project organization design than projects. The role of project owner is fulfilled by a single person rather than a team, and usually only a few sub-teams are required. In small projects the project marketing activities will be less intensive than in projects.

Business processes to be performed by projects are of medium to high strategic importance to the performing organization. Contracting, for example, safeguards the short-term to mid-term financial survival of an organization and is therefore strategically important. In contrast, developing new products or entering a strategic alliance have long-term consequences and are therefore also strategically important.

It is not possible to determine absolutely which business processes are to be performed as small projects, projects or programs. The relevant operationalization must take place within the specific organizational context. For example, what may be categorized as a project in a small company may be considered a routine process for a large company.

The characteristics of business processes, namely their strategic importance, duration, organization involved, etc., make it possible to categorize the organizations for their performance (small project, project or program). Table 1.1 shows an example of such categorization in an Austrian bank. In other companies these specifics will differ, above all for the accruing of external costs.

Table 1.1: Categorizing Organizations for Performing Business Processes (Example)

Criterion	Small Project	Project	Program
Strategic importance	low	medium	medium to high
Duration	at least 2 months	at least 3 months	at least 12 months
Organizations involved	at least 3 organizations (inclusive external partners)	at least 5 organizations (inclusive external partners)	at least 7 organizations (inclusive external partners)
Personnel resources	at least 150 person-days	at least 250 person-days	at least 700 person-days
External costs	at least 0,05 Mio €	at least 0,5 Mio €	at least 2 Mio €

Business processes which don't require projects for their performance are undertaken either by permanent organizational units within the line organization or by working groups. Permanent organizational units are departments, profit centers and service centers. The professionalism of the business processes not performed as projects can be ensured using appropriate business process management. Working groups are groups of limited duration established to fulfill specific assignments. Typical working group objectives are, for example, analyzing weak points in a business process or improving quality in a business process ("quality circle"). Working groups are usually deployed for a short-term and work in a less formal way than projects.

It is not possible to determine absolutely, which business processes are to be performed by small projects, projects or programs. This decision has to be made in the specific organizational context. What may be a process to be performed by the line organization in one case might require a project in another context.

1.4 Projects and Programs: Contexts

Projects and programs are performed within contexts. These must be considered if projects and programs are to be successful.

Key contexts are the strategies, structures and cultures of each project-performing organization, the investment underlying the project or program, and the changes which projects and programs are intended to bring about. This results in relations of project and program managing to strategic managing, to investment planning and investment controlling, to change managing, and to managing the project-oriented organization.

As these contexts are important to the success of projects and programs, they will be examined in detail in following chapters. Strategic managing and investment planning and controlling are examined in Chapter 3, change managing in Chapter 14, and the managing of project-oriented organizations in Chapters 15 and 16.

1.5 Projects and Programs: Benefits

Performing business processes by small projects, projects and programs helps to assure a company's competitiveness. Projects and programs should create the necessary organizational complexity needed to fulfill business processes efficiently, provide adequate organizational structures and methods, and ensure the necessary degree of management attention.

Globalized markets, new technological developments, the need for new cooperative relationships with clients and suppliers, and changing values in society all serve to enhance the complexity of the social environment in which organizations operate. Ashby's law of "Required Variety" states that "only variety can absorb variety".[8]

[8] Ashby, W.R., 1970, p. 94.

Consequently, organizations must establish an appropriate level of inner complexity in order to reflect the complexity of the social environment. The organizational differentiation resulting from the use of projects and programs helps to build up this complexity.

Appropriately organizing creates competitive advantages for organizations.[9] Projects and programs are used and dissolved after the results have been achieved. Project team members are recruited, equipped with the necessary competences to achieve objectives, and then released from the project after project closing. In each case organizations are created which are adequate for the specific need, and used on a temporary basis.

In accordance with Bateson's understanding of information as "a difference which makes a difference"[10], the organizational use of the term "project" makes a difference. By assigning the label "project", an appropriate degree of management attention is assured. Only once a project has been formally defined within the corporate context is professional project management to be applied. This should ensure that the objectives of the project are achieved.

In practice the term "project" is often applied in an inflationary manner—that is, in reference to tasks not worthy of the term. This leads to misunderstandings relating to the use of project management. In daily life the term "program" is used in a variety of ways. Its application as an organizational term still has to be further established. For example, an annual investment program or a company's strategic focus are not programs in the organizational sense. Programs offer new opportunities for differentiation in the management of project-oriented organizations.

In practice, the term "project" is commonly used for temporary organizations which should be managed as programs. In order to highlight the difference in scope and complexity of such organizations, some companies describe these as the "total project" or "major project" and so also manage them as projects. The result is a loss of the organizational potentials which arise through differentiating between projects and programs. The use of programs to perform comprehensive and medium-term business processes ensures higher quality, lower costs, shorter durations and less risk than when these business processes are performed as projects.

Literature

Ashby, W. R.: *An Introduction to Cybernetics,* 5. Auflage, University Paperbacks, London, 1970.

Bateson, G.: *Geist und Natur – Eine notwendige Einheit,* Suhrkamp, Frankfurt am Main, 1990.

[9] Cf. Senge, P., 1994, p. 10 ff.
[10] Bateson, G., 1990, p. 274.

DIN 69901-5:2009-01: *Projektmanagement – Projektmanagementsysteme* – Teil 5: Begriffe, 9. Auflage, Beuth, Berlin, Wien, Zürich, 2009.

Gaitanides, M.: *Prozessmanagement – Konzepte, Umsetzungen und Erfahrungen des Reengineering,* Hanser, München, 1994.

Gareis, R., Stummer, M.: *Prozesse und Projekte,* Manz, Wien, 2007.

Hill, W., Fehlbaum, R., Ulrich, P.: *Organisationslehre 1: Ziele, Instrumente und Bedingungen der Organisation sozialer Systeme,* 5. Auflage, UTB, Stuttgart, 1994.

International Project Management Association (IPMA): ICB. *IPMA-Kompetenzrichtlinie,* Version 3.0, Nijerk, 2006.

Kasper, H.: *Die Handhabung des Neuen in organisierten Sozialsystemen,* Springer, Wien 1990.

Luhmann, N.: *Soziale Systeme: Grundriss einer allgemeinen Theorie,* Suhrkamp, Frankfurt am Main, 1984.

Luhmann, N.: Komplexität, in: Grochla, E. (Hrsg.), *Handwörterbuch der Organisation,* 2. Auflage, Poeschel Verlag, Stuttgart, 1980.

Luhmann, N.: *Funktionen und Folgen formaler Organisation,* Duncker und Humblot, Berlin, 1964.

Project Management Austria (PMA): *pm baseline,* Version 3.0, Wien, 2008.

Project Management Institute (PMI): *A Guide to the Project Management Body of Knowledge (PMBOK® Guide),* 5th edition, Newton Square, PA, 2013.

Senge, P.: *The Fifth Discipline Fieldbook: Strategies and Tools for Building a Learning Organization,* Doubleday, New York, NY, 1994.

Watzlawick, P.: *Wie wirklich ist die Wirklichkeit – Wahn, Täuschung, Verstehen,* Piper, München 1976.

2 Project Management Approaches and New Values

Different perceptions of projects lead to different project management approaches. A mechanistic as well as a systemic project management approach are both represented in the project management literature. According to a mechanistic approach, project management is often reduced to management of the "magic triangle" of project scope, project schedule, and project costs. It is understood as "method toolbox".

Project management—as well as management in general—is based on values. The theories and the values underlying the project management approaches must be presented to obtain clarity about their differentiation. In this chapter, the values of a systemic project management approach are interpreted considering new values, such as agile, resilient, and sustainable development.

2 Project Management Approaches and New Values

2.1 Mechanistic vs. Systemic Management Paradigm

2.2 Mechanistic Project Management

2.3 Systemic Project Management

2.4 New Management Values

2.1 Mechanistic vs. Systemic Management Paradigm

A scientific paradigm can be defined as a coherent bundle of theoretical principles, questions and methods shared by many scientists, which lasts for extended periods in the development of a science.[1] "In management, a paradigm can be viewed as a managerial way of thinking and acting".[2] A distinction can be made between a mechanistic paradigm and a systemic paradigm.[3]

A mechanistic management paradigm is based on the perception of an organization as a trivial system, as a machine. A systemic management paradigm, on the other hand, perceives an organization as a complex system, as a living organism (see Table 2.1).

Table 2.1: Trivial System vs. Complex System

Trivial system: Organization as a machine	Complex system: Organization as an organism
> Understandable > Predictable > Context-independent > Controllable with "residual risk" > Can be influenced directly (in order to improve) > Using standards	> Not (completely) understandable > Unpredictable > Context-dependent > Not controllable, but manageable > Can only be influenced indirectly (by adapting the context) > Accepting differences

A mechanistic management assumes that organizations are understandable, predictable, context-independent, and manageable. This means, for example, that organizations in different cultures "function" in the same way—that management is achieved by the specification of standards and direct influence on employees.

In contrast to this, Malik, for example, sees systemic management as the design and steering of whole institutions in their environment rather than direct management of employees, as the management of many rather than the management of few, as an indirect influence at the meta level instead of direct action at the object level, as action under the criterion of controllability rather than optimality, and acting on the basis of limited information rather than complete knowledge.[4]

The systemic management paradigm is influenced by the models of the "Learning Organization",[5] but also by "Lean Management",[6] and "Total Quality Management"[7] (see Table 2.2).

[1] Asendorpf, J. B., 2009, p. 14.
[2] Němeček, P., Kocmanová, A., 2008, p. 562.
[3] See Kasper, H., 1995.
[4] See Malik, F., 2004.
[5] See Senge, P., 2006.
[6] See Womack, J. et al., 1990.
[7] See Juran, J. M., 1991.

Table 2.2 Influences on the Systemic Management Paradigm

Influence: The Learning Organization
> Differentiation between individual, collective and organizational learning
> Perception of the organization as a competitive factor
> Need to learn and to unlearn
> Continuous and discontinuous learning
Influence: Lean Management
> Process-orientation
> Focus on core competencies
> Flat, lean organizational structure
> Teamwork
> Networking and cooperating
> Continuous development
Influence: Total Quality Management
> Customer-orientation
> Product quality and process quality
> Quality control, quality assurance and quality management.

A management paradigm can be described by presenting the values underlying the management as well as the management objectives, processes, and methods. The values are central, as they are influencing the other dimensions.

2.2 Mechanistic Project Management

The perception of projects as tasks with special characteristics (see Chapter 1) promotes an orientation towards planning and controlling in project management.[8] The main focus is the question of how to carry out a task. Methods for work planning, such as REFA methods[9] or Operations Research methods,[10] form the theoretical basis for this "traditional" mechanistic project management.

[8] Steinle et. al., 1995, p. 354.
[9] See Čamra, J.J., 1976.
[10] See Hillier, F.S., 2001.

For decades, project management was understood as the use of networking techniques for planning and controlling project scope, project schedule, and project costs. Because of the uncertainties associated with unique tasks, risk management methods are also used in traditional project management.

Organizationally, the distribution of formal decision-making powers between project manager, line manager, and project team member appears to have the greatest importance in traditional project management. Pure project organization, matrix project organization, and influence project organization are offered as standards for resolving this area of tension.[11] It is argued that projects require project organization for task management, but that they are not independent temporary organizations. The role of the project owner, which is so important for the success of a project, is not seen as part of this, or is not adequately performed.

The objectives of traditional project management are fulfilling the project scope within the constraints of the project schedule and the project costs. These dimensions are represented as a "magic triangle" (see Fig. 2.1).

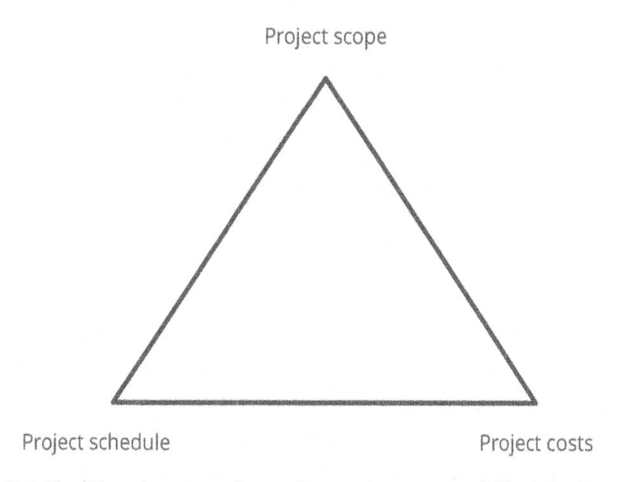

Fig. 2.1: Traditional project dimensions to be managed ("magic triangle").

Project management is understood as a set of methods for managing a project and not as a business process to be designed. Project management methods are used to control and not to structure communications. The focus is not on teamwork. Project managers see themselves above all as "experts" and not as "leaders" with business value responsibility. Project decisions are made according to the hierarchy of the project organization. "Traditional" project management can thus be seen as a mechanistic project management approach.

[11] See Reschke, H., 1989.

2.3 Systemic Project Management

Influence of Organization Theory on Project Management

The perception of projects as temporary organizations promotes the awareness that each project requires a specific organizational design that goes beyond the regulation of decision-making powers for the project manager. An adequate, situational design of the project organization helps to ensure the success of the project.

The organizational design of projects includes the definition of project-specific roles, the development of project organization charts, the establishment of project-specific communication formats, and the agreement of project-specific rules. Concepts such as customer orientation, empowerment, flat organizational structures, teamwork, organizational learning, process orientation, and networking can be implemented in projects. Seeing projects as temporary organizations also promotes project-specific culture development. The targeted selection of a project name, the development of a project-specific language, the formulation of project-specific slogans are relevant project management methods.

Influence of Social Systems Theory on Project Management

The perception of projects as social systems makes it possible to apply concepts and models from the social systems theory for project management. "Systemic" project management is not based on traditional project management, but puts its objectives, processes, methods, and roles into a new context, interprets them, and promotes the development of new concepts and models.

A systemic understanding of project management is derived as a result of the need to manage the boundaries and contexts as well as the complexity and dynamics of projects. Instead of planning and controlling scope, schedule, and costs, what is relevant are the construction of project boundaries and project contexts, the building up and reducing of complexity, and the management of the dynamics of projects.

In constructing project boundaries, a holistic project view is to be ensured. An integrated view of technical, organizational, personnel, and marketing-related solutions is promoted in order to ensure sustainable business value. Project stakeholder analysis, analysis of the pre-project and post-project phases, and analysis of the project's relationships with other projects and with company strategies are used for the management of the project context relations.

Projects require an appropriate degree of complexity to enable connectivity with the (infinitely) complex environment. The building up and reducing of complexity is therefore a project management function. The use of diverse communication formats, such as project workshops, meetings of the project team and of sub-teams, as well as project owner meetings, promotes the building up of complexity. Further organizational possibilities for building up complexity are the differentiation of project roles, the definition of interrelationships between the roles, and the inclusion of different specialist disciplines and representatives of different hierarchical levels in the project team.

The application of different project management methods allows one to construct different project realities. An appropriate project complexity is achieved by relating the different project management methods to each other in a "multi-method approach".

Redundant structures must be created to ensure continuity in the project. Project complexity is also reduced by agreeing on common project objectives, by defining project-specific rules and standards, by developing project plans, and by conducting integrative project team meetings. The different communication formats of a project contribute to the self-reference of the project. Visualizations such as a work breakdown structure, a milestone plan, and a project stakeholder analysis support these communications.

Reflections are meta-communications—that is, communications about communications. Reflections are necessary in order to promote changes in a project. Time, space, and appropriate social know-how are necessary for reflections. The structures necessary for performing a project are therefore formed, questioned, possibly adapted, and newly formed according to new needs in a cyclic process (see Fig. 2.2).

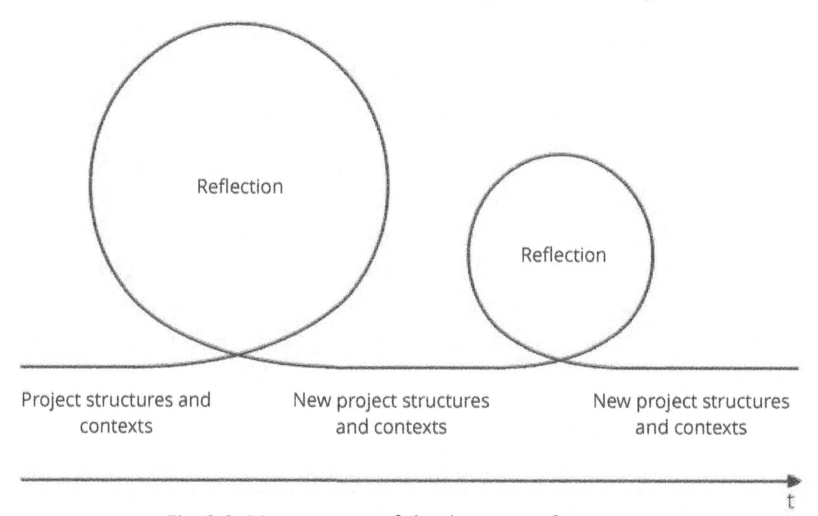

Fig. 2.2: Management of the dynamics of projects.

The dynamics of a project emerge not only from its self-reference but also from interventions of stakeholders of the project. Examples of interventions of stakeholders include new legal requirements imposed by authorities, changes by the customer regarding the scope of services, cancellations by suppliers, an unexpected media echo, lack of motivation in the project team, etc.

The possibility of changing a project depends on its relationships with stakeholders. Momentum of a project can only be achieved when the functionality of the (relative) project autonomy is recognized. The interventions in a project by the stakeholders of the permanent organization therefore should be limited.

Self-referential processes of a project or interventions by project stakeholders can lead to continuous or discontinuous changes of the project. Continuous changes of a project are taken into account by project controlling. Continuous changes are reflected in adapted project structures such as additional project objectives, newly defined project roles, new project schedules, etc., but also in new contextual relationships.

A discontinuous development of a project as a result of a project crisis or a project chance necessitates a change in the project identity. Discontinuous developments are carried out by specific subprocesses of project managing—namely, "transforming a project" or "repositioning a project".

Influence of Constructivism on Project Management

The definition of the boundaries of a project, the assessment of the status of a project at a controlling date, and the definition of a project crisis all are constructions of project realities. A common view of the project status, for instance, provides the basis for a common agreement on measures to direct the project.

Constructions are not right or wrong, but are or are not "viable". A construction is viable if it is possible to function with its help in a specific context. A practicable or usable view is striven for in a specific situation.

Constructions in project meetings are performed through observations and interpretations by the observers. Observers of projects can be members of the project organization but also by representatives of stakeholders. Each observation is carried out by an operation of the observing system, which applies certain observation criteria. In order to assess the project status, one can, for example, look at the progress and adherence to deadlines, but the atmosphere in the project team or the quality of relationships with project stakeholders could also be considered.

Social constructions are the result of a power-influenced negotiation process. This means that the project owner's opinion regarding the project status influences the overall judgment more strongly than that of a project contributor. Constructions can change. For this reason, for example, project controlling meetings are carried out at periodic intervals.

People act on the basis of the importance that events and situations have for them. But there is no "correct" social significance of things and situations. The significance is always context specific. The same situations can have a completely different significance in different contexts and thus lead to different constructs.

According to a mechanistic project management approach, project scope, project schedule, and project costs are the dimensions of a project to be managed. This approach is characterized by an orientation toward planning and controlling.

The systemic project management approach understands the construction of project boundaries and project contexts, the building up and reducing of complexity, and the management of the dynamics of projects as project management objectives.

A description of the development of project management since 1950 can be found in Table 2.3. The traditional mechanistic project management approach continues to be of great importance in practice, even if a newer systemic approach exists.

Table 2.3: Development of Project Management Since 1950

Criterion	Since 1950	Since 1990	Since 2010
Perception of projects	As unique tasks	As temporary organizations and social systems	As temporary organizations and social systems
Management focus	Managing the project scope, schedule and costs	Managing the project scope, schedule and costs, the project organization and the project context relations	Managing the project scope, schedule and costs, the project organization and the project context relations, considering the relations to requirements, programs, and changes
Understanding of project management	Toolbox of project management methods	A business process of the project-oriented organization	A business process of the project-oriented organization
Project success criteria	Performing the project scope under consideration of schedule and costs constraints	Performing the project scope under consideration of schedule and costs constraints; meeting content-related project objectives	Performing the project scope under consideration of schedule and costs constraints; meeting content-related project objectives; contributing to the optimization of the business value of the company
Project types	Major projects with technical objectives	Small, medium and large projects with technical objectives; regional development projects; contracting, marketing, organizational development , etc. projects	Small, medium and large projects with technical objectives; regional development projects; contracting, marketing, organizational development , etc. projects
Role of the project manager	Technical expert, administrator	Content-related expert, administrator	Manager, intrapreneur
Project-performing industries	Military, aviation, construction, engineering, IT	All industries	All industries, public sector

2.4 New Management Values

Values are relatively stable beliefs about the desirable or necessary characteristics of a social system. Values are therefore ideal constructs. They should have a highly binding character, provide meaning, and give orientation to the members of an organization.

Organizations function on the basis of values, which determine the behavior of their members. Values can be observed, for example, in the organization's priorities. An operationalization of values in the corresponding organizational context can be achieved by defining principles and rules.

Definition: Values

Values are relatively stable beliefs about the desirable or necessary characteristics of a social system. They are ideal constructs.

The values of the management paradigm on which the RGC management approaches are based are shown in Figure 2.3.

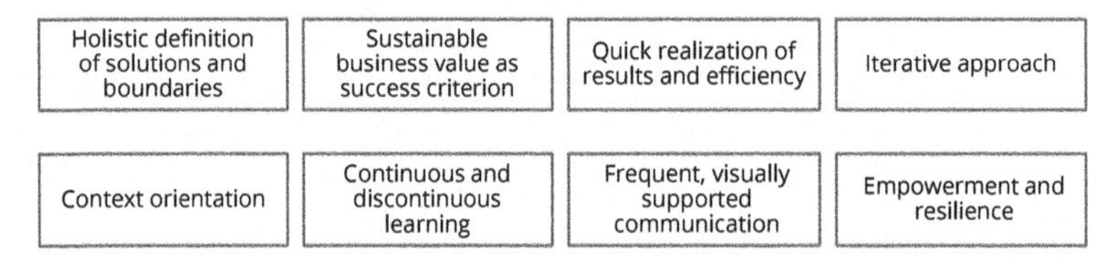

Fig. 2.3: Values on which the RGC management approaches are based.

The RGC management approaches can be described on the basis of the values shown in Figure 2.3. Detailed descriptions can be found for project management in Chapter 6, for program management in Chapter 13, for change management in Chapter 14, and for project portfolio management in Chapter 16.

The values of the concepts of agility, resilience, and sustainable development have been incorporated into the RGC management values. The relatively new values of these concepts are described below.

Values of the Concept "Agility"

Agility can be defined as the ability to change quickly in response to new market conditions. Both permanent and temporary organizations have the ability to be agile. Agility is therefore also relevant for projects and programs.

The agility of an organization expresses itself in its objectives, business processes, methods, roles, and stakeholder relations.

Table 2.4: Agile Manifesto and Agile Principles

Agile Manifesto
> Individuals and interactions over processes and tools > Working software over comprehensive documentation > Customer collaboration over contract negotiation > Responding to change over following a plan
Agile Principles
> Our highest priority is to satisfy the customer through early and continuous delivery of valuable software. > Welcome changing requirements, even late in development. Agile processes harness change for the customer's competitive advantage. > Deliver working software frequently, from a couple of weeks to a couple of months, with a preference to the shorter timescale. > Business people and developers must work together daily throughout the project. > Build projects around motivated individuals. Give them the environment and support they need, and trust them to get the job done. > The most efficient and effective method of conveying information to and within a development team is face-to-face conversation. > Working software is the primary measure of progress. > Agile processes promote sustainable development. The sponsors, developers, and users should be able to maintain a constant pace indefinitely. > Continuous attention to technical excellence and good design enhances agility. > Simplicity – the art of maximizing the amount of work not done – is essential. > The best architectures, requirements, and designs emerge from self-organizing teams. > At regular intervals, the team reflects on how to become more effective, then tunes and adjusts its behavior accordingly.

Definition: Agility

Agility is the ability to change quickly in response to new market conditions.

The agility of projects does not assume the use of agile methods such as Scrum or Kanban. Projects are agile if they take into account agile values and therefore, for example, take an iterative approach, communicate intensively, reflect regularly, etc. An iterative approach means that similar activities are carried out repeatedly in order to approach an overall objective. However, there is also the possibility of using agile methods in projects. Scrum, for example, can be used for the management of specific phases of a project (see Chapter 4).

Definition: Iterative Approach

In an iterative approach, activities are repeated in order to approach an objective. Uncertainty should be adequately managed, and the quality of a solution and the associated business value should be optimized. An iterative approach does not mean that planning is abandoned. Rather, one can only be temporarily sure of one's own approach. Uncertainty is gradually reduced, acceptance is achieved.

One of the strengths of agile methods such as Scrum is the explicit definition of the values underlying the methods. In 2001, a group of software developers published the Agile Manifesto, presented in Table 2.4, along with twelve agile principles.[12]

The central statements of the Agile Manifesto require an interpretation. "Individuals and interactions over processes and tools" means that emphasis is laid on communication, on many (short) meetings, and on the empowerment of employees, rather than on the development of extensive rules. "Working software over comprehensive documentation" means that achieving objectives is more important than documentation. Documentation is still needed, but it is not the top priority. Achievement of objectives is understood as the creation of business value.

"Customer collaboration over contract negotiation" means that an intensive collaboration with the customer is striven for during development, that an understanding of customer requirements and the rapid gathering of customer feedback is important, and that the involvement of additional stakeholders in the cooperation process is also desired. "Responding to change over following a plan" ultimately means that, although defined solution requirements cannot be changed during an iteration, new requirements can be defined and priorities changed after an iteration.

There are similarities between systemic and agile approaches. Thus, for example, self-organization and reflection are characteristics of social systems and they are also agile principles.

Values of the Concept "Resilience"

Resilience can be defined as the robustness of organizations, but also as robustness of teams and individuals. Resilient organizations are characterized by structures that ensure their resilience. Prevention, adaptation, innovation, and culture development can contribute to the resilience of an organization:[13]

> Prevention: A capacity to resist to negative external effects is built up as a precautionary measure.
> Adaptation: A short-term adaptation in order to return to an original position is made.

[12] Beck, K. et al., 2001.
[13] See Wieland, A., Wallenburg, C. M., 2013.

> Innovation: Innovations are implemented in order to exploit benefits.
> Culture development: An organizational culture, which is optimistic, ready to learn, fault-tolerant, but also prepared for confrontation, is assured.

It is possible to distinguish between a proactive form and a reactive form of resilience. The reactive form corresponds to the concept of agility. Agility and resilience are therefore directly related.

Resilience management encompasses all measures with the objective of strengthening the resilience of an organization against external influences. Examples of this are the creation of flexible organizational structures, extensive training of employees, regular feedbacks during everyday business, and frequent communications (in informal networks).

Definition: Resilience

Resilience is the capacity to resist or the robustness of an organization, a team, or an individual to external influences.

Values of the Concept "Sustainable Development"

Sustainable development has received a lot of attention as a result of the "Brundtland Report", published in 1987 by the World Commission on Environment and Development. This report defines sustainable development as "a development that meets the needs of the present without compromising the ability of future generations to meet their own needs".[14]

Sustainable development as a normative concept represents values and ethical considerations.[15] Relevant basic values are justice within and between generations, transparency, fairness, trust, and innovation.[16]

The political concept of sustainable development has been developed for society in general. The values have been defined for society and therefore cannot be transferred to companies without interpretation.

The application of the concept to companies is referred to as corporate sustainability or corporate social responsibility. Over the past few years, companies have committed themselves to implementing the concept of sustainable development. This should go beyond compliance with legal requirements. Critics like Porter and Kramer argue that many of the initiatives of corporate sustainability can be seen as mere lip service.[17] Philanthropic activities are often offered to do "good things" for society. Serious

[14] World Commission on Environment and Development (WCED), 1987, p. 41.
[15] See Adams, W. M., 2006; Davidson, J., 2000; Martens, P., 2006; Meadowcroft, J., 2007; Robinson, J., 2004.
[16] See Global Reporting Initiative, 2011.
[17] See Porter, M. E, Kramer, M. R., 2011.

sustainable development, however, means integrating the principles of sustainable development into the services, products, and processes of organizations, which means rethinking the business.

A process-based understanding of sustainable development is principle based. It takes into account the economic, ecologic, and social; short, medium, and long term; local, regional, and global consequences of developments.[18] The challenge is to "balance" these principles.

Definition: Sustainable Development

Sustainable development is development that takes into account and balances economic, ecologic, and social; short, medium, and long term; as well as local, regional, and global consequences.

These principles of sustainable development are relevant not only to permanent organizations, but also to temporary organizations—namely, projects and programs. Initial studies contribute to an understanding of the link between sustainable development and projects. The focus is on the project content, so-called "green projects", and on the consequences of the project results. So, for example, "impact assessments" are developed for this purpose.[19] Little consideration, however, is given to the principles of sustainable development in project initiation and project management.[20]

Agility, sustainability, and resilience are relatively new values in management. Taking these values into account influences the objectives, processes, methods, and roles of management approaches.

Literature

Adams, W. M.: The Future of Sustainability: Re-Thinking Environment and Development in the Twenty-first Century, *Report of the IUCN Renowned Thinkers Meeting,* Volume 29, 2006.

Asendorpf, J. B.: *Persönlichkeitspsychologie,* 3rd edition, Springer, Heidelberg, 2009.

Beck, K., Beedle, M., van Bennekum, A. et al.: *The Agile Manifesto,* 2001.

Čamra, J. J. (Ed.): *REFA-Lexikon: Betriebsorganisation. Arbeitsstudium, Planung und Steuerung,* 2nd edition, Beuth, Berlin, 1976.

[18] Gareis, R. et al, 2013.
[19] See Martinuzzi, A. and Krumay, B., 2012.
[20] See also Silvius, G. et al, 2012.

Davidson, J.: Sustainable Development: Business as Usual or a New Way of Living? *Environmental Ethics,* 22(1), 45–71, 2000.

Gareis, R., Huemann, M., Martinuzzi, A., Weninger, C., Sedlacko, M.: *Project Management & Sustainable Development Principles,* Project Management Institute (PMI), Newtown Square, PA, 2013.

Global Reporting Initiative: Sustainability Reporting Guidelines, Version 3.1, Amsterdam, 2011.

Hillier, F. S., Lieberman, G. J.: *Introduction to Operations Research,* McGraw-Hill, Boston, MA, 2001.

Juran, J. M.: *Handbuch der Qualitätsplanung,* Moderne Industrie, Landsberg/Lech, 1991.

Kasper, H.: Vom Management der Organisationskulturen zur Handhabung lebender sozialer Systeme, in: Helmut Kasper (Ed.), Post Graduate Management Wissen: *Schwerpunkte des Führungskräfteseminars der Wirtschaftsuniversität Wien,* 189–224, Wirtschaftsverlag Carl Ueberreuter, Wien, 1995.

Malik, F.: *Systemisches Management, Evolution, Selbstorganisation: Grundprobleme, Funktionsmechanismen und Lösungsansätze für komplexe Systeme,* 4th edition, Paul Haupt, Bern, 2004.

Martens, P.: Sustainability: Science or fiction? *Sustainability: Science Practice and Policy,* 2(1), 36–41, 2006.

Martinuzzi, A., Krumay, B.: The Good, the Bad and the Successful—How Corporate Social Responsibility Leads to Competitive Advantage and Organizational Transformation, *Journal of Change Management,* 13(4), 424–443, 2012.

Meadowcroft, J.: Who Is in Charge Here? Governance for Sustainable Development in a Complex World, *Journal of Environmental Policy and Planning,* 9(3), 299–314, 2007.

Němeček, P., Kocmanová, A.: *Management Paradigm, 5th International Scientific Conference "Business and Management",* 559–564, Vilnius, 2008.

Porter, M. E., Kramer, M. R.: The Big Idea: Creating Shared Value. *Harvard Business Review,* 89, 1–2, 2011.

Reschke, H.: Formen der Aufbauorganisation in Projekten, in: Reschke, H., Schelle, H., Schnopp, R. (Eds.), *Handbuch Projektmanagement,* Volume 2, TÜV Rheinland, Cologne, 1989.

Robinson, J.: Squaring the Circle? Some Thoughts on the Idea of Sustainable Development, *Ecological Economics,* 48(4), 369–384, 2004.

Senge, P.: *The Fifth Discipline: Art & Practice of the Learning Organization,* Doubleday, New York, NY, 2006.

Silvius, G., Schipper, R., Planko, J., van den Brink, J., Köhler, A.: *Sustainability in Project Management,* Gower, Surrey, Burlington, VT, 2012.

Steinle, H., Bruch, H., Lawa, D. (Eds.): *Projektmanagement: Instrument moderner Dienstleistung,* edition Blickbuch Wirtschaft, Frankfurt am Main, 1995.

Wieland, A., Wallenburg, C. M.: The Influence of Relational Competencies on Supply Chain. Resilience: A Relational View, *International Journal of Physical Distribution & Logistics Management,* 43(4), 300–320, 2013.

Womack, J. P., Jones D. T., Roos, D.: *The Machine That Changed the World,* Simon and Schuster, New York, NY, 1990.

World Commission on Environment and Development (WCED): *Our Common Future,* Oxford, 19.

3 Strategic Managing and Investing

The objective of strategic managing an organization is to ensure its sustainable development. Strategic managing includes the planning and controlling of the objectives, strategies, structures and cultures of an organization within its context. In order to realize an organization's strategic objectives, the investment portfolio and project portfolio will also be planned and controlled. Investments and projects derive from an organization's strategic objectives.

Strategic managing and investing are key contexts of projects and programs. One outcome of strategic managing is an implementation plan which contains the measures, projects, and programs for implementing an organization's strategies. Thus projects and programs always have a strategic context.

The life cycle of an investment object includes the planning, implementing, use, and potential decommissioning of an investment object. The objective of the business process "planning an investment object" is to decide to implement, or to decide not to implement an investment object.

The success of an investment can be ensured by performing the business process "controlling the benefits realization" during its implementation, use or operation. The basis for controlling the benefits realization is created while planning an investment object. The application of methods for planning an investment object and for controlling the benefits realization is shown in the RGC case study "Values4Business Value" (see Sections 3.3 beginning on page 39 and 3.4 on page 45).

The business processes described in this chapter are shown in the following overview. For a better understanding of the relations to other processes, the business processes described in later chapters are also shown.

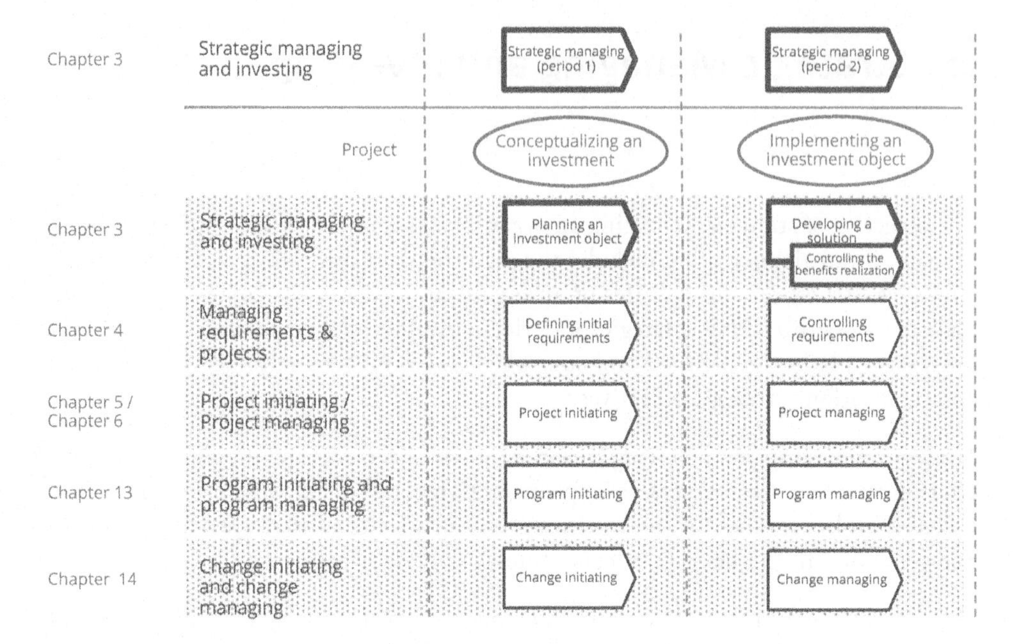

Overview: "Strategic managing" and "investing" in context.

3 Strategic Managing and Investing

3.1 Strategic Managing an Organization

3.2 Investment Definition and Investment Types

3.3 Planning an Investment Object

3.4 Controlling the Benefits Realization of an Investment

3.1 Strategic Managing an Organization

Strategic Managing: Objectives

The objective of the strategic managing of an organization is to ensure the sustainable development of this organization within its context. Managing the overall objectives, strategies, structures, and cultures provides sense and orientation for the employees of a particular organization. It provides a basis for operative managing. In contrast to operative managing, strategic managing has a medium-term rather than a short-term focus.

In accordance with a systemic management approach, strategic managing perceives organizations holistically within their particular contexts. A holistic perception implies that the following structural dimensions of an organization are to be considered:

> services and products,
> organizational structures (processes, roles, rules, etc.) and cultures,
> personnel structures (number of employees and their qualifications),
> infrastructures (building infrastructure, information, and communication technology), and
> budget and financing.

Relevant contextual dimensions of an organization are:

> history and expectations about the future,
> stakeholders such as markets and client segments, competitors, partners, and suppliers, and
> the superordinate social system to which the organization contributes.

These structural and contextual dimensions determine an organization's identity.

Definition: Strategic Managing

Strategic managing is the planning and controlling of the overall medium-term objectives, strategies, structures, and cultures of an organization within its context.

Strategic Managing: Process

Strategic managing includes the strategic planning and controlling of an organization. It includes tasks such as analyzing the strengths and weaknesses of the organization and its competitors, analyzing the organization's strategic position, defining strategic scenarios for developing the organization, planning and controlling its strategic objectives and strategies for their realization, as well as planning implementation.

The organization's investment portfolio must also be considered in planning and controlling the strategic objectives. The objective of planning the investment portfolio is to select those investments which will make an optimal contribution to realizing the organization's strategic objectives. Only investments which are realized by projects or programs will be considered. Comprehensive investment planning does not need to be performed for small investments such as equipment or furnishing. In such cases, investment decisions usually fall within the competence of specific managers of profit or cost centers. These small investments do not need to be considered in the investment portfolio.

Strategic managing also involves taking the decisions needed to implement the strategic objectives. These are listed in the implementation plan. "Strategy implementation [. . .] refers to decisions that are made to install new strategy or reinforce existing strategy. The basic strategy implementation activities are establishing annual objectives, devising policies, and allocating resources. Strategy implementation also includes the making of decisions with regard to matching strategy and organizational structure, developing budgets, and motivational systems".[1] The measures, projects, and programs for implementing the strategic objectives will be decided upon and the project portfolio structured within the framework of implementation planning for "project-oriented organizations" (see Chapter 15). Consequently, project portfolio management is a task of strategic managing.

An example of the process of "Strategic managing" is shown in Table 3.1 in the form of a responsibility chart.

The implementation of planned investments or the implementation of planned measures, projects, and programs takes place within the framework of operative managing and is not part of strategic managing.

The process of strategic managing is undertaken periodically. The frequency of the processes and the scope of the tasks to be fulfilled depend on the dynamics of the particular organization. "All strategies are subject to future modification because internal and external factors are constantly changing. In the strategy evaluation and control process managers determine whether the chosen strategy is achieving the organization's objectives".[2]

The planning periods considered in strategic managing also depend upon the particular industry. For example, the IT and telecom industries have shorter planning cycles than the construction or engineering industries. However, planning cycles are becoming perceptibly shorter. For example, an Austrian telecom company no longer plans its investment portfolio once a year, but rather every three months. The results of a cycle of strategic managing are the basis for the following cycle.

Strategic Managing: Methods

Strategic analyses form the basis for strategic planning and controlling. SWOT analysis, Porter's value chain analysis,[3] Kaplan and Norton's balanced scorecard,[4] the BCG

[1] Barnat, R., 2005.
[2] Barnat, R., 2005.
[3] Cf. Porter, M. E., 1985.
[4] Cf. Kaplan, R., Norton, D., 1992.

Table 3.1: Business Process "Strategic Managing"—Responsibility Chart

Business Process: Strategic Managing - Example							
	Roles						
Legend P ... Performing C ... Contributing I ... Being informed Co ... Coordinating Process tasks	Coordinator Strategy team	Strategy team	Strategic decision makers	Project Portfolio Group	Employees	Stakeholders	Tool/Document
Collecting information	P				C	C	
Checking strategic plans	Co	P					
Performing SWOT-Analysis	Co	P					
Developing strategic scenarios	Co	P					
Strategic positioning	Co	C	P				
Adapting values, mission (if required)	Co	C	P				
Planning strategic objectives	Co	P					1
Developing structures for realizing the strategic objectives	Co	P					
Managing the investment portfolio	Co	C	P				2
Developing an implementation plan	Co	P		C			3
Managing the project portfolio	Co	C		P			4
Deciding about objectives, strategies, investment portfolio, implementation plan, project portfolio	Co		P	C			
Informing about results of the strategic managing	Co	P	I	I	I	I	

Tool/Document:
1 ... Objectives plan
2 ... Investment portfolio data base
3 ... Implementation plan
4 ... Project portfolio data base

product portfolio matrix,[5] and investment analyses can all be used to analyze an organization's structural dimensions.

Methods for analyzing the contextual dimensions include stakeholder analyses, Porter's model of the "five competitive forces", or the PEST model (politics, economics, social/demographics, technology).[6]

5 Cf. Henderson, B., 1973.

6 Cf. Kendall, N., 2016., without year

Developing strategic scenarios for the organization within its contexts supports the process of deciding upon future strategic positioning. The strategic position forms the basis for developing a vision, for defining the values and the mission, for planning the objectives and strategies, and for developing an implementation plan for realizing these objectives.

An organization's vision describes in relatively basic terms what should be achieved over the medium term—for instance, in the format of: "In five years we are . . .". A vision is often communicated using slogans, images and metaphors. The mission describes and organization's ideal actual status—for example, in the format of: "This is how we are today and how we differentiate ourselves from others . . .". An organization's objectives plan describes the concrete objectives for the next one to two years—for example, in the format: "Next year we are . . .". The organization's strategies describe how the objectives should be achieved—for instance, in the format: "We want to achieve our organizational objectives by . . .". The analyses of the investments to be implemented and the agreed investment portfolio are a result of the planning of the objectives. An implementation plan can be prepared on the basis of the objectives and strategies. It includes measures, projects, and programs, as well as the overall project portfolio.

The investments to be implemented can be selected either by a separate consideration of individual investments, or by taking an integrated approach by considering an investment portfolio. An organization's investment portfolio can be defined as the investments being implemented and planned, together considering the relations between these investments.

Definition: Investment Portfolio

An organization's investment portfolio can be defined as the investments being implemented and planned together, considering the relations between these investments.

For investments considered separately, the contribution of each investment towards realizing the organization's strategic objectives is measured. The resulting contributions are compared and the "best" investments selected. A scoring method based on the criteria of the balanced scorecard model of Kaplan and Norton can be used in this selection process (see Fig. 3.1).[7] For each investment, the contribution to realizing financial objectives, stakeholder-related objectives, innovation objectives, and business process and resource objectives can be considered.

When applying an integrated approach, investment portfolios are analyzed. Alternative investment portfolios can be compiled and compared with one another. The relationships between investments can be complimentary, neutral, or competing.

[7] Cf. Kaplan, R.; Norton, D., 1992.

Investment	Realization of investment objectives	Score	Approved/Denied
A		90	Approved
B		86	Approved
C		81	Approved
D		70	Denied
E		65	Denied

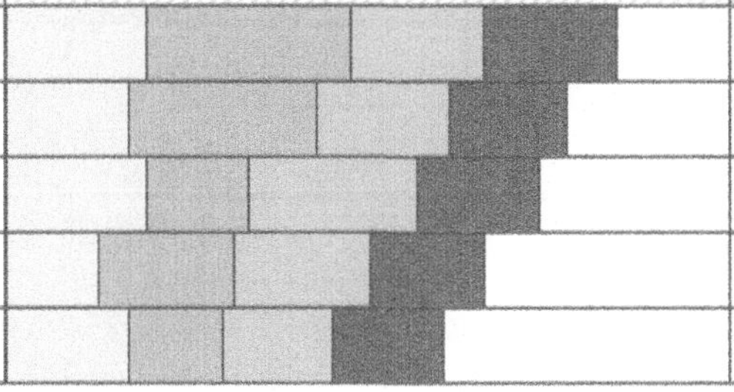

Financial objectives
Stakeholder-related objectives
Business process and resource objectives
Innovation objectives

Fig. 3.1: Scoring and selecting investments (example).

Rather than considering and selecting individual investments, investment portfolios are analyzed and the optimal portfolio is selected. The optimal investment portfolio is the one which best fulfills the strategic objectives of the organization.

The project portfolio is managed as part of strategic managing within the framework of implementation planning. The business process "project portfolio managing" is usually performed more frequently than the business process "strategic managing" (see Chapter 16).

A summary of the results of strategic managing is shown in Figure 3.2.

The results of strategic managing should relate to each other. The results should provide medium-term orientation and, consequently, be relatively stable. Therefore, attention should be paid to the complexity and dynamics of organizations when formulating the vision and objectives.

Strategic Managing: Organization

In principle, strategic managing is a (top) management responsibility. Strategic decisions should be developed by a strategy team (see the roles in Table 3.1). The project portfolio group (see Chapter 15), employees, and stakeholders of the organization should be involved in the communication processes of analyzing, planning, coordinating, and deciding. This broad involvement generates diversity which, on the one hand, increases creativity in the process of strategic managing, and, on the other, contributes to ensuring acceptance of the results. All employees and selected stakeholders must be informed of the results of strategic managing in an appropriate format.

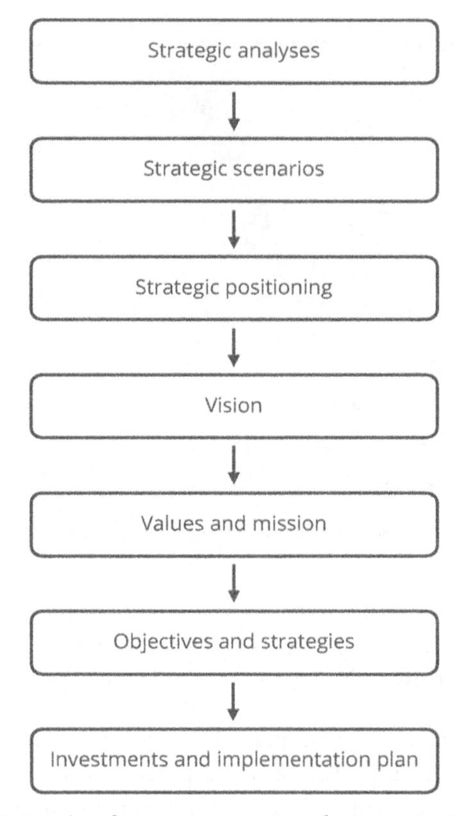

Fig. 3.2: Results of strategic managing of an organization.

If in larger organizations fundamental strategic changes are required, then the business process "strategic managing" may be so comprehensive, complex, and above all strategically important, that it is advisable to perform it as a project.

Strategic Managing and Projects

The results of strategic managing also determine an organization's project portfolio (see Chapter 16). New projects and programs are identified and documented in the implementation plan, and restructuring of the project portfolio is required.

Projects are derived from an organization's objectives and strategies. Projects which make no contribution to fulfilling objectives should not be performed. An ad hoc "bottom up" identification of projects is, however, possible. In this case, the projects should be appropriately linked back to the strategic objectives.

The contributions made by projects to fulfilling the strategic objectives of an organization must be communicated to the members of the project organizations and to the project stakeholders. This is a project management responsibility. This creates "sense" and contributes to motivating the members of the project organizations.

On the other hand, project results also influence an organization's strategic objectives. The relationships between an organization's strategic objectives and projects are shown in Figure 3.3. Managing these relationships as part of project portfolio management is also a task of strategic managing.

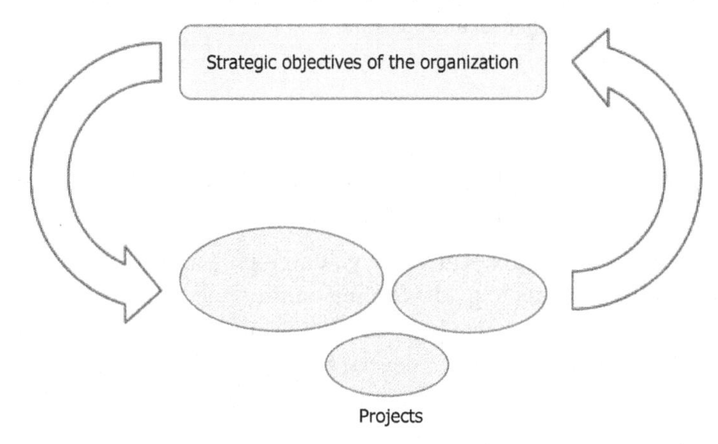

Fig. 3.3: Relations between strategic objectives of an organization and projects.

3.2 Investment Definition and Investment Types

Investment Definition

In the literature, an investment is defined as the transfer of capital into fixed and current assets.[8] In fiscal terms, an investment is understood as a process which is characterized by a series of cash outflows and a series of cash inflows.[9] The predilection for the fiscal definition can be explained by its operationality. It allows traditional investment analysis methods to be used as a basis for decision making.

Here an investment is understood as a chain of business processes required to ensure an organization's capacity to provide services. For this not only fixed and current assets are necessary, but personnel and organizational competencies must also be created, and medium-term to long-term collaborative relationships with stakeholders must be established.[10]

Investments may relate to various objects—namely, to a product or service, a market, the organization, the personnel, the infrastructure, the financing, and a stakeholder relation. Therefore, unlike in accounting, the definition of an investment used here is not limited to depreciable fixed assets.

[8] Cf. Albach, H. 1962.

[9] Cf. Boulding, K. E. 1936.

[10] In practice, the terms "initiative" or "task" are often used instead of the investment terms applied here. Although this reduces the risk of the term being confused with the fiscal definition of investment, this cannot be justified in business terms. For example, no business case analysis or cost–benefit analysis is prepared for an initiative or a task.

Definition: Investment

An investment is defined as a chain of business processes required to ensure an organization's capacity to provide services. A differentiation may be made between product or service-related investments, market-related, organization-related, personnel-related, infrastructure-related, financing-related, and stakeholder-related investments.

Investment Types

In accordance with the structural and contextual dimensions of organizations defined above, investment types can be differentiated as follows:

> product or service-related investments (e.g., developing and selling a product),
> market-related investments (e.g., developing and serving a market),
> organization-related investments (e.g., developing and using an organization),
> personnel-related investments (e.g., developing and deploying personnel),
> infrastructure-related investments (e.g., developing and operating ICT infrastructure),
> finance-related investments (e.g., an initial public offering), and
> stakeholder-related investments (e.g., developing and using a supplier relation).

The general life cycle of an investment object includes the phases of planning, implementing, operating or using, and any subsequent decommissioning of an investment object. The life cycle of an investment object begins once the investment decision is taken. The life cycle of an industrial plant, for example, includes the following chain of business processes: "constructing the industrial plant", "operating the industrial plant", "maintaining the industrial plant", and "decommissioning the industrial plant". The upstream business process "conceptualizing the industrial plant investment" provides a temporal context for this investment.

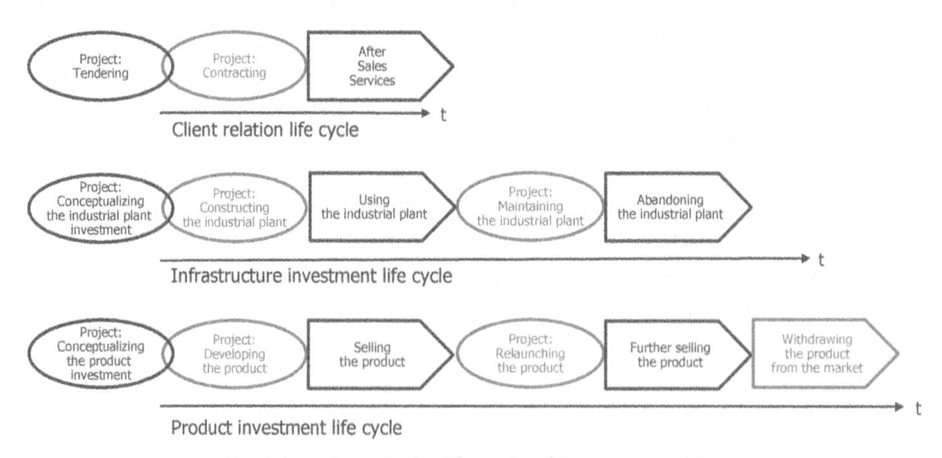

Fig. 3.4: Projects in the life cycle of investment objects

Investment, Projects and Programs

Projects and programs can be applied to perform business processes within the life cycle of an investment object, or prior to this for conceptualizing an investment. Figure 3.4 presents in the form of ellipses those processes which, when of appropriate scope and appropriate complexity, are performed as projects or programs. Where two or more projects are performed sequentially, a chain of projects results.

3.3 Planning an Investment Object

Planning an Investment Object: Objectives

The planning of an investment object takes place during the conception of an investment. The objectives of planning an investment object are to prepare for and reach an investment decision. A decision shall be reached with respect to implementing an investment, but not with respect to the form of the organization for implementing the investment. This later decision is made during project or program initiating. Planning an investment object produces a basis for strategic managing (see Fig. 3.5).

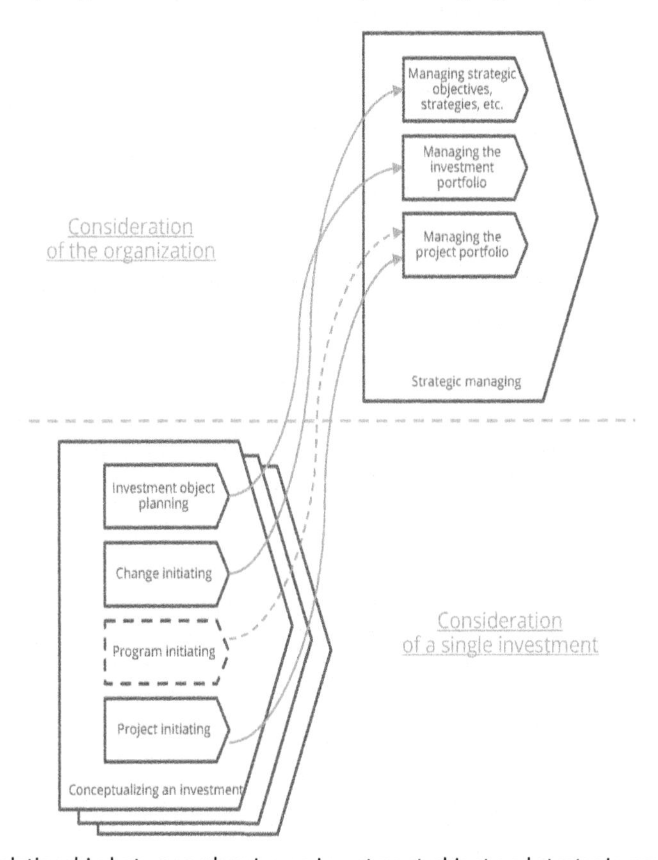

Fig. 3.5: Relationship between planning an investment object and strategic managing.

Figure 3.5 shows that investment object planning, change initiating, and program or project initiating are sub-processes of conceptualizing an investment. The results of these sub-processes provide input for managing the strategic objectives, strategies, etc. of an organization, for managing the investment portfolio, and for managing the project portfolio. As symbolized by stacking in Figure 3.6, several investments are usually conceived at the same time. The results from conceptualizing different investments are then the basis for strategic managing.

Planning an Investment Object: Process

An idea for an investment is the basis for performing the business process "Planning an investment object". This process includes analyzing the actual situation, defining investment objectives, defining and describing possible solutions, selecting a solution, defining initial requirements for the selected solution, and performing an investment analysis (see Fig 3.6). This information can be summarized in an investment proposal which forms the basis for an investment decision. An example of the business process "conceptualizing an investment", which shows the tasks involved in planning an investment object, is given in Table 3.5 for the case study "Values4Business Value" (page 47).

Planning an Investment Object: Investment Analysis Methods

The key methods for analyzing an investment are business case analysis and cost–benefit analysis. The basis for these analyses is the definition of the boundaries of the considered investment.

"Business case" is another frequently used name for investment. Consequently, the business case analysis is a form of investment analysis. The business case analysis includes the description of the effects of an investment, its monetary assessment in the form of cash outflows and cash inflows, the results of investment calculations, and possibly also simulation calculations. The results of investment calculations are financial ratios such as net present value, return on investment (ROI), and amortization period.

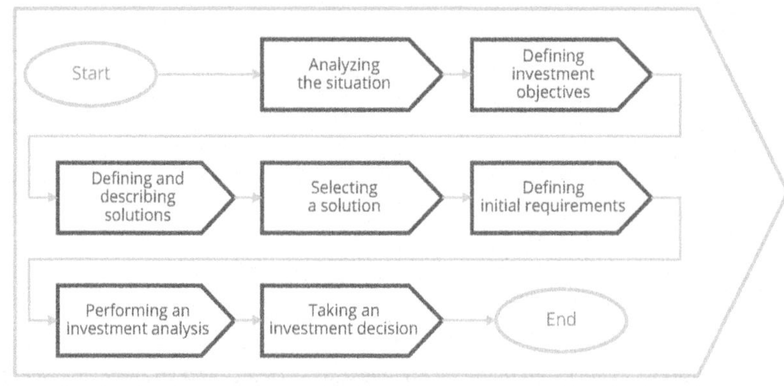

Fig. 3.6: Business process "Planning an investment object"—flow chart.

The OGC investment appraisal template for performing a business case analysis is shown in Figure 3.7.[11]

Project name:			Business sponsor:		
Management summary					
Financial appraisal summary					
Cash flow	Expenditure			Savings	Cumulative cash flow
	Capital	Non-capital	Total		
Year 0					
Year 1					
Year 2					
Year 3					
Totals					
Investment ratios					
Benefits (£)					
Payback period					
Net present value (NPV)					

Fig. 3.7: The OGC Investment Appraisal Template.

The basic rules and methods of business case analysis are described in the Excursus "Business Case Analysis". An Austrian telecom company's rules for developing a business case analysis are shown in Table 3.2.

Table 3.2 Rules for Developing Business Case Analyses—Example of a Telecom Company

Rules for preparing business case analyses
> Methods used: Application of the net present value method and calculation of the amortization period based on the cumulative discounted cash flow
> Form of presentation: Negative sign for cash outflows
> Period of consideration: Period for which the investment object can be used economically; the standard is 5 years; any reinvestment must be taken into account.
> Current assets: Inventory and receivables to be considered
> Interest rate to be applied: A mixed rate is assumed, which also includes interest on equity

Excursus: Business Case Analysis

According to the fiscal definition of an investment, the advantages represented by an investment are determined by its cash outflows and cash inflows. At the start of an investment, the cash outflows usually exceed the inflows, while later the inflows dominate.

[11] Cf. Jenner, S., Kilford, C., 2011

The cash flows relevant for evaluating an investment are its future cash flows. Historic cash flows—that is, past outflows and inflows—are not to be considered in the investment analysis.

Cash flows can be forecast using certain or uncertain assumptions—that is, deterministic or stochastic cash flows may be used. Due to the different timing of the individual cash flows, two or more cash flows cannot be directly compared. However, due to the principle of equivalence, it is possible to calculate a value for a particular point in time which is equivalent to a cash flow. The value of a cash flow which is calculated for the beginning of the investment life cycle is called "net present value".

The choice between two or more investments then becomes a choice between net present values of investments. The internal rate of return is the interest rate which sets the net present value of a cash flow as zero. The following assumptions for defining an interest rate for calculating the net present value are possible:

- the opportunity costs reflecting the yields of the best unrealized investment opportunity,
- a subjective minimum interest rate (minimum attractive rate of return), which is the result of a management decision, and
- the average capital costs calculated as the weighted costs of financing with equity and borrowed capital.

All the above interest rates usually include a risk premium and an inflation premium.

Dynamic investment analysis methods are briefly described in Table 3.3.

Table 3.3: Methods for Investment Analysis

Method	Calculation	Rules for decisionmaking
Net present value	$K = \sum_{t=0}^{T} Z_t (1+i)^{-t}$ (t=0, ... , T) i = required interest rate T = Life cycle K = Net present value Z = cash flow	An investment is beneficial if its net present value is zero or positive. In the case of two or several alternative investments, the one with the largest net present value is the most beneficial.
Internal rate of return	$K = \sum_{t=1}^{T} Z_t (1+r)^{-t} = 0$ r = Internal interest rate T = Life cycle K = Net present value Z = cash flow	An investment is beneficial if the internal rate of return is the same as or greater than the minimum rate of return required by the company. In the case of two or several alternative investments, the one with the highest internal interest rate is the most beneficial.
Dynamic amortization period	$-Z_0 = \sum_{t=1}^{n} Z_t (1+i)^{-t}$ (t=1, ... , n) Z_0 = Initial payment Z_t = Cash flow n = Amortization period	In the case of two or several alternative investment, the one with the shortest amortization period is the most beneficial.

The cost–benefit analysis shows the consequences of investments in the form of costs and benefits. These are described and, as far as possible, monetary assessments are performed. Where market prices are available for specific cost and benefit types, these are considered in the analysis. Where there is no perfect competition—for example, where prices are subsidized—or there are no prices—for instance, for evaluating a job loss—then shadow prices may be used. Shadow prices are fixed on the basis of assumptions.

In contrast to business case analysis, in a cost–benefit analysis the costs of an investment which do not result in cash outflows are also considered. Non-monetary costs and benefits are described verbally and by using indicators.

The principles of sustainable development can be considered in the cost–benefit analysis. That means that:

> economic, ecological, and social costs and benefits of an investment are considered,
> costs and benefits of an investment for the investor as well as for selected stakeholders are considered,
> short-, medium-, and long-term effects of the investment are considered,
> local, regional, and global effects of the investment are considered, and
> a social discount rate for discounting future costs and benefits is applied in order to take into account the interests of future generations.

Due to stakeholder orientation as a principle of sustainable development, considerations are not limited to the effects of an investment for the investor, but also consider the effects on stakeholders.

Cost–benefit analysis enables a more holistic approach to analyzing an investment than business case analysis. An example of a cost–benefit analysis which considers the principles of sustainable development and the different perspectives of the investor and the stakeholders is shown in Table 3.6 for the RGC case study "Values4Business Value" (page 49).

Due to their long-term nature, investment decisions are subject to more uncertainty than other organizational decisions. There are two ways of considering the risks associated with an investment. On the one hand, a sensitivity analysis can be conducted to determine the parameters upon which the benefits of an investment depend. In the simplest form of sensitivity analysis the extent to which deviations adversely impact an investment is examined for one parameter.

A sensitivity analysis can also be performed for several parameters. In this case, the result of the sensitivity analysis is a series of critical points which indicate the boundaries within which parameters may vary without making an investment no longer attractive. On the other hand, simulation calculations may be performed in which uncertain parameters are defined as random variables for which probability distributions are adopted. Investment analysis tips are given in Table 3.4.

Table 3.4: Tips for Investment Analysis

Tips for investment analysis
> Investment analysis has a high practical relevance for product and infrastructure investments. It is becoming increasingly important for marketing and organizational investments.
> It is of little relevance for contracting projects, since the main focus here is on profits achievable in the short-term. However, the company performing a contract can perform an investment analysis from the customer's perspective as a marketing tool.
> Investments should be described as holistically as possible, i.e., not only technical investment objectives but also marketing, organization or personnel-related objectives must be considered.
> If possible, monetary values for costs and benefits should be assumed. These provide an important basis for communication in the investment decision making process.
> The monetary assumptions of costs and benefits should be supplemented by non-monetary indicators and qualitative statements.
> The investment analysis should be subject to periodic controlling.

Investment decisions usually involve a relatively high level of capital and have medium- to long-term economic, ecological, and social consequences. After investment decisions are reached, they are regarded as relatively fixed and unchangeable.

However, the objective must be the ability to react flexibly to changes in the context of an organization within the life cycle of the investment. The theory of real options offers a model permitting flexibility in the investment life cycle.[12] Specific courses of action within the investment life cycle are identified and considered in planning. Growth and abandonment options in particular are regarded as possible courses of action.

When applying the model of real options, a flexible investment is valued more highly than a rigid investment. Although the formal use of this model is theoretically demanding, it is certainly relevant as a basic "way of thinking".

Planning an Investment Object: Methods for Proposing an Investment

An investment proposal is a means of describing a proposed investment. An investment proposal includes a summary of the problem situation and the reason for the investment, the investment objectives, the description of the investment object,

[12] Cf. Hommel, U. et al., 2013.

investment ratios, and the contribution of the investment to realizing the ecological and social objectives of the project-oriented organization.

An example of an investment proposal for the case study "Values4Business Value" is shown in Table 3.7 (page 53).

The investment proposal is the basis upon which the decision makers decide whether to implement an investment. The description of the problem situation shall ensure a common view of the actual status. The description of the investment objectives serves to show the ideal status to which the investment is intended to contribute. Taking into account the principles of sustainability, the contributions towards realizing the objectives of the organization can be differentiated into economic, ecological, and social objectives, and also regarding the timing. The investment analysis and the relevant investment plans must be included with the investment proposal.

3.4 Controlling the Benefits Realization of an Investment

Controlling Benefits Realization: Definitions

Controlling benefits realization is called benefits realization management (BRM) in the literature and is usually covered together with program management. The "Managing Successful Programs" standard lays out key fundamentals,[13] and the PMI program management standard shows the connection with BRM.[14] BRM methods are comprehensively described by Levin and Green.[15] APM defines BRM as "the identification, definition, planning, tracking, and realization of business benefits"[16]; Bradley defines it as "the process of organising and managing, so that potential benefits, arising from investment in change, are actually achieved".[17]

In the "Managing Successful Programs" standard benefit is defined as "a measurable outcome, which is perceived as an advantage by one or more stakeholders". The benefits include cost reductions, efficiency increases, and image improvements. Benefits should be formulated as improvement objectives—for example, less, more, higher, or better. Costs are defined as the negative effect of an investment.

Controlling benefits realization is a business process whose objective is to optimize the cost–benefit difference of an investment. Therefore the process is not limited to considering the benefits of an investment; it also considers their costs. Benefits realization controlling is not concerned with projects or programs. Instead, the considered costs and benefits relate more to investments, although these are implemented in the form of projects or programs. This relation is shown in Figure 3.9 (on page 54).

[13] Cf. Sowden, R. et al., 2011, S. 73 ff.

[14] Cf. PMI, 2013, p. 33 ff.

[15] Cf. Levin, G., Green, A.R., 2013.

[16] Cf. APM, 2012.

[17] Cf. Bradley, G., 2010.

(text continues on page 54)

Case Study: Values4Business Value—Planning an Investment Object

Values4Business Value: Relations to RGC Strategic Managing

In the course of RGC strategic managing, it was determined in 2014 that the RGC management approaches required a fundamental enhancement.

Although new content-related developments such as agile approaches, principles of sustainable development, managing requirements, and controlling benefits realization were consistently included in training and consulting activities, this did not lead to an integrated consideration of their consequences. Growing criticism of "traditional" project management in the literature and in practice necessitated a clear theoretical positioning of the RGC management approaches. Only after the developments were documented in a publication was it possible to communicate them adequately to customers and to the management community.

These strategic management results were the basis for establishing a working group in 2015 for conceptualizing the investment "Values4Business Value". The investment object was planned and the implementation project initiated as part of the conception process. An overview of the business processes related to the investment "Values4Business Value" is shown in Fig. 3.8.

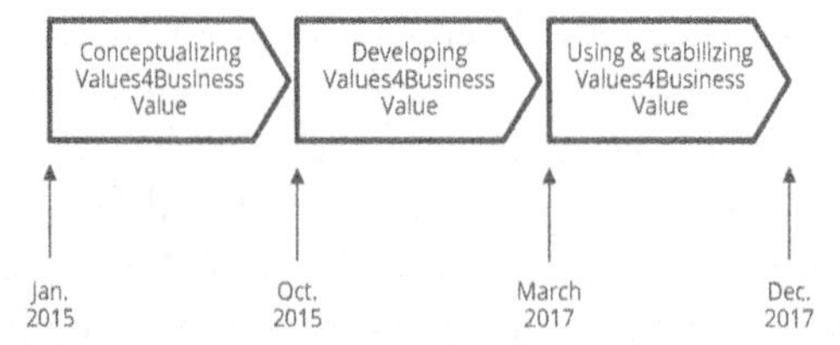

Fig. 3.8: Chain of business processes related to the investment "Values4Business Value".

Although no project was defined to perform the conception, with the performance instead being undertaken by a working group, several project management methods were used. Table 3.5 shows the structure of tasks in the form of a responsibility chart.

Selected results of the conception—namely, the cost–benefit analysis and the investment proposal—are shown and interpreted below. The results of defining the "initial" business requirements have been incorporated into the cost–benefit analysis. The results of defining the "initial" solution requirements have been summarised in an "initial" product backlog. The process "defining initial requirements", the methods used in this process, and examples are described in Chapter 4.

The cost–benefit analysis of the investment "Values4Business Value" is shown in Table 3.6 (beginning on page 49). The first part shows the perspective of RGC as the investor, with the perspectives of key stakeholders presented in the second part.

The cost–benefit analysis was developed as part of the conception of "Values4Business Value". During the initial development, the effects of the investment were considered from the perspective of the investor RGC. It was clear that the investment was advantageous. In a later phase, where the marketing considerations were of primary importance, the cost–benefit analysis was expanded to consider the costs and benefits of other stakeholders.

The monetary evaluations of the costs and benefits were undertaken by the working group which performed the conception. There was no discounting or compounding of future costs

Table 3.5: Conceptualizing the Investment "Values4Business Value"—Responsibility Chart

Legend
P...Performing
C...Contributing
I...Being informed
Co...Coordinating

Tools/documents:
1) ..... Documentation of the stakeholder requirements
2) ..... Documentation of the business requirements
3) ..... Documentation of Epics and of User Stories
4) ..... Backlog
5) ..... Cost-benefit analysis
6) ..... Initial project plans
7) ..... Communication tools

WBS-Code	Phase / Work package	Roles										
		Initiator	Initiation team	Investment decision makers	Requirements manager	Marketing experts	Scrum expert	Project management expert	Controller	Representative Publisher	Other stakeholders	Tool/Document
1	Peforming situation analysis											
1.1	Analyzing the management approaches		P									
1.2	Analyzing publications		P									
1.3	Analyzing RGC services		P									
1.4	Cross-checking the situation analysis	I	P									
1.5	Finalizing the situation analysis		P									
2	Defining stakeholder requirements											
2.1	Defining RGC's requirements		P		C							1)
2.2	Defining the publisher's requirements		P		C					C		
2.3	Defining the RGC clients' requirements		P		C	C						
2.4	Defining the readers' requirements		P		C	C						
2.5	Defining the management community's requirements		P		C						C	
2.6	Cross-checking the stakeholder requirements	I	P		C							
2.7	Finalizing the stakeholder requirements		P		C	C						
3	Describing solutions and selecting a solution											
3.1	Defining solutions		P		C	C						
3.2	Describing the solution 1-x		P		C	C				C		
3.3	Selecting a solution	P	C		C					I		
4	Analyzing and planning management innovations											
4.1	Analyzing project, program and change mangement approaches		P									
4.2	Analyzing management values		P									
4.3	Analyzing the sustainable development approach		P									
4.4	Analyzing agile approaches		P									
4.5	Analyzing requirements management		P									
4.6	Analyzing benefits realization management		P									
4.7	Planning management innovations		P									
4.8	Cross-checking management innovations	I	P									
4.9	Finalizing management innovations		P									

(continues on next page)

Table 3.5: Conceptualizing the Investment "Values4Business Value"—Responsibility Chart *(cont.)*

Legend: P...Performing; C...Contributing; I...Being informed; Co...Coordinating

Tools/documents:
1) ..... Documentation of the stakeholder requirements
2) ..... Documentation of the business requirements
3) ..... Documentation of Epics and of User Stories
4) ..... Backlog
5) ..... Cost-benefit analysis
6) ..... Initial project plans
7) ..... Communication tools

WBS-Code	Phase / Work package	Initiator	Initiation team	Investment decision makers	Requirements manager	Marketing experts	Scrum expert	Project management expert	Controller	Representative Publisher	Other stakeholders	Tool/Document
		Roles										
5	Defining initial requirements											
5.1	Defining initial business requirements	C	P		C	C			C			2)
5.2	Cross-checking initial business requirements	C	P		C				C			
5.3	Finalizing initial business requirements	C	P		C				C			
5.4	Defining initial solution requirements		P		C		C			C	C	3)
5.5	Cross-checking initial solution requirements	I	P		C		C					
5.6	Finalizing initial solution requirements		P		C		C					
5.7	Defining initial transition requirements		P		C		C					
5.8	Developing the initial product backlog		P		C	C	C		C	I		4)
6	Developing a cost-benefit analysis											
6.1	Drafting the cost-benefit analysis		C		C	C			P			5)
6.2	Cross-checking the cost-benefit analysis	I	C		C	C			P	C		
6.3	Finalizing the cost-benefit analysis		C		C	C			P			
7	Initiating an implementation project											
7.1	Developing initial project plans		C		C	C	C	P		C		6)
7.2	Analyzing the project portfolio		C		C	C	C	P				
7.3	Defining project strategies	I	C		C	C	C	P		C	C	
7.4	Developing the project proposal		C		C	C	C	P				
8	Deciding regarding the investment and the implementation project											
8.1	Developing the documents for decision making		P		C	C						
8.2	Presenting the proposal	C	P		C							
8.3	Providing additional information	I	P		C	C		C		C		
8.4	Deciding regarding the investment	C	C	D								
8.5	Deciding regarding the implementation project	C	C	D								
8.6	Communicating the decisions	C	P	M	C					I	I	7)

V. 1.001 v. S. Füreder per 5.10.2015

or benefits. This accords with the principles for sustainable development in which future benefit surpluses are considered just as important as current benefits.

From the perspective of the investor as well as from the perspectives of the other stakeholders considered, the results indicated that the benefits hugely outweighed the costs, and that it would be possible to amortize project costs in the short term (after two years). As the results were so unequivocal there was no need for a sensitivity calculation and formal risk analysis for the investment. On the other hand, from the investor's perspective, the implementation costs were surprisingly high.

Case Study continues on page 53

Table 3.6: Cost–Benefit Analysis for the Investment "Values4Business Value"

Cost-Benefit Analysis
Values4Business Value

V. 1.001 v. S. Füreder per 5.10.2015

Investor's perspective: Developing								
Types of costs/benefits	Description of the costs/benefits	Monetary assessment	Impact					
			economic	ecological	social	local	regional	global
Phase: Developing - Costs								
Personnel costs: Project managing (project owner, manager, team)	Project starting, project controlling, project transforming, project closing (600 h à 75.-)	€ 45,000.00	x			x		
Personnel costs: Authors	Researching and writing (1800 h à €100,-)	€ 180,000.00	x			x		
Personnel costs: Editor	Research, revision, design (800 h à 75,-)	€ 60,000.00	x			x		
Personnel costs: Assistance	Design of figures and directories (200 h à 50.-)	€ 10,000.00	x			x		
Personnel costs: Developer	Development of prototypes (3 seminars, 2 lectures, 1 event, each 30 h à € 100.-)	€ 18,000.00						
Personnel costs: Consultants, network partners	Content-related development by informing and reflexion workshops (5 persons à 40 h à 100.-)	€ 20,000.00	x			x		
Personnel costs: Marketeer	Events, mailing, initial marketing (80 h à 75.-)	€ 6,000.00	x			x		
Personnel costs: E-Book designer	Developing the e-book (200h à 75.-)	€ 15,000.00	x	x		x		
Material costs	100 books for initial marketing (à 70.-)	€ 7,000.00	x			x		
Costs for services: Peer review workshops	Catering, room and materials for peer review workshops (lump sum)	€ 3,000.00	x		x	x		
Costs for services: Book design	Layout and design of the book (lump sum)	€ 3,000.00	x			x		
Costs for services: Design of RGC documents	Layout and design of the documents (lump sum)	€ 2,000.00	x			x		
Costs for services: Book presentations	Catering, room and materials for 3 book presentations (lump sum)	€ 12,000.00	x		x		x	
Phase: Developing Total monetary costs		€ 381,000.00						
Qualitative Costs: Family and friends	Less time for family and friends				x	x		
Qualitative Costs: RGC employees	Less time for leadership of RGC		x		x	x		
Phase: Developing - Benefits								
Monetary benefits: Prototypes	Management innovations applied in prototypes (3 seminars, 2 lectures, 1 event, each € 5000.-)	€ 30,000.00	x		x	x		
Qualitative benefits: RGC USP	RGC USP made more distinctive by management innovations		x			x		
Qualitative benefits: Management approaches, services/products	Management approaches and RGC services/documents optimized		x			x		
Qualitative benefits: Book	Management innovations documented in the book		x				x	
Phase: Developing Total monetary benefits		€ 30,000.00						

(continues on next page)

Table 3.6: Cost–Benefit Analysis for the Investment "Values4Business Value" ***(cont.)***

Investor's perspective: Applying								
			Impact					
Types of costs/benefits	Description of the costs/benefits	Monetary assessment	economic	ecological	social	local	regional	global
Phase: Applying - Costs (Purchasing costs of the books not considered. The contributions are calculated as sales revenues minus the purchasing costs.)								
Personnel costs: Authors per year	Costs for additional marketing; participation in 1 event / year	€ 2,000.00	x				x	
Personnel costs: Marketeer per year	Costs for regular marketing	€ 3,000.00	x				x	
Phase: Applying Total monetary costs per year		€ 5,000.00						
Phase: Applying - Benefits								
Monetary benefits: Consulting services per year	Contributions of consulting services increased in the DACH-region	€ 100,000.00	x				x	
Monetary benefits: Trainings and RGC events per year	Contribution of tainings and RGC events increased	€ 50,000.00	x			x		
Monetary benefits: Book sales (hardcover) per year	Contribution of book sales (hardcover) increased in the DACH-region (600 à € 15.-)	€ 9,000.00	x				x	
Monetary benefits: Book sales (e-book) per year	Contribution of book sales (e-book) realized in the DACH-region (400 à € 10.-)	€ 4,000.00	x	x			x	
Monetary benefits: Savings of personnel costs per year	Personnel costs saved by higher efficiency of the RGC management	€ 20,000.00	x		x	x		
Monetary benefits: Royalties for authors per year	Royalties: 1000 books à 6 €	€ 6,000.00	x			x		
Phase: Applying Total monetary benefits per year		€ 189,000.00						
Qualitative benefits: Managing director per year	New RGC managing director Lorenz Gareis positioned		x		x		x	
Qualitative benefits: RGC image by innovations per year	RGC USP made more distinctive by management innovations		x				x	
Qualitative benefits: Cooperations with publishers	Potential for further cooperations with publishers created		x		x	x		
Qualitative benefits: Binding personnel per year	Personnel bound by the RGC USP		x		x	x		

(continues on next page)

Table 3.6: Cost–Benefit Analysis for the Investment "Values4Business Value" ***(cont.)***

Stakeholders' perspective: Applying									
			Impact						
Types of costs/benefits	Description of the costs/benefits		economic	ecological	social	local	regional	global
Cooperating publishers - Costs (no figures were made available from the publishers)									
Material costs per year	Production costs of the book		x			x		
Personnel costs per year	Marketing/PR for the book		x			x		
Costs for services per year	Royalties for authors		x			x		
Cooperating publishers - Benefits (no figures were made available from the publishers)									
Monetary benefits: Book sales per year	Revenues of book sales		x				x	
Qualitative benefits: Improved publisher's image per year	Image improvement through innovative management publications		x		x		x	
Readers: Practitioners (project managers, consultants, trainers, teachers) - Costs									
Material costs: Purchasing the book	Purchasing of the book, 700 units per year à € 70.-	€ 49,000.00	x			x		
Personnel costs: Reading	Time of reading of 700 persons à 30 hours à € 50.-	€ 1,050,000.00	x			x		
Total monetary costs		€ 1,099,000.00						
Readers: Practitioners (project managers, consultants, trainer, teachers) - Benefits									
Monetary benefits: Improved efficiency	More efficient and effective management, (saving 10% of working time per year = 700 persons à 160 h à € 200.-)	€ 22,400,000.00	x		x	x		
Qualitative benefits: More appreciation	Appreciation of cooperation partners and supervisors				x	x		
Total monetary benefits		€ 22,400,000.00						
RGC Clients (Training/Consulting) - Costs									
Costs of training and consulting services	Unchanged, no additional costs per year							
RGC Clients (Training/Consulting) - Benefits									
Qualitative benefits: Business Value created	Increase in efficiency and effectivity through further developed RGC management approaches		x		x	x		
Readers: Students - Costs									
Material costs: Purchasing the book	Purchasing of the book (300 units à € 60.-)	€ 18,000.00	x			x		
Personnel costs: Reading	Time of reading (300 students à 40 hours à € 15.-)	€ 180,000.00	x			x		
Total monetary costs		**€ 198,000.00**						
Readers: Students - Benefits									
Monetary benefits: Better skills	Contribution to skills development (300 students, higher annual salary, starting salary of € 37.000.- instead of € 35.000.-)	€ 600,000.00	x			x		
Qualitative benefits: Success in exams	Contribution to the success in exams of 300 students per year		x		x	x		
Total monetary benefits		**€ 600,000.00**						

(continues on next page)

Table 3.6: Cost–Benefit Analysis for the Investment "Values4Business Value" *(cont.)*

Investment ratios for 3 years of application			Cost-benefit ratio of the investment	Interpretation
Monetary cost-benefit ratio from the investor's perspective	Phase: Developing	-€ 351,000.00	€ 201,000.00	Cost-benefit ratio of the investor for 3 years = Cost-benefit ratio of applying (3 x 184.000 €) minus cost-benefit ratio of developing (351.000 €)
	Costs:	-€ 381,000.00		
	Benefits:	€ 30,000.00		
	Phase: Applying	€ 184,000.00		
	Costs per year:	-€ 5,000.00		
	Benefits per year:	€ 189,000.00		
Cost-benefit ratio from the reader's perspective: Practitioners	Costs per year:	-€ 1,099,000.00	€ 63,903,000.00	Cost-benefit ratio of practitioners for 3 years = Benefits for 3 years (3 x 22.400.000 €) minus costs for 3 years (3 x 1.099.000)
	Benefits per year:	€ 22,400,000.00		
Cost-benefit ratio from the reader's perspective: Students	Costs per year:	-€ 198,000.00	€ 1,206,000.00	Cost-benefit ratio of students for 3 years = Benefits for 3 years (3 x 600.000 €) minus costs for 3 years (3 x 198.000)
	Benefits per year:	€ 600,000.00		

Table 3.7: Investment Proposal "Values4Business Value"

Title of the investment:	Values4Business Value
Type of investment:	☐ Organization ☐ Marketing ☐ Infrastructure ☐ Personnel ☐ Stakeholder relationships X Services
Reasons for the investment	
Due to the strong competition on the consulting market the evolution of the RGC management approaches can contribute to making the RGC USP more distinctive.	
The book "Happy Projects!", in which the RGC management approaches are documented, is still regarded a "classic" in the project managment community. However, new developments like agility, requirements management, benefits realization management, etc. are missing. Connections between project, program and change management are insufficiently described. Therefore an innovation of the management approaches is necessary.	
The marketing of the book "Happy Projects!" was good in Austria, but insufficient in Germany and Switzerland. The evolution of the management approaches can establish a basis for new markeing measures in the DACH region.	
An e-book can be developed as a new format. This can contribute to digitalization objectives.	
Investment objectives	
Mangement innovations applied in prototypes	Contribution of e-book sales realized in DACH-region
USP made more distinctive by management innovations	Personnel costs reduced by higher efficiency of the RGC management
RGC Management approaches and services/products optimized	Royalties for authors ensured
Mangement innovations documented in a book	New RGC managing director Lorenz Gareis positioned
Contribution of consulting services increased in the DACH-region	RGC USP in the DACH-region made more distinctive by management innovations
Contribution of trainings and RGC events increased	Potential for further cooperations with publishers created
Contribution of book sales (hardcover) increased in the DACH-region	Personnel bound by the RGC USP
Description of the investment object	
RGC management approaches: Process management, project management, program management, management of the project-oriented organization and change management	
Materials (designs, Power Point slides)	
Book: PROJECT.PROGRAM.CHANGE	
Consultants, personnel and network partners	
Contribution to the realization of economic objectives	
Cost-benefit ratio of the investor: € 201.000.-	
Amortization period: 2 years	
Risk: Little interest in the management innovations in the DACH region due to resistance to implement new approaches.	
Initiator	Initiation team

Values4Business Value: Investment Proposal

Table 3.7 shows as an example the investment proposal for the investment "Values4-Business Value". It was possible to adopt key information for developing the investment proposal from the cost–benefit analysis. "Values4Business Value" is a service-related investment, as the RGC management approaches are regarded as a "technology" for fulfilling services. The reason for the investment has already been interpreted in the description of the relations with strategic managing. The key objects considered within the framework of the investment were:

> the RGC management approaches for process, project, program, and change management,
> the services and products,
> the book *PROJECT.PROGRAM.CHANGE,*
> the documents for providing RGC services (e.g., designs, PowerPoint presentations, etc.),
> the consultants, employees, and networking partners.

The investment ratios were taken from the results of the cost–benefit analysis. A key factor in deciding to perform the investment "Values4Business Value" was the strategic objective of strengthening the position of RGC as a management innovator in the DACH region.

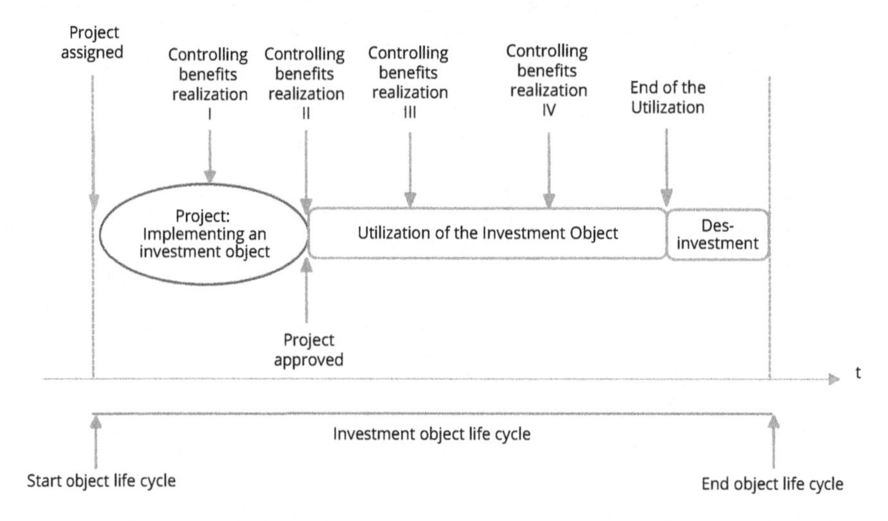

Fig. 3.9: Relation between implementation project and investment.
(Figure is described on page 45.)

Controlling Benefits Realization: Objectives

The success of an investment can be controlled using the business process "controlling the benefits realization". The objectives of the business process "controlling the benefits realization" are:

> the optimization of the cost–benefit difference of an investment,
> the assurance that an investment contributes to the sustainable business value of the organization,
> the rapid realization of investment benefits by defining quick wins,
> the prompt abandonment of an unsuccessful investment,
> the organization of learning for other investments, and
> the involvement of stakeholders in controlling the benefits realization.

From the point of view of managing requirements, benefits realization controlling should ensure that both the business requirements and the solution requirements are fulfilled. "Start with the end in view" is the motto of benefits realization controlling.

Controlling the benefits realization of investments creates the basis for controlling the investment portfolio within the framework of an organization's strategic management. In practice, benefits realization controlling is (still) rarely applied, although it represents a significant communication and optimization opportunity for investments.

Definition: Controlling Benefits Realization

Controlling the benefits realization of an investment is a business process whose objective is to optimize the cost–benefits difference of the investment. The process is not limited to considering the benefits of an investment, but also considers its costs.

Controlling the benefits realization of investments creates the basis for controlling the investment portfolio within the framework of an organization's strategic management.

Controlling Benefits Realization: Structure

Benefits realization controlling must be performed with reference to the results of planning the investment object. Benefits realization controlling begins after the investments decision, with the implementation of an investment. During the process of implementing an investment object, the following tasks must be fulfilled:

- communicating the expected costs and benefits,
- prioritizing and selecting the costs and benefits to be controlled,
- defining metrices for the costs and benefits to be controlled,
- periodically controlling the costs and benefits,
- planning and approving measures for optimizing costs and benefits,
- controlling potential risks of the costs and benefits of the investment,
- preparing periodic benefits realization reports, and
- communicating the realized benefits.

During the operation or use of an investment object, the following tasks must be fulfilled:

- periodically controlling the costs and benefits,
- planning and approving measures for optimizing costs and benefits,
- controlling potential risks of the costs and benefits of the investment,
- preparing periodic benefits realization reports,
- communicating the cost–benefit difference realized.

In each case the extent of controlling during the operation or use of the investment object must be agreed upon. If, during this phase, the investment appears to offer no further optimization potential, then controlling of the benefits realization may be performed until only shortly after the start of operation or use. This is the general rule in practice. For example, an Austrian bank uses a value capturing process for investments with a project budget larger than €250,000. The investment implemented is controlled by the Project Management Office for a maximum period of two years from project closing. It is the responsibility of the defined benefit owners to realize the benefits.

Controlling Benefits Realization: Methods

Specific methods can be used for controlling the benefits realization. The cost–benefit analysis drawn up during the planning of an investment object is the basis for controlling benefits realization. Key methods for controlling benefits realization include a benefits profile, a list of the benefits to be controlled, a benefits realization plan, a benefits relations analysis, and a benefits realization report.

The objectives of the methods for controlling benefits realization are described in Table 3.8.

Controlling Benefits Realization: Organization

Controlling benefits realization allows an investment to be controlled. In principle, it is the initiator of an investment who is responsible for the controlling. Investment initiators are usually the organizational units that enjoy the key benefits of an investment. As controlling benefits realization is also important during the time in which an investment object is implemented, the members of the project organization of an implementation project must also be involved in this process.

Table 3.8: Objectives of the Methods for Controlling Benefits Realization

Method for controlling benefits realization	Objectives
Performing a cost-benefit analysis	> Providing a basis for deciding whether to implement an investment > Analyzing the economic, environmental and social costs and benefits of an investment for the investor and for stakeholders > Holistic analysis of an investment > Using the analysis as a tool for communication and controlling
Developing a benefit profile	> Describes the benefit, the form of the evaluating the benefit, the organizations that benefit, other benefits to which the benefit contributes, the measurement standard for the benefit, and the responsibility for the benefit
Preparing a list of benefits to be controlled	> Listing the benefits that are to be controlled
Preparing a benefits realization plan	> Showing what benefits are to be realized to what extent and when > Describing the standard of measurement for benefits, the objectives to be achieved, and the actual benefits achieved for each period
Analyzing benefits relations	> Showing the relations between benefits and interpreting the relations
Preparing a benefits realization report	> Periodic reporting of the benefits and/or costs realized as of the controlling date > Reports on the modified benefits-costs-difference

Case study: Values4Business Value—Controlling Benefits Realization

Values4Business Value: Basis for Controlling Benefits Realization

In 2015, RGC decided to implement the investment "Values4Business Value". The basis for controlling the benefits realization was provided by the cost–benefit analysis. The benefits to be subject to controlling were selected. These benefits are shown in Table 3.9. The focus was placed on quantifiable benefits. Benefit profiles were drawn up for each benefit to be controlled, as shown in Table 3.10. It was decided that the benefits realization would be controlled through to the end of 2017.

Table 3.9: List of Benefits to Be Controlled

List of Benefits to be Controlled

Developing Values4Business Value

Benefit	To be controlled
01 Management innovations applied in prototypes	
02 RGC USP made more distinctive by management innovations	
03 Management approaches and RGC services/products optimized	✔
04 Management innovations documented in a book	✔
05 Contributions of consulting increased in the DACH-region	✔
06 Contributions of trainings and RGC events increased in the DACH-region	✔
07 Contributions of book sales (hardcover) increased in the DACH-region	✔
08 Contributions of e-book sales realized in DACH-region	✔
09 Personnel costs saved by higher efficiency of the RGC management	
10 Royalties for authors ensured	
11 New RGC managing director positioned	
12 RGC USP in the DACH-region made more distinctive by management innovations	✔
13 Potential for further cooperations with publishers created	✔
14 Personnel bound by the RGC USP	✔

V. 1.001 v. S. Füreder per 5.10.2015

Table 3.10: Benefit Profile "Contribution from Book Sales (Hardcover) in the DACH-Region Increased"

Benefit Profile

Developing Values4Business Value

Benefit number:	07	Type of benefit:	Quantitative benefit
Benefit:		Contribution from book sales (hardcover) in the DACH-region increased	
Importance of the benefit:		High importance	
Organization, that benefits:		Division: RGC Products	
Project, that brings benefit:		Developing Values4Business Value	
Benefits, that the benefit contributes to:		12 RGC USP in the DACH-region made more distinctive by management innovations	
		13 Potential for further cooperations with publishers created	
Ratio for measuring the benefit:		Contribution of book sales (hardcover) per country in €	
Responsibility for the benefit:		Head of the division: RGC Products	

V. 1.001 v. S. Füreder per 5.10.2015

Table 3.11 provides the example of the planned realization for two benefits. The benefits realization plan shows how the benefits can be realized, by when, and to what extent. As benefits may be interdependent, an analysis of the benefits relations was also undertaken (see Fig. 3.10).

Table 3.11: Benefits Realization Plan

Benefits Realization Plan
Developing Values4Business Value

#	Benefit	Ratio	Objective per year	Actual: 2017	Actual: 2018
7	Contribution of book sales (hardcover) increased in the DACH-region	Contribution of book sales (hardcover) in the DACH-region per country per year	Cont. in GER: € 2.000,- Cont. in AUT: € 5.000,- Cont. in CH: € 2.000,-	Cont. in GER: € ... Cont. in AUT: € ... Cont. in CH: € ...	Cont. in GER: € ... Cont. in AUT: € ... Cont. in CH: € ...
12	RGC USP in the DACH-region made more distinctive by management innovations	Survey results on the RGC image in the DACH-region	Perception as a provider of an integrated management approach	Perception: ...	Perception: ...

V. 1.001 v. S. Füreder per 5.10.2015

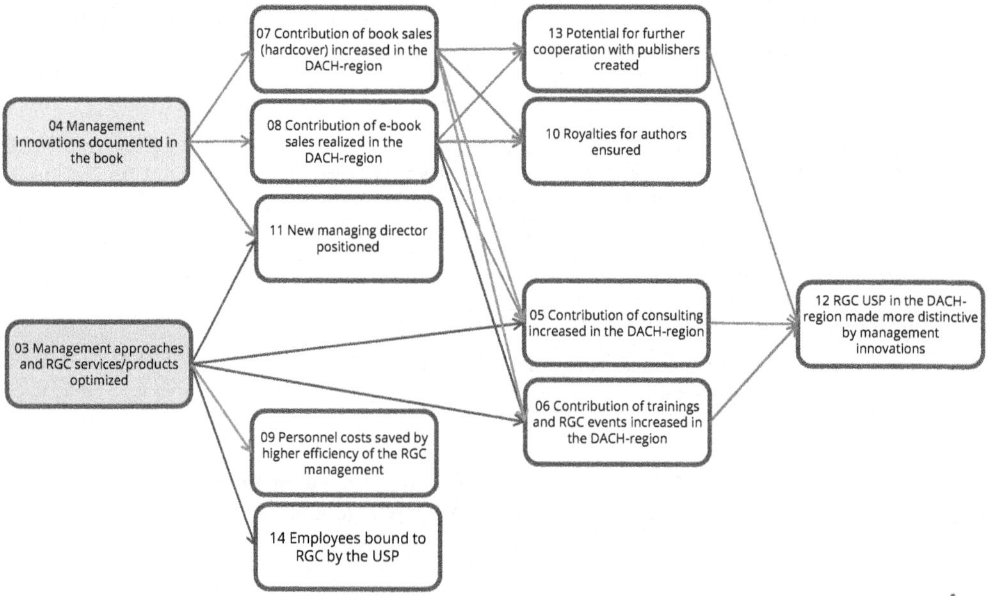

Fig. 3.10: Benefits relations analysis.

A fundamental challenge in controlling the benefits realization lies in structuring the project so that benefits can be identified and measured. A differentiation can be made between "enablers" and other benefits. In the case study "Values4Business Value", the benefits "optimizing management approaches and RGC documentation" and "documenting management innovations in the book" are the central enablers.

In agreement with the initiator of an investment, the project owner, the project manager, and the project team members of an implementation project may record all of the current investment costs and benefits, adapt future expected investment costs and benefits as the result of more information becoming available during implementation, and possibly take measures to optimize the cost–benefit difference.

Therefore, the roles responsible for controlling the benefits realization investment during the implementation project may be the initiator of the investment, the project owner, and the project manager. As shown above, the responsibility for realizing benefits can also be transferred to the benefits owners. In this case, the benefits owners should be integrated into the project organization. The strategy team or the Project Portfolio Group (see also Chapter 15) must provide periodic benefits realization reports in order for the results of the controlling of the benefits realization to be considered as part of strategic managing.

The prerequisite for the project manager taking on responsibility for controlling the benefits realization during the performance of an implementation project is a new self-understanding of project managers as "intrapreneurs". Project managers not only "administer" projects, they also contribute to ensuring the realization of investment benefits.

During the operation or use of the investment object, related information can be prepared and relayed by the facility management or product management.

Literature

Albach, H.: *Investition und Liquidität: die Planung des optimalen Investitionsbudgets,* Betriebswirtschaftlicher Verlag Dr. Th. Gabler, Wiesbaden, 1962.

Association for Project Management (APM): *APM Body of Knowledge,* 6th edition, APM, Buckinghamshire, 2012.

Barnat, R.: *The Nature of Strategy Implementation,* abgerufen von http://www.introduction-to-management.24xls.com/en201 (30.9.2016), 2005.

Boulding, K.E.: Time and Investment, *Economica,* Volume 3, London, 1936.

Bradley, G.: *Benefit Realisation Management: A Practical Guide to Achieving Benefits Through Change,* 2nd edition, Gower, Surrey, Burlington, VT, 2010.

Henderson, B.: *The Experience Curve—Reviewed IV. The Growth Share Matrix or the Product Portfolio,* The Boston Consulting Group, Boston, MA, 1973.

Hommel, U., Scholich, M., Vollrath, R. (Hrsg.): *Realoptionen in der Unternehmenspraxis: Wert schaffen durch Flexibilität,* Berlin, Heidelberg, Springer, 2013.

Jenner, S., Kilford, C.: *Management of Portfolios,* TSO (The Stationary Office), Norwich, 2011.

Kaplan R., Norton D.: The Balanced Scorecard—Measures that Drive Performance, *Harvard Business Review* (1–2), 1992.

Kendall, N.: *What is Strategic Management?* abgerufen von http://www.applied-corporate-governance.com/what-is-strategic-management.html (25.10.2016).

Levin, G., Green, A. R.: *Implementing Program Management: Templates and Forms Aligned with the Standard for Program Management,* 3rd edition, CRC Press, 2013.

Porter, M. E.: *Competitive Advantage: Creating and Sustaining Superior Performance,* Free Press, New York, NY, 1985.

Project Management Institute (PMI): *The Standard for Program Management,* 3rd edition, PMI, Newton Square, PA, 2013.

Sowden, R., Wolf, M., Ingram, G.: *Managing Successful Programmes (MSP),* 4th edition, The Stationary Office (TSO), Norwich, 2011.

4 Managing Requirements and Projects

Requirements related to a solution to be developed can be distinguished into business, stakeholder, solution, and transition requirements. These different types of requirement can be managed using either sequential or iterative approaches.

In a sequential approach, all requirements are defined before implementing a solution. The defined business requirements then form the basis for an investment analysis, while the solution requirements form the basis for the formulation of project objectives and the development of project plans. During an implementation project, the solution requirements are considered when the requirements are changed, when testing, and when accepting the solution.

In an iterative approach, the initial solution requirements are described in the business process "Defining initial requirements" and documented in an "initial backlog". The definition of additional requirements, prioritization of the requirements, etc., is performed during the implementation of a solution in the process "Controlling requirements". In this case, the requirements are managed in parallel with the development of the solution. As an example, methods for managing requirements are applied in the RGC case study "Values4Business Value" (see Section 4.6 on page 72).

The relations between the above-mentioned business processes for managing requirements and other business processes discussed in this publication are shown in the following overview.

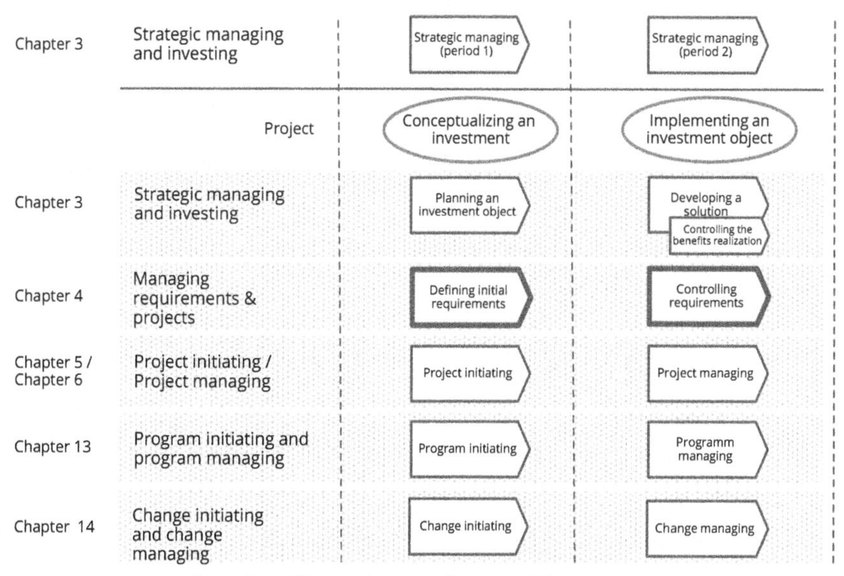

Overview: "Managing requirements" in context.

4 Managing Requirements and Projects

4.1 Requirements and Projects: Critical View from Practice

The objectives of managing requirements are the definition of requirements for an intended solution and the assurance that the defined requirements are fulfilled. A distinction is made between requirements that describe the solution ("solution requirements"), and requirements describing the benefits of the solution for the organization ("business requirements").

In managing requirements, reference is made to solutions, which correspond to investment objects (see Chapter 3). A "solution" can, for example, be an adapted building, a new product, a developed organization, developed personnel, a managed stakeholder relation, or a combination of these. Managing requirements is therefore object oriented. Project management is process oriented. The focus is on the process of fulfilling the defined requirements.

Mechanistic project management is based on the so-called "magic triangle" of project scope, project schedule, and project costs. Project objectives, as an aggregated description of the solution requirements, are not considered in this magic triangle. Project objectives are often described in formal project assignments (e.g., project charters). These are often "frozen", because project assignments are only changed in exceptional cases. There is usually no separate project objectives plan. There is therefore no active controlling of the project objectives, and therefore also no controlling of the solution requirements.

The objectives of an organization regarding an investment are usually defined in a business case analysis or cost–benefit analysis (see Chapter 3). The defined costs and benefits accord with the business requirements. If costs and benefits to stakeholders are also considered in a cost–benefit analysis, this accords with the definition of stakeholder requirements. The results of a business case analysis or cost–benefit analysis form the basis for an investment decision but are generally not considered further in project management. Because of a lack of controlling of the project objectives (or the aggregated solution requirements) and the investment objectives (or the business requirements), there is little results orientation in project management.

Managing requirements is not seen as an integral part of projects. Following a sequential approach, requirements are understood as the basis for project management. They are defined before the project begins. During the performance of projects, requirements are considered, on the one hand, for processing change requests and, on the other hand, during the testing and the acceptance of the solution by external or internal customers. Changes of requirements are managed by means of change requests. This application of the concept of changing to a change of solution requirements corresponds to a very narrow change understanding. A change of solution requirements may require adaptation of the project scope, the project schedule and the project costs. These adaptations are performed during the execution of the change request process. Associated changes in the project organization, project stakeholder relations, relations with other projects, etc. are usually not explicitly considered.

The problems for managing requirements are different for external and internal projects. As a basis for the award of contracts which are to be performed in the form of projects, customers often describe the requirements to be fulfilled in detailed specifications. In the preparation of these specifications, however, often only the solution requirements to be fulfilled by the supplier are specified. Other solution components which are to be fulfilled by the customer are usually not considered. If these additional requirements are not considered, no holistic project view can be obtained. Interfaces are created that often are not managed, or not adequately managed, by the project. Suppliers are only responsible for their own scope of services. There is no holistic management, project complexity is not appropriately built up, and the project plans cover only parts of the actual project.

In internal projects, solution requirements are often only roughly defined before the start of the project, and it is expected that the requirements will be clarified during the project performance. As a result, realistic planning of the project scope, project schedule, project costs, etc. is not possible, because the project objectives are unclear at the start of the project.

Responsibility for fulfilling the requirements lies with the project owner. This is not adequate, as the project owner is usually positioned too high in the hierarchy of the project-oriented organization and therefore cannot assume operational tasks and responsibilities. On the other hand, the project owner does not generally cover the interests and competencies for securing all components of a solution. Iterative approaches, such as Scrum, define the product owner as the role for performing these tasks. In the case of external projects, such as construction or engineering projects, the customer often employs an architect or an "engineer" as a representative of his interests.

As this analysis shows, in practice there is a need for an adequate managing of requirements. This need can be met by integrating requirements management into projects and project management. An iterative approach supports this integration, because, in this case, managing requirements is combined with the development of content-related solutions.

4.2 Requirement Definition and Requirement Types

Requirement Definition

Different methods for managing requirements use different terms and techniques. For example, the terms "specification", "epic", "user story", and "feature" are all used to refer to requirements. The sometimes-necessary simultaneous use of different methods is thereby made more difficult. In the following, the similarities and differences of the methods are analyzed, and an integrative approach for managing requirements is presented.

In recent years important principles for managing requirements have been made available in the *Business Analysis Body of Knowledge® (BABOK® Guide)* of the International Institute of Business Analysis™ (IIBA®)[1] and in publications by the International Requirements Engineering Board (IREB) and the Project Management Institute (PMI).[2]

Common to the different sources is a basic requirement definition. A "requirement" is understood as a necessary quality of a solution or the ability of a solution to achieve an objective.

Requirement Types

It is possible to differentiate between business, stakeholder, solution and transition requirements (see Fig. 4.1).

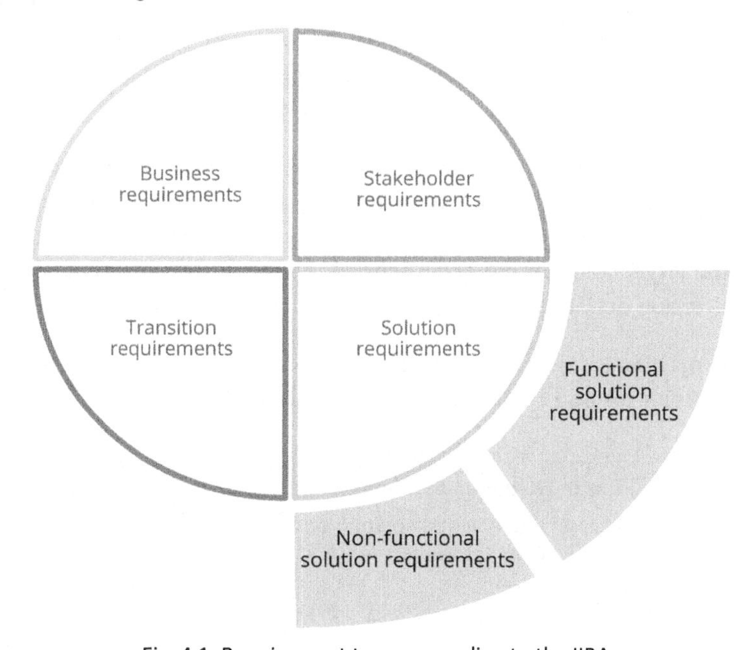

Fig. 4.1: Requirement types according to the IIBA.

Business requirements describe the benefits of a solution for an organization. They explain why an investment is to be made, what benefits for the organization will result

[1] See International Institute of Business Analysis, 2015

[2] See Project Management Institute, 2016

from the solution, and how the success of the solution will be measured. From the point of view of requirements management, the costs and benefits of an investment (see Chapter 3) therefore represent business requirements.

Stakeholder requirements describe the different requirements of different stakeholders regarding a solution. If costs and benefits to stakeholders are also described in a cost-benefit analysis, this approach accords with the description of stakeholder requirements. There is therefore a strong link between the methods of investment theory and the management of requirements.

Solution requirements can be distinguished into functional and non-functional requirements. The functional requirements describe the concrete functions of a solution, while the non-functional requirements describe the conditions under which a solution must function. Non-functional requirements are, for example, capacity, speed, security, and availability requirements.

Solution requirements describe functions and not the tasks for developing or implementing a solution. However, information required for planning the development or implementation appropriately is provided by the defined requirements. The solution requirements provide a basis for the definition of project objectives. Project objectives correspond to aggregated solution requirements. Defining project objectives without a description of solution requirements is not sufficient.

Transition requirements describe the features of a solution which enable a transition from an actual state to a target state, but which are no longer required after the transition. They are temporary and cannot be developed until the actual state has been analyzed and the target solution has been defined. Examples of transition requirements are data conversion or data migration as part of the development of a new IT solution.

In requirements management, constraints are also defined which restrict possible solutions. A constraint for an IT solution could be: "Do not use a public cloud solution".

Requirement Definition and Requirement Types

A requirement is a necessary quality of a solution or the ability of a solution to achieve an objective. A distinction can be made between business, stakeholder, solution, and transition requirements.

Requirement Types and Project Objectives or Investment Objectives

Requirements must not be equated with project objectives or investment objectives. Project objectives correspond to aggregated functional solution requirements. Investment objectives, such as the business case objectives or cost–benefit differences of an investment, accord with the business requirements. Stakeholder requirements can be incorporated into both the project objectives and the investment objectives (see Fig. 4.2).

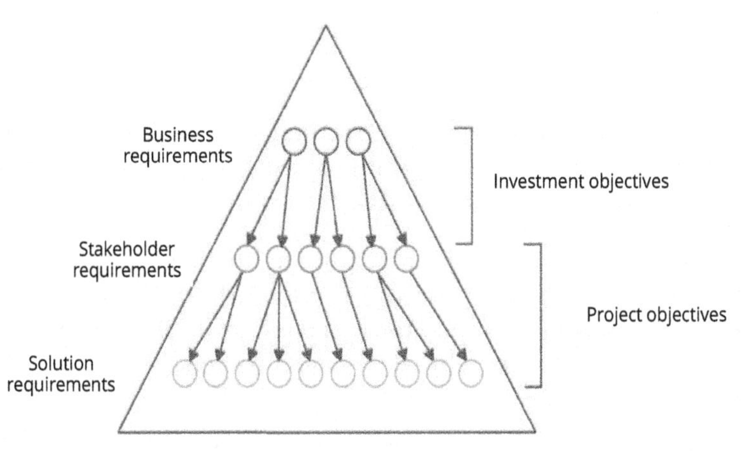

Fig. 4.2: Relation between requirements, project objectives, and investment objectives.

4.3 Managing Requirements Using a Sequential Approach

The management of requirements can be performed using either a sequential or an iterative approach. Figure 4.3 shows two sequential models—namely, the "waterfall model" and the V-model.[3]

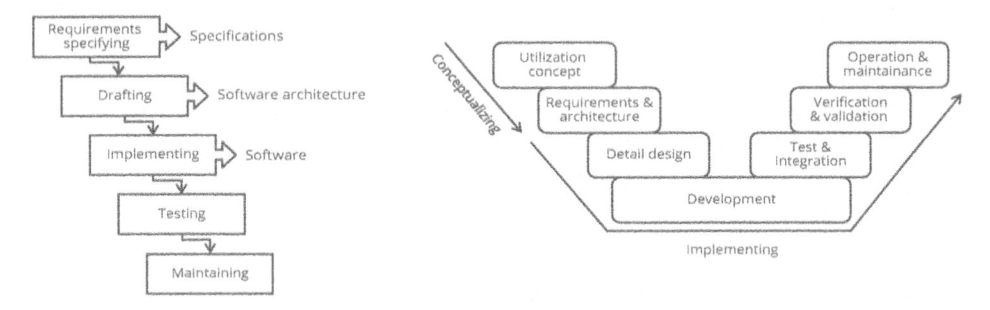

Fig. 4.3: Sequential models: Waterfall model and V-model.

The waterfall model is the best-known sequential model, and it is used primarily in IT and construction. The waterfall model separates the phases defining requirements, designing, implementing, verifying, and maintaining.[4] At each phase transition, it is assumed that the respective preceding phase has been completed. Sequential models are based on defining the requirements for a solution before the implementation of the solution. The definition of requirements is the basis for planning the implementation project. After requirements have been defined, there is a "design freeze". A

[3] These two models are life cycle models for a solution or an investment object, but they are not project models, because the phases presented do not fall within the boundaries of projects. They include, for example, maintenance for the solutions considered.

[4] The water flowing over the multiple cascades of a waterfall, which cannot flow back because of gravity, symbolizes the (intermediate) results achieved.

similar approach is used in the models "Big Design Up Front (BDUF)" and "Upfront Engineering".

In the phases of defining requirements and designing (see Fig. 4.3) documents such as a product specification and a software architecture, etc., in which requirements are described in an increasing degree of detail, are produced.[5]

If additional requirements are identified during the implementation of a solution, they must be handled by a change request process. In this process the project plans are also adapted. Deviations from the original project plans are usually seen as errors, which could have been avoided by better project planning.

The waterfall model, the V model, Big Design Up Front (BDUF) and Upfront Engineering are all based on precise defining of requirements before implementing a solution. The defined requirements form the basis for project planning.

4.4 Managing Requirements Using the Scrum Model

Scrum: Definition

Scrum is an "agile method"—that is, an iterative model. On the one hand, it supports an iterative management of requirements, and on the other hand, it supports the development of solutions for meeting the defined requirements. The use of Scrum should contribute to ensuring the quality of the solution developed and to make it possible to react quickly to market changes. A more detailed description of the objectives, roles, and processes of Scrum is provided in Chapter 5.3. Here the focus is on describing the managing of requirements when applying Scrum.

Scrum Terms: User Story, Epic, Product Backlog

Scrum differentiates between solution requirements of different granularity—namely, user stories, epics, and product backlogs.[6] A user story corresponds to a functional solution requirement.[7] An epic is a set of related user stories, and a product backlog includes all user stories. User stories serve as a basis for discussions of the solution requirements between the product owner (or product team) and the Scrum team. This discussion serves to concretize the solution requirements.[8]

[5] The resulting documents are described differently in different approaches to requirements management. The *BABOK® Guide* describes documentation for this purpose—for instance, a vision statement and a requirements document.

[6] Some authors use the term "feature" instead of the term "user story".

[7] The distinction between user story and solution requirement is not made clear in the literature (see Gloger, B., 2016, p. 129f). A user story is a need which is defined by discussions between the product team and the Scrum team and thus becomes a requirement. However, there is no statement about the concrete difference or any different form of documentation.

[8] Scrum roles, communication formats, and artefacts are defined in Chapter E.

Definitions: User story, epic, product backlog

A user story corresponds to a functional solution requirement. An epic is a set of related user stories. The product backlog includes all user stories.

User stories can be described in different formats. A common format is: "As a (persona) I can (do something) so that I (get some benefit)". An example of a user story for the "Values4Business Value" solution is: "As a project manager, by reading the book *PROJECT.PROGRAM.CHANGE,* I get information on agility in projects that I can use in my management practice".

User stories can be documented on Post-Its or story cards. For example, a story card can include a user story ID, a story name, a textual description from a stakeholder's perspective, the estimated workload, the iteration planned for processing the story, the benefit of the story to the stakeholder, and acceptance criteria. Similar cards can also be used to describe epics. To describe wide-ranging, complex user stories, methods such as process descriptions, mock-ups and prototypes can be used in addition to the textual description.

A user story is testable and user-oriented. It should therefore be largely independent of other user stories. Multiple user stories that are processed in one "sprint"—that is, in an iteration—should produce a Minimum Viable Product (MVP) or Minimum Marketable Feature (MMF). An MVP represents a clearly defined, deliverable solution component, which already secures a customer benefit.

User stories correspond to functional requirements. In the Scrum model the "product vision" defined by the product owner (or the product team) corresponds at least partially to business requirements.

Managing Requirements with Scrum: Process

In the Extended Scrum model (see Fig. 4.4, on next page), the management of requirements is combined with the development of the content-related solution.

The basis for requirements management with Scrum is the development of an initial product backlog by the product owner or the product team. This is done, for example, as part of a feasibility study or conception. It can be produced, for example, from use cases (see below), epics and user stories. In a prioritization meeting, these epics and user stories are prioritized.

The time and budget available for the development of the solution are also determined. It is checked which user stories can be implemented within this defined framework. For this purpose, the cost of implementing the individual user stories is estimated. Prioritization criteria are the business value and risk of each user story as well as dependencies between user stories.

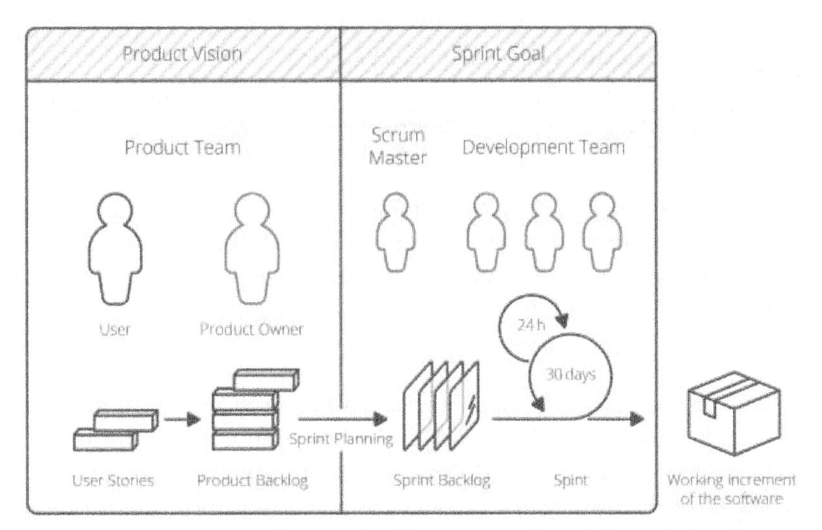

Fig. 4.4: Extended Scrum model (further development based on Highsmith, 2010).

In contrast to a planning-oriented approach, in which the objectives determine the scope, schedule, and costs, in this change-oriented approach the schedule and costs determine the objectives and scope. The detailed objectives and thus also the scope are variable, but the schedule and costs are fixed by the number of sprints with defined durations and resources.

The initial product backlog is created by means of the concretization and reduction of user stories achieved by prioritization. Though the solution is iteratively specified in Scrum, this does not mean that the results to be achieved are unclear at the beginning of the project. The initial product backlog is the basis for a definition of the desired objectives in the "product vision",[9] and for the development of project plans. For this purpose, the defined initial requirements must be available to a sufficient extent and with a sufficient level of detail.

The initial product backlog enables the planning of the necessary sprints and the selection of the first user stories to be handed over to the Scrum team. The user stories are concretized by the product team and the Scrum team in the process of sprint planning. This produces a sprint backlog. The tasks to be fulfilled for the concrete user stories are defined. The product backlog is processed by fulfilling these defined tasks.

The product team further develops the product backlog in parallel with the content-related work of the Scrum team. On the basis of the initial product backlog, additional user stories are identified during implementation by the product team. New information and needs are considered, and the defined epics and user stories are prioritized

[9] The product vision describes the basic idea of the solution and answers the questions "What?", "Why?", and "For whom?"

once again. This produces an updated product backlog. The repeated reorientation of the product backlog during the project implementation prevents the fulfillment of non-beneficial requirements. "While useful as a guide, excessive detail in the early stages of a project may be problematic and misleading in a dynamic environment".[10]

As an iterative model, Scrum defines functional solution requirements in parallel with the development of the content-related solution. This requires organizational integration of the person responsible for the solution ("the product owner") into the Scrum organization.

Because it is possible to define new user stories iteratively, Scrum does not use change requests. In Scrum, requirements are defined initially and further developed and concretized on an ongoing basis. Only requirements which can be realized within a reasonable period of time are considered. The definition of new requirements is seen as part of the Scrum process. In contrast to sequential models, recourse to previous activities is therefore permitted.

4.5 Strengths and Weaknesses of the Models for Managing Requirements

Managing Requirements: Strengths and Weaknesses of Sequential Models

Sequential models have strengths when the requirements for a solution are fundamentally clear and the technologies to be used are already known. They are also advantageous when contractual security is important and documentation and transparency are essential.

The "up-front" identification and description of requirements is seen as a weakness. The requirements for the solution are defined before the project begins. The project scope, project schedule and project costs planned on this basis are insufficient if there are extensive changes to the requirements. This leads to stress, a lack of efficiency and only partial fulfillment of the customer's needs. Another disadvantage is a low level of consideration of the changing project environment.[11] It is difficult to react to changes such as new legislation or new technologies.

Managing Requirements: Strengths and Weaknesses of the Scrum Model

The Scrum model has strengths when the requirements for a solution are partly unclear, the technologies to be used are new, and when there is a lot of time pressure. The advantages of using the Scrum model for requirements management are the prioritization of requirements based on business value, the proactive risk management by the inclusion of stakeholders, and the ongoing re-prioritization of requirements by

[10] See Serrador, P. et al., 2015
[11] See Hüsselmann, C., 2014

the product team. The agility of this model of requirements management results from the fact that the insights gained during the development of the solution are used to concretize and prioritize requirements.

Weaknesses of the Scrum model for requirements management include the high organizational complexity that results from including the product team or stakeholders in the project organization. The uncertainty about the concrete solution to be expected is also seen in practical terms as a weakness. The lack of familiarity of the project management community with the methodology and with many of its new terms is another disadvantage.

Selecting a Requirements Management Model

One challenge in projects is to find a suitable balance between the earliest possible definition of requirements and the securing of sufficient flexibility to take account of changing business conditions.

Choosing a suitable model for requirements management (sequential or iterative) is a strategic decision to be made as part of the project initiation (see Chapter 5). This decision is made in the context of the management paradigm which is being used and may also consider the tendering and award procedure, especially in the case of public administration projects.

4.6 Managing Requirements: Sub-Processes

Managing Requirements: Objectives

The objectives of managing requirements are the definition and prioritization of requirements for an intended solution, and the assurance that the defined requirements are fulfilled. The risk that the solution developed does not meet stakeholder requirements is to be minimized.

Sub-Processes: Initial Requirements Defining and Requirements Controlling

In managing requirements, one can differentiate between the sub-process "Initial requirements defining", which is performed as part of the business process "Developing a feasibility study" or "Conceptualizing an investment", and the sub-process "Requirements controlling", which is performed within the framework of implementing a solution.

Tasks for defining the initial requirements are above all:

> identifying the need (the problem or the opportunity),
> analyzing the actual state,

- defining the target state, for example, by means of (initial) stakeholder requirements,
- defining the (initial) business requirements,
- defining possible solutions,
- selecting a solution, and
- defining the (initial) solution requirements.

When using sequential models, the fulfillment of these tasks leads to detailed descriptions of the requirements. When using iterative models such as Scrum, the fulfillment of these tasks results in an "initial product backlog".

The tasks of the sub-process "Initial requirements defining" are shown in Table 3.5 (on page 47) in an excerpt from a responsibility chart for the conceptualizing of an investment. The tasks depicted are to be fulfilled regardless of which model is used. However, the level of detailing of the requirements documentation produced is significantly higher in sequential models than in the Scrum model.

The role names selected in the responsibility chart in Table 3.5 assume the use of the waterfall model. The tasks to be fulfilled by requirements managers when applying the waterfall model are presented in the context of the other tasks and make visible the need for cooperation between the various roles.

The main tasks of controlling requirements during the implementation of a solution are:

- controlling the business requirements and the stakeholder requirements,
- controlling the solution requirements,
- testing the solution components or the solution, and
- accepting the solution.

The tasks of controlling the requirements can be fulfilled either iteratively in accordance with the Scrum model, or in the form of change requests in accordance with sequential models. Iterative controlling also includes defining new requirements and prioritizing requirements.

Key challenges in requirements management are identifying the relevant requirements and establishing consensus among the stakeholders regarding the requirements. Systematically controlling the requirements by using appropriate methods and tools and communicating the requirements, is also challenging.

Managing Requirements: Methods

Methods for managing business requirements include stakeholder analysis, use cases, business case analysis, cost–benefit analysis, and methods for controlling benefits realization (see Chapter 3).

The most important methods for identifying and defining solution requirements are use cases, document analysis, interviews, observations, surveys, SWOT analysis, requirements workshops, prototyping, focus groups, and an interface analysis.[12] Prioritization of requirements can be achieved by using the MoSCoW (Must-Should-Could-Won't) method.

Texts and/or models can be used to identify requirements. A common model for identifying requirements is the use case.[13] Use cases are described in the format <Role><Activity><Object>, for example: "Project Managers select information in the book *PROJECT.PROGRAM.CHANGE*".

Use cases can be represented graphically. For this, Unified Modeling Language (UML) notation can be used. Examples of the representation of use cases in the UML can be found in Figures 4.5 and 4.6 for the RGC case study "Values4Business Value—Defining Initial Requirements". In the examples, a solution is represented as a box, roles as stick figures and use cases as ellipses. To provide as complete a representation as possible, use cases can be differentiated, for example by stakeholders or by features.

In sequential models, it is possible to create requirements documentation of different levels of detail based on use cases. When using Scrum, user stories, epics, and product backlogs are derived from use cases. These can be visualized on a Scrum board. Examples of an initial product backlog, an epic card, a user story card, and a Scrum board for the case study can be found in Tables 4.1 to 4.3 (page 65–80) and Figure 4.7 (page 80).

Managing Requirements: Organization

The tasks of managing requirements can either be performed only by a user who has an investment need, or the user can be supported by requirements managers. When using Scrum, requirements management is performed by the product owner and the Scrum team, whereby the product owner is a representative of the business area of the user.

According to IIBA and IREB, specific roles for requirements management are the business analyst and the requirements engineer. A business analyst cooperates with stakeholders to collect, analyze, communicate, and validate requirements. A requirements engineer receives and manages solution requirements, represents requirements using models, ensures the transparency of requirements, and creates test cases. Generally, both business analysts and requirements engineers can be considered and called requirements managers.

Requirements management can be seen as a service provided by requirements managers for internal or external customers (see Excursus: Managing Requirements as a Service).

[12] See International Institute of Business Analysis, 2015, p. 140 f.
[13] See International Institute of Business Analysis, 2015, p. 356.

Excursus: Managing Requirements as a Service

Managing requirements can be seen as a service. The following services can be differentiated:

- Contributing to optimizing an existing solution,
- Contributing to implementing a new solution,
- Contributing to developing a feasibility study or to conceptualizing an investment.

The advantage of seeing requirements management as a service lies in the distinction between possible services and in adequate communication of requirements management services to customers.

The distinction between the services and their categorization as services of small, medium, or large scope enables appropriate organizational design and adequate planning of personnel for performing the requirements management.

Case Study: Values4Business Value – Defining Initial Requirements

Defining the Needs of Stakeholders by Means of Use Cases

A breakdown of the use cases by features and by stakeholders for the solution "Values4Business Value" was performed. The breakdown by features includes, for example, use cases for RGC services, for the book, the marketing and the e-book. The breakdown by stakeholders, includes, for example, use cases for customers of RGC services, for various book readers (project managers, students, etc.), for RGC, and for the publishing companies involved. The different examples of use cases in Figures 4.5 and 4.6 focus particularly on the stakeholders "Reader: Project manager".

It can be seen from Figure 4.5 that the "Reader: Project manager" has a need to select information in the book on the basis of the table of contents and index, to gather information and to exchange it with other readers. User stories could be formulated on the basis of these use cases.

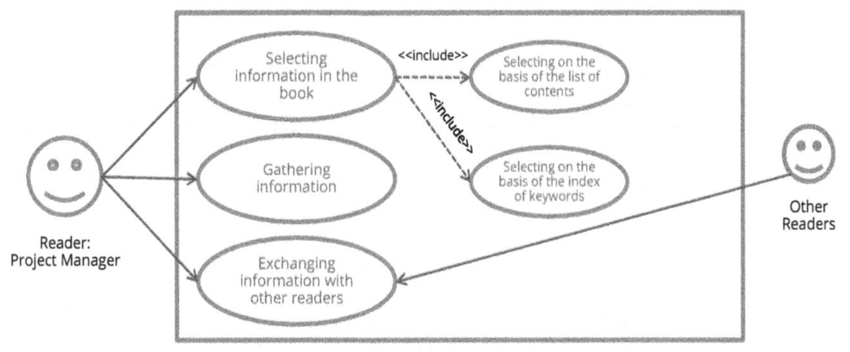

Fig. 4.5: Use cases for book readers.

The exchange of information between "Reader: Project manager" and "Other readers" seemed to be of fundamental interest. An epic for the development of a suitable IT platform for knowledge management was defined (see initial product backlog in Table 4.1), but because of its low priority, the project did not provide a solution component for this.

In Figure 4.6, use cases for different stakeholders are considered in relation to each other. For example, the relations between readers and booksellers and between readers and users (namely "projects") are presented. This increases complexity, but also makes additional needs visible.

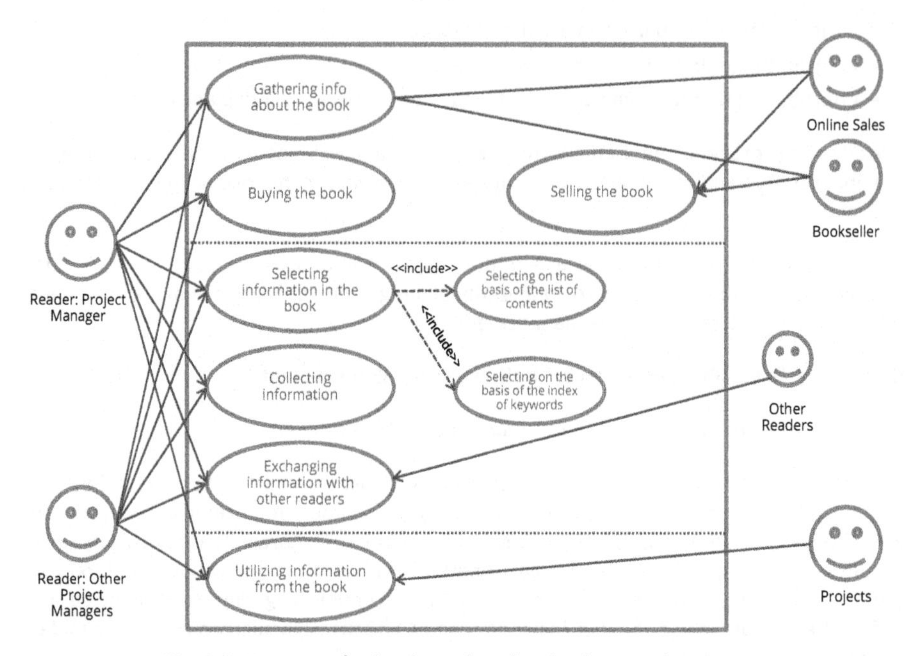

Fig. 4.6: Use cases for book readers, booksellers, and projects.

Defining Business Requirements

The use cases for the stakeholders RGC, the publishers, customers and readers of the book were the basis for the definition of the business requirements for the solution "Values4Business Value" during conceptualization.

In developing the cost–benefit analysis, the business requirements from the viewpoint of the investor RGC were described as benefits. Further impacts of the investment "Values4Business Value" were also taken into account in the cost–benefit analysis by considering costs and benefits for other stakeholders (see Chapter 3).

Defining Initial Solution Requirements: Initial Product Backlog

The initial product backlog for the solution "Values4Business Value" is shown in Table 4.1.

For the solution "Values4Business Value", backlog parts were defined for the different solution components. These backlog parts, namely prototyping, book, marketing, e-book, RGC services and products, as well as an IT platform for knowledge management, are shown in Table 4.1.

The definition of the initial solution requirements and the "initial product backlog" took place as part of the conceptualization of the investment "Values 4Business Value". Not all solution components were considered in the initial product backlog. The focus was on the

Table 4.1: Initial Product Backlog for the Solution "Values4Business Value"

Initial Product Backlog

Developing Values4Business Value

Backlog & EPIC Nr.	EPIC	User Story Nr.	User Story	Person days, resources	Prioritiy 1–3
1.00.00	Backlog: Prototypes				
1.01.00	Sustainable development				
		1.01.01	Prototype: Lecture at the forum "Sustainable development"	3 PD, 1 person	1
		1.01.02	Prototype: Seminar "Principles of a sustainable development and project management"	2 PD, 1 person	1
1.02.00	Agile approaches				
		1.02.01	Protoype: Seminar "Agility & Projects"	10 PD, 2 persons	1
		1.02.02	Prototype: Lecture " Agility & Processes"	3 PD, 1 person	1
1.03.00	Benefits realization management				
		1.03.01	Prototype: Lecture at HP 16 "Only benefit counts"	2 PD, 1 person	2
		1.03.02	Prototype: Developing a case study with RBI	4 PD, 2 persons	2
1.04.00	Business modelling				
		1.04.01	Prototype: Consulting digital transformation	4 PD, 1 person	3
		1.04.02	Prototype: Business model for "Mobility Point"	4 PD, 1 person	3
2.00.00	Backlog: Book				
2.01.00	Integrative information				
		2.01.01	Integrative information about values and management approaches	6 PD, 1 person	1
		2.01.02	Integrative information about sustainable development	4 PD, 1 person	1
		2.01.03	Integrative information about agility in projects	10 PD, 2 persons	1
		2.01.04	Integrative information about benefits realization management	6 PD, 1 person	1

(continues on next page)

Table 4.1: Initial Product Backlog for the Solution "Values4Business Value" ***(cont.)***

Initial Product Backlog

Developing Values4Business Value

2.03.00	Book chapters				
		2.03.01	Chapter on projects and programs	10 PD, 2 persons	1
		2.03.02	Chapter on project management approaches and new values	10 PD, 2 persons	1
		2.03.02	Chapter on strategic managing and investing	15 PD, 2 persons	1
		2.03.03	Chapter on requirements management and projects	15 PD, 2 persons	1
		...	...		
		2.03.17	Chapter on business modelling	20 PD, 2 persons	3
2.25.00	Additional book contents (directories, figures, case studies, etc.)				
		...	...	...	...
3.00.00	Backlog: Marketing				
4.00.00	Backlog: E-Book				
5.00.00	Backlog: Services and products				
6.00.00	Backlog: IT-Platform - knowledge management				
V. 1.001 v. S. Füreder per 5.10.2015					

identification of the basic solution components and the definition of user stories for the "prototyping" and "book" backlog parts, which have been worked on first. The other backlog parts were only defined later in the various phases of the implementation project (see the work breakdown structure of the project "Developing Values4Business Value" in Chapter 7).

Defining Initial Solution Requirements: Epic Card and User Story Card

The epics for the initial backlog parts are shown in Table 4.1. As an example of the description of an epic, the epic card "Integrative information" is shown in Table 4.2.

Table 4.2: Epic Card "Integrative Information" for the Solution "Values4Business Value"

Epic Card

Developing Values4Business Value

V. 1.001 v. S. Füreder per 5.10.2015

Epic ID:	2.01.00	Epic name:	Integrative information
Description:		Gathering information on management innovations, such as values in management, agility, sustainable development, requirements management, which are relevenat for several chapters	
Effort of work and resource allocation:		30 person days, 2 persons	
Plannend iteration:		Iteration 1 and 2	
Benefits for developer/author:		Basis for the further development of the RGC management approaches, for the development of chapters	
Acceptance criteria:			
Values of a systemic management paradigm described as a basis for the RGC management approaches			
Information (texts, examples, references, etc.) on agility, sustainable development, requirements management gathered			
Information on management innovations assigned to certain book chapters			

The epic "Integrative information" included the user stories "Integrative information about values in management", "Integrative information about agility", "Integrative information about sustainable development", etc. For each user story, the tasks "developing a draft", "additional analyzing", "completing", and "assigning to chapters" were to be fulfilled. The user story "Integrative information about agility" was chosen as an example of a user story for the epic "Integrative information" (see Table 4.3).

Project managers are interested in the topic of "agility". There is little clarity about this, because the community of software developers has relatively little project management background, and the project management community has little background with agile methods. A clarification of the relations between these topics in the chapters of the book should therefore be useful for all stakeholders, but above all for project managers.

Communicating Solution Requirements: Scrum Board

A photo of the Scrum board for the epic "Integrative information" is shown in Figure 4.7. The possible states of tasks for each user story was defined as "to do", "in progress" or "done". A "definition of done" was produced for each user story in order to establish a consensus on the results to be achieved.

Table 4.3: User Story Card "Integrative Information About Agility" (Requirements of Developers/Authors)

User Story Card

Developing Values4Business Value

V. 1.001 v. S. Füreder per 5.10.2015

User Story ID: ⍰	2.01.03	User Story Name:	Integrative information about agility
Description:		Information on agility in projects, programs and changes, which are relevant for several chapters, and their assignment to certain chapters	
Effort of work and resource allocation:		10 person days, 2 persons	
Planned iteration:		Iteration 1	
Benefits for developer/author:		Basis for the development of chapters relating to agility in projects, programs and changes	
Acceptance criteria:			
Information on agility in projects, programs and changes gathered			
Information on agility in projects, programs and changes assigned to certain chapters			

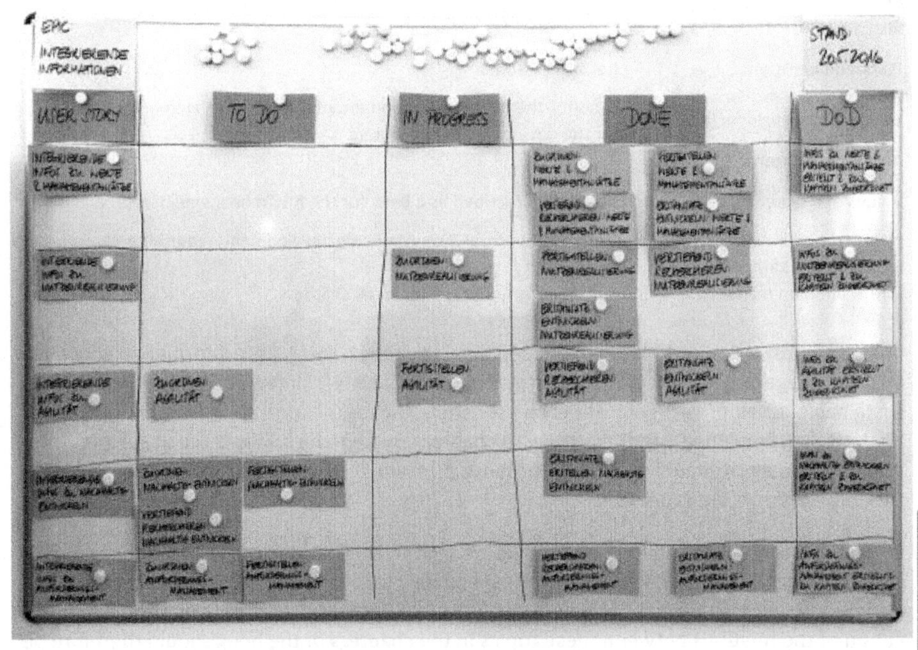

Fig. 4.7: Scrum board.

Literature

Gloger, B.: *Scrum Produkte zuverlässig und schnell entwickeln,* 5th edition, Carl Hanser, München, 2016.

Hüsselmann, C.: Agilität im Auftraggeber-Auftragnehmer-Spannungsfeld: Mit hybridem Projektansatz zur Win-Win-Situation, *Projekt Management aktuell,* 25(1), 38–42, 2014.

International Institute of Business Analysis (IIBA): *A Guide to the Business Analysis Body of Knowledge® (BABOK® Guide)* 3.0, IIBA, Toronto, 2015.

Project Management Institute (PMI): *PMI Professional in Business Analysis (PMI-PBA®) Handbook,* PMI, Newton Square, PA, 2016.

Serrador, P., Pinto, J. K.: Does Agile Work? A Quantitative Analysis of Agile Project Success, *International Journal of Project Management,* 33(5), 1040–1051, 2015.

5 Project Initiating

The objectives of project initiating are the development of initial project plans and the selection of an appropriate organization to carry out a relatively unique and extensive business process. Appropriate temporary organizations could be a small project, a project, or a program. The decision to implement an investment object is a prerequisite for this organizational decision. In practice, the differentiation between the investment decision and the organizational decision is often not made, which can lead to a lack of transparency and incorrect decisions.

As part of project initiating, project boundaries are to be defined, initial project plans are to be developed, and decisions about project strategies are to be made. The use of agile methods in a project is a possible project strategy. In practice, the most commonly used agile method is "Scrum"—it is described in an excursus (page 91).

Project initiating is always part of a higher-level business process such as conceptualizing an investment or making a tender offer. The application of methods for project initiating is illustrated by the example of the RGC case study "Values4Business Value" (see Section 5.4 starting on page 93).

Relations between the business process "Project initiating" and other business processes discussed in this publication are shown in the following overview.

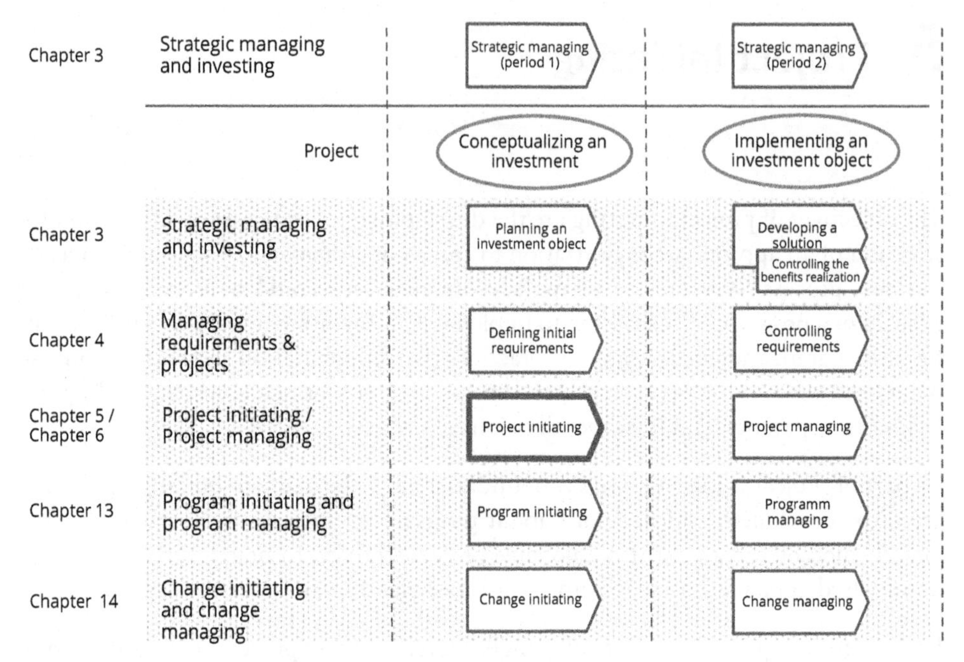

Overview "Project initiating" in context.

5 Project Initiating

5.1 Project Initiating: Objectives

5.2 Project Initiating: Process

5.3 Applying Agile Methods: A Possible Project Strategy

5.4 Project Initiating: Methods

5.5 Project Initiating: Organization

5.1 Project Initiating: Objectives

Project initiating is a business process of the project-oriented organization. The objectives of project initiating are:

> a basis for project starting established,
> an appropriate organization for performing a comprehensive business process selected,
> project assigned, and
> selected stakeholders in the initiating process involved.

A non-objective of project initiating is the development of detailed project plans. These are developed by the project team in the project starting sub-process.

The start event for project initiating is the decision of the initiator of an investment to initiate a project. The end event is the issued (or not issued) project assignment.

Definition: Project Initiating

Project initiating is a business process of the project-oriented organization with the objective of establishing the basis for project starting, selecting the appropriate organization for performing a comprehensive business process, and possibly issuing a project assignment.

5.2 Project Initiating: Process

The process of "Project initiating" is shown in a flow diagram in Figure 5.1 and in a responsibility chart in Table 5.1. The responsibility charts show the responsibilities for performing the process tasks.

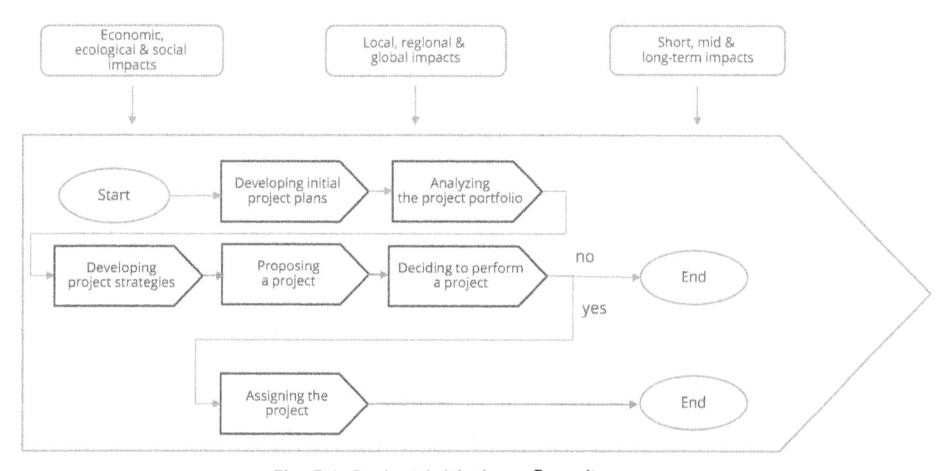

Fig. 5.1: Project initiating—flow diagram.

Table 5.1: Project Initiating—Responsibility Chart

Sub-process: Project initiating

Tool/Document:
1 ... Initial project plans
2 ... Project portfolio data base
3 ... Documentation of the project strategies
4 ... Project proposal
5 ... Minutes of meeting
6 ... Project assignment

Legend

P ... Performing
C ... Contributing
I ... Being informed
Co ... Coordinating

	Roles								
Process tasks	Initiator	Initiation team	PM Office	Project Portfolio Group	Expert pool managers	Project owner	Project manager	Stakeholders	Tool/Document
Developing initial project plans		P		C	C				1
Analyzing the project portfolio		C	D						2
Developing project strategies	I	P	C						3
Proposing a project	I	P	C						4
Deciding to perform a project	I	C	C	P	I			I	5
Assigning the project			I	I	I	P	C		6

Project initiating involves the following tasks:

> developing initial project plans based on a definition of appropriate project boundaries,
> analyzing the project portfolio,
> developing project strategies,
> proposing a project,
> deciding to perform the project, and
> assigning the project.

Developing Initial Project Plans

The basis for developing initial project plans is the definition of the project boundaries. Project boundaries are adequately defined if the project objectives and the project scope make it possible to fulfill the requirements of the investor and key stakeholders. A "holistic" project definition considers all closely linked objectives and avoids sub-optimizations by partial solutions.

The boundaries of a project as a social system are defined in terms of time, content, and social relations. Timing requires the definition of a start event and an end event of the project. In order to define the project content, the structural and contextual dimensions of organizations, as discussed in Chapter 3, have to be considered. It is necessary to define whether project objectives are being pursued in relation to services and products, markets, organizational structures, personnel, infrastructure, or, for example, the stakeholder relations of an organization. Social definition is performed by defining the roles responsible for the performance of a project.

Defining boundaries defines what belongs to a project. Clarity about these boundaries is also achieved by exclusion of what does not belong to the project, which is project context.[1] This includes tasks and decisions which are performed in other projects or in the line organization, as well as tasks and decisions of the pre-project and post-project phases. In these, like the project stakeholders, the projects implemented at the same time and the objectives or strategies of the organization represent project contexts.

Project boundaries cannot be defined correctly or incorrectly, but are the result of agreements between the project initiator and the initiation team. They are social constructs. Table 5.2 provides tips for constructing holistic project boundaries.

Table 5.2: Tips for Constructing Holistic Project Boundaries

Tips for defining holistic project boundaries
• Project boundaries must be established so as to create a "self-contained whole".
• The project owner-project manager relation should remain the same for the duration of a project.
• Project boundaries should enable to agree on operational project objectives.
• The definition of project boundaries makes it possible to plan a project. Only what can be given boundaries can be planned!
• Projects in project chains, such as a tendering project and a contracting project or a conceptual project and an implementation project must be distinguished from one another.

Results of developing initial project plans are drafts of the project objectives plan, the work breakdown structure, the project schedule, the project cost plan, the project stakeholder analysis, and the description of the project organization. If necessary, a draft of a project risk analysis may also be developed.

"Initial" means, for example, that the work breakdown structure is broken down only to the third level, that a project milestone plan may be sufficient as a scheduling method, and that key project stakeholders are only identified, without stakeholder

[1] Exclusions can be "included" in the project or program at a later date.

strategies yet being planned. The local, regional, and possibly global ecological and social impacts of a project can be considered in the initial project plans. A non-objective of project initiating is the development of detailed project plans. These are developed by the project team during project starting.

Analyzing the Project Portfolio

The decision of whether or not to initiate a project also depends on the relations of the project under consideration to other projects. These relations, which can involve either synergy or conflict, must be analyzed. The consequences of "including" a new project in the project portfolio of the project-oriented organization must be considered in decision making.

Developing Project Strategies

"Project strategies" are strategies for the realization of the project objectives which relate to the project as a whole and which are relevant for the duration of the project. The following types of project strategy can be distinguished:

- Procurement strategies: Examples are the use of a main contractor or of individual companies; domestic or international procurement; different tender procedures.
- Strategies for the use of technology: An example is using or not using a cloud solution in an IT project.
- Strategies for the use of methods: Examples are the requirements management approach to be used, the project or program management approach to be used, using agile methods or not.
- Project personnel strategies: Examples are using a project-specific incentive system or not; performing project-related personnel development or not.
- Strategies for the planning of stakeholder relations: Examples are different contract types with customers and suppliers; different types of relations with authorities; continuous communication of project objectives and interim results, or only communication of the project results at the project end.

Proposing a Project Proposal and Deciding to Perform the Project

The project proposal is the basis for deciding which organizational form should be used to perform a business process. An example of a project proposal for the case study "Values4Business Value" is given in Table 5.5 (on page 95) in Section 5.4.

The following possible organizational forms are available for performing a business process: the line organization, a working group, a small project, a project, a network of projects, a program, or several programs. In project initiating, it is assumed that a project is the appropriate organizational form. However, this has to be challenged in the course of project initiating, and it is necessary to decide between assigning a small project or a project (see Chapter 1).

Assigning the Project

The project owner assumes an important management task in the project. The selection of an appropriate project manager is one of his decisions. As part of project initiating, the nominated project owner assigns the project to the project manager and the project team.

The main results of project initiating are the initial project plans, the documented project strategies, the project proposal, and the possible project assignment. The decisions taken must be documented in minutes.[2]

5.3 Applying Agile Methods: A Possible Project Strategy

The use of agile methods in projects is gaining in importance. The decision to use agile methods is a project strategy defined during project initiating. In the following, criteria and prerequisites for the use of Scrum in projects are analyzed. Scrum objectives, processes, roles, and communication formats are described in the excursus on Scrum (on page 91).

Agile Methods

"Agile methods" are all methods which refer to the Agile Manifesto (see Chapter 2). Different agile methods exist, such as adaptive software development, extreme programming, feature-driven development, usability-driven development, Scrum, Kanban, design thinking, and rapid prototyping. Scrum[3] is the best-known agile method for which well-defined descriptions are available.[4] The study "Status Quo Agile" by Ayelt Komus shows that Scrum and Kanban are the most frequently used agile methods.[5] The application of these methods is for 90% in software development.

Use of Scrum vs. Applying Agile Values in Projects

The use of Scrum as an agile method must be distinguished from applying agile values in projects. Agile values can be applied in any project. In order to be able to use Scrum in a project, the project must be suitable, and an adequate understanding of Scrum must be available in the organization performing the project.

In principle, Scrum can be used not only for software development but also for product and organizational development, research, marketing, etc. The use of Scrum and thus the execution of sprints requires repeatability of tasks (in a project phase). Repetitions enable improvements in processes as a result of reflections. This repeatability

[2] The tasks of project management are described in Chapter 7 "project management", the relevant methods in Chapters 9 to 12.

[3] Scrum is a copyright for the Iterative Software Development Framework of Scrum Alliance.

[4] See Gloger, B., 2016.

[5] See Komus, A., 2012.

and the relative uniformity of tasks makes it possible to work in the desired dedicated Scrum teams with a maximum of five to seven people.

Scrum is a form of micro-management, as tasks are planned, executed, and controlled in detail on a daily or even hourly basis. Project management can be seen in relation to this as macro-management, since planning and controlling are less detailed.

Scrum is a form of micro-management, since tasks are planned, executed, and controlled in detail on a daily or even hourly basis. Project management can be understood in relation to this as macro-management, since planning and controlling are less detailed.

Scrum can be applied either by permanent teams in the company's permanent organization or by temporary teams in projects. Permanent teams represent *de facto* an organization with different (development) departments. If Scrum is applied by permanent teams, these teams can fulfill either ongoing tasks which are not suitable for projects or parts of projects. The tasks to be fulfilled as parts of projects represent epics for the team. In this case, the epics to be fulfilled must be prioritized by a decision-making committee in which the corresponding project managers are represented.

Scrum can be used for one or many project phases in a project. Temporary Scrum teams form project sub-teams (see Chapter 7).

A holistic project understanding excludes the management of a project with Scrum only. A project always includes tasks that are unique and not repetitive (e.g., procurement, commissioning, training, etc.) Scrum can therefore not be seen as a substitute for project management, but as a supplement.

In addition to Scrum, there exist other methods in project management, such as project stakeholder analysis and project risk management, for dealing with the complexity and dynamics of projects, and there are other roles and communication formats which can be used. This is also recognized by representatives of agile methods. One speaks of hybrid approaches in projects.

Scrum and Hybrid Approaches in Projects

The term "hybrid" can be interpreted in various ways when using Scrum in projects: Firstly, "hybrid" can mean that Scrum is used for some projects or project types and not for others. For example, a German pharmaceutical group uses Scrum for conceptual projects, but not for certified processes of development projects. "Hybrid" can also mean that Scrum is used in some business areas, but not in others. The term "hybrid" is most often used for the use of Scrum in a project in combination with the waterfall model.

A combination of the waterfall model (see Chapter 4) with Scrum is not an appropriate form for managing a project, because in this case two different management paradigms are combined in one project. This combination corresponds to an "add-on" model without additional benefits. It probably even incurs additional costs. An appropriate form of the use of Scrum in a project should be combined with a systemic project management approach in which agile values are applied (see Chapter 2). When Scrum is used in a project, it is necessary to integrate it appropriately into the project structures to secure the added value of Scrum.

A combination of the waterfall model and Scrum is not an appropriate form for managing a project, because in this case two different management paradigms are combined in one project. The use of Scrum in a project should be combined with a systemic project management approach in which agile values are applied.

Excursus: Scrum

Scrum: Name

Takeuchi and Nonaka have characterized the connection between Scrum and rugby by means of a picture of a team pressing forward with rapid passing of the ball.[6]

Scrum: Objectives

The following objectives are to be realized by the use of Scrum:

> a strong solution orientation by clear definition of results assured,
> intermediate results defined as Minimum Viable Products (MVP),
> a lean solution assured by prioritization,
> rapid implementation of requirements in sprints performed,
> rapid response to changes performed, thereby promoting innovation,
> competitive advantages created through speed and flexibility,
> learning organized by reflection after each sprint,
> employees motivated through the possibility of self-organization.

"Fail early, release frequently!" and "Stop starting, start finishing!" are rules that promote the realization of these Scrum objectives.

Scrum: Processes, Communication Formats, and Roles

On the one hand, Scrum supports an iterative management of requirements (see Chapter 4) and, on the other hand, the development of solutions for fulfilling the defined requirements. Therefore, two parallel processes are running in Scrum (see Fig. 4.4 on page 70): on the one hand, the process of identifying and defining requirements in the form of epics and user stories, which are combined in backlogs; and on the other hand, the process of planning and implementing sprints for the development of solutions for fulfilling requirements.

[6] See Takeuchi, H., Nonaka, I., 1986.

The process for each sprint in Scrum is shown in Figure 5.2. Each sprint starts with sprint planning. In the sprint planning meeting, the requirements to be selected from the product backlog are clarified and the objectives of the sprint are defined. The requirements selected are summarized in the sprint backlog, and the approach for realizing the sprint objectives is planned. Then the development team starts work. A product increment as Minimum Viable Product (MVP) should result at the end of the sprint.

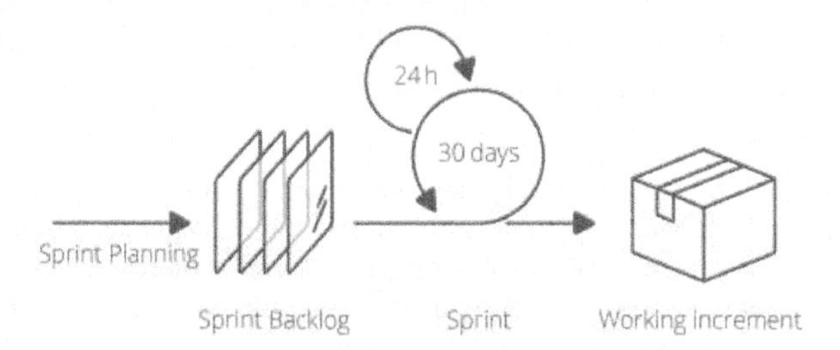

Fig. 5.2: Process for each sprint in Scrum.

Communication is important for performing the Scrum processes. The following communication formats are distinguished in Scrum: Sprint planning, daily scrum, sprint review, and sprint retrospective. The participants of the sprint planning meeting are the product owner, the development team, possibly the user, and the Scrum master.

The objective of the sprint planning meeting is to clarify the requirements, the objectives, and the tasks of the sprint. The participants of daily Scrum meetings are the development team and the Scrum master. The daily Scrum meetings take place daily and are short. Progress is announced, the tasks to be carried out on this day are planned, and any current problems are identified.

The participants of the sprint review meeting are the development team, the product owner, the Scrum master, and the user. The objective is to present the results achieved and to obtain feedback from the user. The development team, the product owner, and the Scrum master participate in the sprint retrospective. The objective is to reflect and to identify possible improvements in the work process.

Visualizations, such as Scrum boards or burn-down charts, are important for supporting communications. The tasks to be fulfilled can be seen on a Scrum board. The tasks are therefore not an amorphous concept, but are given a shape and a flow by visualization. Decisions such as prioritizations are thereby facilitated.

The product owner (or the product owner team), a Scrum master, and a development team are roles for the execution of Scrum processes. Stakeholders such as users or external customers can also be included in Scrum processes.

The product owner fulfills the tasks of analyzing requirements, prioritizing requirements and accepting the MVP. He or she is responsible for the business value of the created solution. The multi-disciplinary development team develops the solution. It cooperates closely with the product owner and possibly with users. It is responsible for meeting the schedule and for delivering good quality. The Scrum master directs the development team and tries to solve possible problems in order to assure the productivity of the process.

5.4 Project Initiating: Methods

Methods for project initiating are basically all project management methods which are also used in project starting (see Table 6.2, page 106). However, as described above, in project initiating, these methods are used to develop initial project plans. Other methods for project initiating are analyzing the project portfolio and developing a project proposal.

The project proposal form and the project assignment form must be designed in the same way. The project proposal shows the structures and contexts of the proposed project. The initial project plans are appended to the project proposal. This information enables the project portfolio group to make a decision about whether to perform the business process as a small project or a project. The initial project objectives plan, the initial project strategies plan, and the project proposal of the RGC case study "Values4Business Value" are presented in Tables 5.3 to 5.5 as examples of the use of methods in project initiating (starting on next page).

Canvases are an alternative form for representing project proposals and assignments. Canvases can be used to present a project idea, proposal or assignment, but also the project status.[7] With the help of a canvas, as popularized by Osterwalder/Pigneur, a complex situation is divided into components.[8] Their contents are recorded in the relevant fields. The common framework and the visualization used provide additional information benefits. An example of a project canvas is shown in Figure 5.3. The fields used in a project canvas depend on the underlying project management approach. Again, there is a need for clear terms and methods. The objectives of the use of project canvases are similar to the objectives of agile concepts—namely, visualization, interdisciplinary teamwork in the development process, involvement of stakeholders, etc.

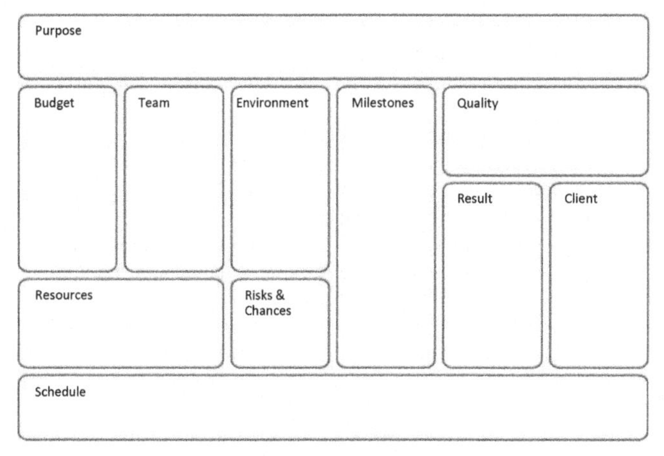

Fig. 5.3: Project canvas.[9]

[7] See Habermann, C., 2016.

[8] See Habermann, C., 2016, p. 36.

[9] See Habermann, F., Schmidt, K., 2016.

Case Study: Values4Business Value—Project Initiating

Table 5.3: Initial project objectives plan for the project "Developing Values4Business Value"

Initial Project Objectives Plan Developing Values4Business Value
Main objectives
RGC management approaches (process management, project management, program management, management of the project-oriented organization, change management) further developed by innovations
RGC management approaches documented in a book (hardcover and e-book)
Services (consulting, training, events, lectures) further developed
Optimized services and book marketed
Non-objectives
English version of the book developed

V. 1.001 v. S. Füreder per 5.10.2015

Table 5.4: Initial Project Strategies Plan for the Project "Developing Values4Business Value"

Initial Project Strategies Plan Developing Values4Business Value
Project strategies regarding procurement
Peer Review Group: Cooperation with management experts to ensure the practical orientation of the new RGC management approaches
Project strategies regarding the use of technology
E-book: Designing the book in a way that enables the realization of an e-book
Project management documentation: Using the RGC tool sPROJECT
Project strategies regarding the use of methods
Project managing: Iterative approach in the project "Developing Values4Business Values"
Researching: Literature analysis, case studies, interviews, reflection workshops, ...
Visualizing: Using a Scrum-board, project plans and change plans
Project strategies regarding the project personnel
Project personnel development: Organize the learning of the project personnel by working on case studies and reflecting, etc.
Project strategies for project stakeholders' relations
Key accounts: Involvement of the Peer Review Group members, information about the benefits of the management innovations
RGC consultants and networking partners: involvement in the development process

V. 1.001 v. S. Füreder per 5.10.2015

Table 5.5: Project Proposal for the Project "Developing Values4Business Value"

Project Proposal
Developing Values4Business Value

Type of project:	Small project X Project		
Project start date:	10/5/2015	Project end date:	5/31/2017
Project objectives:		Project non-objectives:	
RGC management approaches further developed by innovations		English version of the book developed	
RGC management approaches documented in a book (hardcover and e-book)			
Services (consulting, training, events, lectures) further developed			
Optimized services and book marketed			
Project phases:		Project costs:	Project earnings:
1.1 Project managing		381.000.-	30.000.-
1.2 Prototyping and planning			
1.3 Developing Chapters: Cycle 1			
1.4 Developing Chapters: Cycle 2			
1.5 Further developing services and products			
1.6 Planning the marketing activities, further developments			
1.7 Producing the hardcover and e-book			
1.8 Initial marketing			
Project owner:	R. Gareis, L. Gareis	Project manager:	S. Füreder
Project team members:			
PTM Analysis: R. Gareis		PTM Production: Representative Manz Verlag	
PTM Development: L. Gareis		PTM E-book: Representative Manz Verlag	
PTM Marketing: V. Riedling			
Decisions and documents of the pre-project phase:			
Cost-benefit analysis, concept „Values4Business Value", initial project plans			
Expectations of the post-project phase:			
Costs and benefits realized according to the cost-benefit analysis, "Values4Business Value" stabilized			
Relationships to other projects:			
HAPPYPROJECTS 16, Symposium "Project Audit 2016", Symposium "Project Audit 2017", HAPPYPROJECTS 17, sPROJECT-adaptation, RGC digitalization, customer projects			
Project stakeholder:			
Management, consultants and trainer, office staff, publisher Manz, media, readers			
Annex:	Initial project plans		

Project owner team — Project manager

V. 1.001 v. S. Füreder per 5.10.2015

5.5 Project Initiating: Organization

The development of initial project plans and a project proposal is performed by an initiation team in which the designated project manager and future project team members may be represented.

Project initiating provides a basis for project portfolio management. The decision to start a project is not an isolated one but is made in the context of an organization's project portfolio, which is changed by adding a new project. Thus project-oriented organizations use an investment decision-making committee to make the investment decision, and assign to a project portfolio group the task of making an appropriate organizational decision for implementing an investment. Departmental managers or managers of expert pools are to be involved in clarifying the availability of the required project resources (see Chapter 15).

The basis for issuing a project assignment is the decision to implement an investment by a small project or project. The project assignment should be issued in writing. The issuing of the project assignment is the formal start event for a project.

In the project starting process, detailed planning and the inclusion of additional experts as team members can lead to an adaptation of the project assignment. This cyclical formulation of the project assignment increases the quality of the agreed-upon structures.

In assessing the quality of the process "project initiating", the design of the business process—that is, its duration, costs, and organization—must be considered in addition to the results of the process. By defining project initiating as a business process, the objectives, process, and responsibilities are formalized. The clear differentiation between the investment decision and the organizational decision helps to ensure decision-making quality and reduces the risk of bad investments and of inadequate organizational forms for implementing investments. In practice, this process often takes a long time, and the responsibilities and decision-making competencies are not made clear.

Literature

Gloger, B.: *Scrum. Produkte zuverlässig und schnell entwickeln,* 5th edition, Carl Hanser, München, 2016.

Habermann, F.: Der Project Canvas – Projekte interdisziplinär definieren, *Projekt Management aktuell,* 1, 36–42, 2016.

Habermann, F., Schmidt, K.: *The Project Canvas. A Visual Tool to Jointly Understand, Design, and Initiate Projects, and Have More Fun at Work,* Gumroad E-Book, Berlin, 2014.

Komus, A.: *Studie: Status Quo Agile. Verbreitung und Nutzen agiler Methoden – Ergebnisbericht (Langfassung),* Hochschule Koblenz, Koblenz, 2012.

Takeuchi, H., Nonaka, I.: The New New Product Development Game, *Harvard Business Review,* 64 (1-2), 1986.

6 Business Process: Project Managing

"Project managing" is a business process of project-oriented organizations that should contribute to the successful performance of projects. The business process includes the sub-processes project starting, project coordinating, project controlling, project transforming or repositioning, and project closing. Project marketing is a task to be performed in all sub-processes of project management. The business process "Project administrating" is to be carried out in parallel with project managing.

From the values underlying a systemic project management approach, the objectives to be achieved, the tasks to be performed, the methods to be used, and the communication formats and project roles are derived. The project management approach can be adapted according to specific project demands and project types. The benefits of project management are in appropriate handling of the complexity and dynamics of projects.

The relations between the business process "Project managing" and other relevant business processes are shown in the following overview.

Chapter 3	Strategic managing and investing	Strategic managing (period 1)	Strategic managing (period 2)
	Project	Conceptualizing an investment	Implementing an investment object
Chapter 3	Strategic managing and investing	Planning an investment object	Developing a solution Controlling the benefits realization
Chapter 4	Managing requirements & projects	Defining initial requirements	Controlling requirements
Chapter 5 / Chapter 6	Project initiating / Project managing	Project initiating	Project managing
Chapter 13	Program initiating and program managing	Program initiating	Programm managing
Chapter 14	Change initiating and change managing	Change initiating	Change managing

Overview: "Project managing" in context

6 Business Process: Project Managing

6.1 Project Managing: Objectives

"Project managing" is a business process of project-oriented organizations that should contribute to the successful performance of projects. It includes the sub-processes project starting, project coordinating, project controlling, (possibly) project transforming or repositioning, and project closing.

The objectives of project managing are dependent on the project management approach used. The objectives described below assume a systemic project management approach and a process orientation of project managing (see also the Excursus "Method Orientation vs. Process Orientation in Project Managing").

Excursus: Method Orientation vs. Process Orientation in Project Managing

A mechanistic project management approach (see Chapter 2) is method oriented. Project management is understood as the application of methods for planning and controlling project scope, project schedule, project resources, project costs, etc. The quality of project management is assessed on the basis of the existence and quality of the project plans. It is assumed that a good knowledge of methods ensures good project management.

A systemic project management approach is process oriented. Project managing is understood as a business process, which includes the sub-processes project starting, project coordinating, project controlling, (possibly) project transforming or repositioning, and project closing. These sub-processes are related.

Appropriate project management methods are used to perform the sub-processes of project managing. They therefore do not lose their importance in a process-oriented approach. Rather, there is an integrative application of project management methods. It cannot be an objective of project management to produce a good schedule. Rather, the objective is, for example, to start a project optimally. All methods used in the starting process, including scheduling, must be considered in an integrative manner in order to achieve an optimized overall solution.

Individual methods, such as project scheduling or project cost planning, cannot be considered as project management processes.[1] If the application of individual project management methods is perceived as processes, the objective of setting process boundaries in a holistic manner is not achieved. The process "project scheduling" is not to be assessed on the basis of the process results achieved, process duration, and process costs. Rather, the project starting process as a whole must be considered and optimized.

The basic objective of the business process "Project managing" is to contribute to the successful performance of a project by means of professional management. It is possible to differentiate between economic, ecological, and social project management objectives. The objectives of project management are shown in Table 6.1.

[1] See Habermann, C., 2016, p. 47.

Table 6.1: Objectives of the Process "Project Managing"

Objectives of the process: Project managing
Economic objectives > Project complexity, project dynamics and relations to project contexts managed > Project starting, project coordinating, project controlling and project closing professionally performed; possibly also project transforming and repositioning > Economic impacts optimized possibly for a program, for the change, for the implemented investment
Ecological objectives > Local, regional and global ecological impacts of the project considered > Ecological impacts optimized possibly for a program, for the change, for the implemented investment
Social objectives > Project personnel recruited and allocated, incentive systems applied, project personnel assessed, developed and released > Local, regional and global social impacts of the project considered > Social impacts optimized possibly for a program, for the change, for the implemented investment > Stakeholders involved in project managing

Specific for the systemic approach presented here, the objective is to optimize the economic, ecological, and social impact of a project on a program that the project may be part of, and on the change and investment implemented by the project. These contexts are described in Chapter 13, "Program Initiating and Program Managing", and in Chapter 14, "Change Initiating and Change Managing".

In order to operationalize the project management objectives defined in Table 6.1, the objectives of the sub-processes of project management are defined below. From a systemic point of view, the objective of project starting is to establish a project as a social system. The objective of project controlling is to support the evolution of a project, and the objective of project closing is to dissolve a project as a social system. The objective of project transforming and project repositioning is to manage a project discontinuity and create a new project identity. The objective of project coordinating is to ensure internal and external project communication in order to assure project progress.

A description of the sub-processes of project management and a detailed definition of their objectives are provided in Chapters 9 to 12.

The following structural dimensions of projects must be managed:

> the solution requirements, the objects of consideration of a project, the project objectives, the project strategies,
> the project scope, the project schedule, the project resources, the project budget, the project risks,
> the project organization, the project culture, the project personnel, and
> the project infrastructure.

The following contextual dimensions of projects must be managed:

> the decisions, measures, and documents of the pre-project and post-project phases,
> the project stakeholders,
> the other projects in the project portfolio of the project-oriented organization,
> the organizational objectives and organizational strategies, and
> the investment implemented by a project.

These structural and contextual dimensions of projects to be managed are shown in Figure 6.1. Exclusive consideration of the dimensions of the "magic triangle" is obviously not enough.

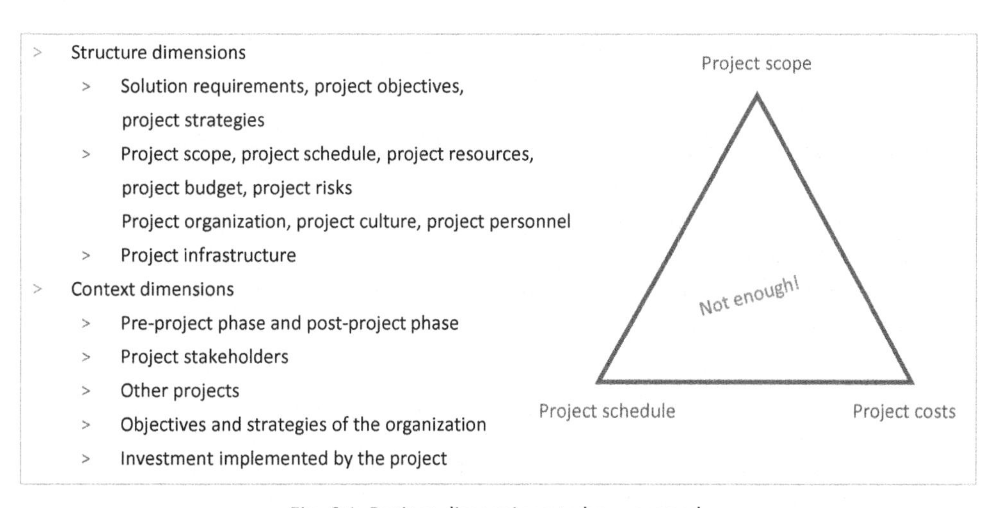

Fig. 6.1: Project dimensions to be managed.

6.2 Project Managing: Process

The advantage of perceiving project managing as a business process is that its objectives, tasks, responsibilities, and key performance indicators (KPIs) are described in order to assure the process quality. The quality of project managing can be assessed on the basis of the quality of the design of the process and on the project management results achieved. The project management personnel therefore require both social skills for designing the process and methodological competencies for performing the process.

An overview of the process of project managing is shown in Figure 6.2. It also illustrates how the principles of sustainable development are considered.

Project managing begins when the project is assigned to the project manager and to the project team by the project owner, and it ends when the project owner approves the project. The sub-processes can be seen in the representation of the project managing

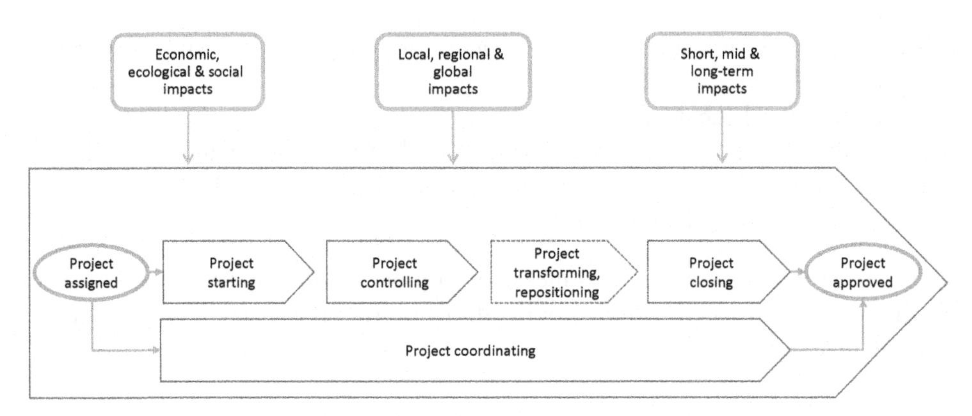

Fig. 6.2: Overview of the business process "Project managing"—flow diagram.

process. Project coordinating is carried out on a continuous basis. The performance of the other sub-processes occurs within defined time limits. By definition, project starting and project closing are each carried out once. Project controlling is to be carried out several times in a project and usually takes place periodically. The need to manage a project discontinuity by project transforming or project repositioning is situation dependent.

Relations Between the Sub-Processes of Project Managing

A consistent performance of project managing is based on a consistent application of terms and methods in all sub-processes. Project management results can be optimized by considering the relations between the sub-processes. The following relations exist:

- In project starting, the structures for project coordinating, project controlling, and project closing are to be planned.
- In project starting, the criteria for measuring the project success in project closing are determined by defining the project objectives.
- In project starting, working forms are established to be applied in project coordinating, project controlling, and project closing (e.g., project meetings, project workshops, reporting, reflections).
- In project starting, by applying scenario techniques and by developing alternative project plans, measures for avoiding project crises, promoting project chances, and providing for project discontinuities can be planned.
- In project starting, the management of structurally determined project changes can be planned.
- In project coordinating, the plans developed in project starting (and adapted in project controlling) are used for communicating.
- In project controlling, the project plans developed in project starting are used for communicating and are possibly updated.

> In project transforming or repositioning, alternative project plans which may have been developed in project starting and current project plans from the last project controlling cycle are used.
> In project closing, project plans developed in project starting and adapted in project controlling form the basis for assessing the project performance and for securing organizational learning.
> Project marketing takes place in all sub-processes, using a uniform project marketing strategy. The project marketing strategy is defined in project initiating and verified in project starting.

Excursus: Project Marketing as a Project Management Task

Many projects are characterized by a strong orientation towards content but a weak orientation towards communication. Members of the project organizations concentrate primarily on fulfilling work packages and do not recognize the need to communicate the objectives, content, and organization of the project to its stakeholders in order to contribute to the success of the project.

Project marketing is defined as project-related communication with project stakeholders. It is not only the quality of a project's results, but also their acceptance which must be ensured.

The success (S) of a project can be defined as the product of the quality (Q) of the results of a project and their acceptance (A): $S = Q \times A$. If the acceptance of the project results is low despite good quality, the project success is also low.

Project marketing is a success factor for projects. The objectives of project marketing are:

> securing appropriate management attention for the project,
> securing the necessary resources for the project,
> securing acceptance of the (interim) results of the project,
> minimizing conflicts in the project, and
> encouraging the identification of members of the project organization with the project.

Conflicts in the project and in relations with the project stakeholders are minimized by appropriate project information. Expectations about project results are also to be developed. Feedback on the project should be gathered, and a dialog with stakeholders should be established. The personal self-marketing interests of members of the project organization are taken care of.

Project marketing is a task to be performed in all sub-processes of project managing. It therefore does not usually constitute a separate sub-process of project managing. Project marketing is particularly important in the project starting process. Initial information about the project and the project identity can be communicated. After this, project marketing measures should be maintained at an appropriate level intensity. In project controlling, it is possible to communicate the interim results of the project and changes in the project structures. In project closing, not only the project results, but also the process of project work and the contributions of the members of the project organization to project success should be communicated.

The basis for project marketing is the project stakeholder analysis and the project culture development. Stakeholder analysis makes it possible to define differentiated marketing strategies and measures for different stakeholders. Within the framework of project culture development, the project name, a project logo, project slogans, and project-specific values are defined, which are used in project marketing. The central tools of project marketing are printed media such as a project folder or a project newsletter, project-related events such as a "project vernissage" or a project presentation, a project homepage, and also the project management documentation such as a project manual, project progress reports, or project scorecards. All members of the project organization are responsible for project marketing. Not just the project manager, but also the project owner and project team members are responsible for appropriate communications with stakeholders. Project collaborators who contribute only to specific parts of the project do not have any responsibility for project marketing.

6.3 Project Managing: Contexts

The business processes of investment planning (see Chapter 3) and project initiating (see Chapter 5) take place before starting a project to implement an investment. Controlling the benefits realization to ensure a successful investment (see Chapter 3) takes place during an implementation project and after its end during the use of the investment object. The content-related business processes for developing a solution and the process "Project administrating" are performed in parallel with project managing (see the overview at the beginning of this chapter, page 97).

The content-related business processes of a project depend on the project type and are therefore not dealt with here. In the case of an IT conception project, for example, the following content-related phases are to be performed: gathering information, analyzing the actual-state, defining and describing possible solutions, planning the implementation, reporting, and decision-making. Engineering, procuring, producing and transporting, constructing and assembling, and training and commissioning are essential business processes in an engineering project.

The business process "Project administrating" includes the tasks of administrating project personnel, project-related customer and supplier contracts, and project infrastructure, as well as the filing of project correspondence and project documents. The tasks of project administrating are not part of project managing; they belong to a separate process. Project administrating work packages are to be included in the work breakdown structure. This contributes to a clear distinction between project managing and project administrating, and thus to the professionalization of project managing.

The objectives of project administrating are to ensure the transparency of personal, customer, and supplier documents and correspondence, as well as to provide quick access to personnel, customer, and supplier data. For large projects, project administrating can be performed by a project administrator (e.g., a contract expert) or a project assistant. In the case of small and medium-sized projects, the role of project administrator is usually fulfilled by the same person as the project manager role.

6.4 Project Managing: Designing the Business Process

The business process "Project managing" is to be designed according to the specific needs of a project. The use of project management methods, standard project plans and checklists, project communication formats, project management infrastructure, and possibly project management consultants must be planned.

Project Managing: Use of Project Management Methods

To define basic rules regarding the use of project management methods in projects and the structuring of project organizations is a corporate governance task. In organizational guidelines of project-oriented organizations, the methods to be used for small projects and projects are specified. Methods that are used for project starting, project coordinating, project controlling, project transforming, project repositioning, and project closing are differentiated. A "must" and a "can" use of each method can be defined. A recommendation for this is shown in Table 6.2 (on next pages). The use of "can" methods and the required level of detail are to be decided project specifically.

Each project plan resulting from the use of a project management method is a model of the project and contributes to constructing the project reality. By developing various project plans, it is possible to create a management complexity that is appropriate for dealing with the project complexity. The achieved level of detail of individual project plans contributes to this.

Project plans such as a work breakdown structure, a project organization chart, or a project stakeholder analysis are project models and contribute to the construction of the project reality by the project organization. By developing various project plans, it is possible to create an appropriate management complexity that corresponds to the complexity of the project.

The quality of the project plans can be optimized by a multi-method approach. The completeness of the project plans can be ensured by relating them to each other. So, for example, the results of a project stakeholder analysis are to be considered in work breakdown structuring and/or project cost planning. Project plans are therefore developed iteratively.

In the sub-process "Project starting", the project team should develop detailed project plans based on the initial project plans created in project initiating. This allows the creativity of the team to be used and encourages the identification of the project team members with the results. Preparatory work for the development can be performed by a smaller group of selected project team members. Efficient and goal-oriented teamwork can be ensured by using moderation techniques. Visualization techniques promote communication in project managing and support the documentation of the results.

Table 6.2: Methods for Managing Small Projects and Projects (Recommendation)

Methods for the sub-processes of project managing	Small Project	Project
Methods for project starting		
Planning project objectives	Must	Must
Planning project strategies	Can	Can
Planning objects of consideration	Can	Can
Developing a work breakdown structure	Must	Must
Specifying a work package	Can	Can
Planning project milestones	Must	Must
Developing a project bar chart	Can	Must
Developing a project network (e.g.. CPM schedule)	Can	Can
Planning project resources	Can	Can
Planning a project budget	Must	Must
Planning project cash flows	Can	Can
Analyzing and planning project stakeholder relations	Must	Must
Analyzing project risks	Can	Must
Analyzing project scenarios	Can	Can
Developing alternative project plans	Can	Can
Performing a cost-benefit analysis	Can	Must
Developing a project assignment	Must	Must
Developing a project organization chart, listing project roles	Must	Must
Developing a project responsibility chart	Can	Can
Planning project communication formats	Can	Must
Defining project rules	Can	Must
Defining a project name	Must	Must
Developing a project logo	Can	Can
Agreeing a project specific incentive system	Can	Can
Planning project-related personnel development	Can	Can

(continues on next page)

Project plans are often understood as tools for documentation and for controlling. In fact, project plans are also tools for decision-making, leadership, and communication. Target group-specific project information can be designed and communicated through the appropriate use of information and communication technologies.

Table 6.2: Methods for Managing Small Projects and Projects (Recommendation) ***(cont.)***

Methods for sub-processes of project managing	Small Project	Project
Methods for project coordinating		
Keeping a to-do list	Must	Must
Developing minutes of meetings	Must	Must
Developing work package approval certificates	Can	Can
Methods for project controlling		
Reporting the project status (e.g. project score card)	Must	Must
Analyzing the earned value	Can	Can
Developing a project trend analysis	Can	Can
Adapting project plans	Must	Must
Methods for project transforming or project repositioning		
Planning ad-hoc measures and additional measures	Must	Must
Analyzing causes of a project discontinuity	Must	Must
Planning strategies to cope with a project discontinuity	Must	Must
Methods for project closing		
Listing remaining tasks and tasks for the post-project phase	Must	Must
Adapting the project stakeholder analysis	Can	Can
Adapting the cost-benefit analysis	Can	Must
Developing the project closing report	Must	Must
Performing an exchange of experience workshop	Can	Can
Assessing members of the project organization	Can	Can
Approving a project	Must	Must

Chapters 9 to 12 describe the sub-processes of project managing and the methods listed above for the individual sub-processes.

Project Managing: Use of Checklists and Standard Project Plans

The efficiency of project managing can be increased by the use of checklists and standard project plans. The development and provision of checklists and standard project plans are also corporate governance tasks of the project-oriented organization. These tasks can be performed, for example, by a PMO (Project Management Office).

For example, the following checklists can be used for project managing:

> Checklist: Agenda of a project start workshop
> Checklist: Agenda of a project controlling workshop
> Checklist: Agenda of a project owner meeting
> Checklist: Table of contents of the project manual

In a project manual, it is necessary to distinguish between project management documentation and project results documentation.

Standard project plans can be used to manage repetitive projects. Standardization represents a tool for organizational learning and/or knowledge management for the project-oriented organization. Project plans that can be standardized are, for example, work breakdown structures, work package specifications, objects of consideration plans, milestone lists, project organization charts, and project responsibility matrices. If standard project plans are used in a project, these are to be suitably adapted project specifically by the members of the project organization.

Project Managing: Use of Different Project Communication Formats

In project managing, the communication formats individual meetings, project meetings, and project workshops can be combined. As can be seen in Figure 6.3, project workshops have the highest resource requirements, but they are nevertheless recommended to ensure project management quality.

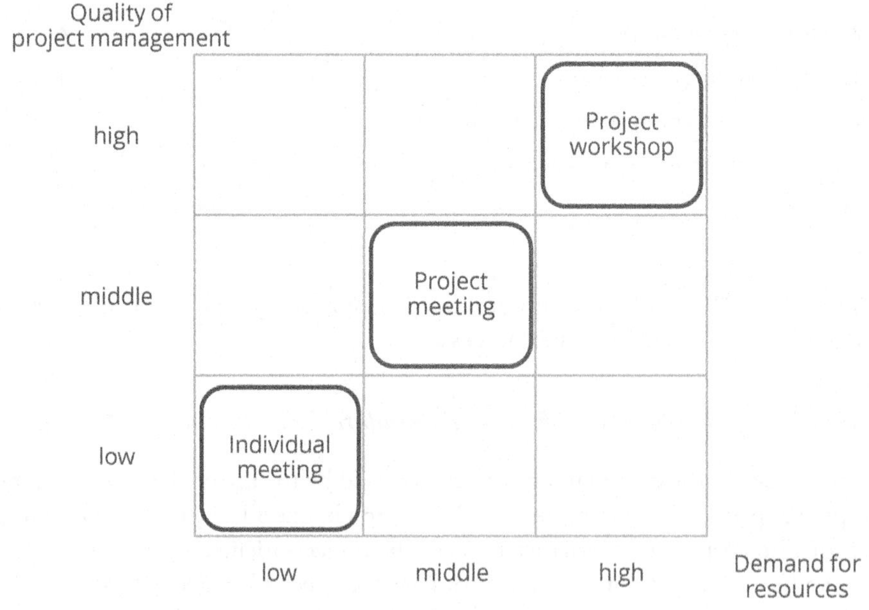

Fig. 6.3: Communication formats in a project.

Different communication formats can be combined in the individual sub-processes of project managing. By combining different communication formats in project starting, different objectives are pursued. The objective of the project manager's meetings with individual project team members is to provide basic information about the project and to exchange mutual expectations about cooperation. This initial orientation provides a basis for participation in project meetings and the project start workshop.

A common communication format in project starting is the kick-off meeting. The objective of a kick-off meeting is for the project owner and the project manager to inform the project team about the project. This usually involves "one-way communication" for a period of two to three hours with little opportunity for interaction.

The objective of a project start workshop is for the project team to develop a common "big project picture". The interaction between team members in the workshop makes a significant contribution to project culture development. A project start workshop can last from one to three days and usually takes place in a moderated form outside the daily workplace. The number of participants in a project workshop should not exceed 15 persons. The project team members should participate in the whole workshop.

Representatives of project stakeholders can participate as guests in the workshop. At the end of the workshop, the main results should be presented to the project owner. In larger projects, it may be necessary to have several kick-off meetings and project start workshops in different locations with different target groups.

The objectives of regular project meetings are the exchange of information between members of the project organization about the project status and agreement on the further approach. The objectives of regular project owner meetings are for the project manager to inform the project owner about the project status and to ensure strategic project decisions.

Project Managing: Design of the Project-Related Infrastructure

Professional project management requires the use of an appropriate infrastructure of information and communication technologies (ICT) as well as an appropriate office infrastructure.

The planning of the ICT infrastructure to be used in a project poses a particular challenge for virtual project organizations with project team members who work in different locations. The use of common project management and office software must be ensured, and the required hardware has to be made available. Decisions have to be made about the use of appropriate communication tools such as project management portals, collaboration software, telephone conferencing, and video conferencing.

A project-specific office infrastructure is to be provided for holding project meetings, giving project presentations, and creating a work space for a possible Project Office.

Project Managing: Project-Specific Use of Project Management Consultants

Consulting can be provided not only for permanent, but also for temporary organizations—that is, projects and programs. Consultants can be used to support content-related work as well as to support the management of a project.

In management consulting of a project, the project as a social system is the client of the consultant. It is also possible for individuals or teams to be clients of project-related consulting. The consulting of an individual (e.g., a project manager) or a team (e.g., a project team) is considered as "coaching".

Through project management consulting or project management coaching, the quality of the managing of a project is to be assured or improved. The use of a project management consultant or coach is particularly recommended in the project starting process. The use of an external expert can also be useful for dealing with a project discontinuity.

The decision about using a project management consultant or coach should be made jointly by the members of the project organization. The consulting role can either be carried out by a competent member of the project-oriented organization or by an external consultant (for more information, see Chapter 16).

6.5 Project Initiating and Project Managing: Values

The values of agility, resilience, and sustainable development described in Chapter 2 are of increasing importance in management and also influence project initiating and project managing. Authors such as Radatz and Hanisch criticize "traditional" project management and say it is already dead.[2,3]

Greater speed, more networking, and more mobility are called for; location independence and simultaneity are required; self-organization through charisma, competence, dialog, and reflection are success factors for projects. "Agile project management" and collaboration with "digital natives" based on rough specifications instead of the use of project plans are proposed as alternatives.

In fact, it is the mechanistic project management approach which is being criticized. The authors are not familiar with the systemic project management approach. However, further development of management approaches is necessary because of the increasing complexity and dynamics of permanent and temporary organizations. It is

[2] See Radatz, S., 2013.
[3] See Hanisch, R., 2013.

possible to learn from agile methods, such as Scrum and design thinking (see Excursus: Design Thinking).

Excursus: Design Thinking

The design-thinking approach is used to deal with complex problems—for example, for optimizing existing products or creating new products, services, and processes. The objective is to develop solutions that best serve the needs and wishes of users and to ensure their feasibility and marketability. Various principles and methods from the design world are used.

As with Scrum, design thinking is characterized by an iterative approach, the involvement of stakeholders, the use of prototypes, and work in multidisciplinary teams supported by visualization techniques for value creation.

Lockwood defines design thinking as "a human centered innovation process that emphasizes observation, collaboration, fast learning, visualization of ideas, rapid concept prototyping, and concurrent business analysis [. . .]".[4]

The basis for the application of a design-thinking approach is the process type depicted in Figure 6.4. This consists of six iteratively processed phases.

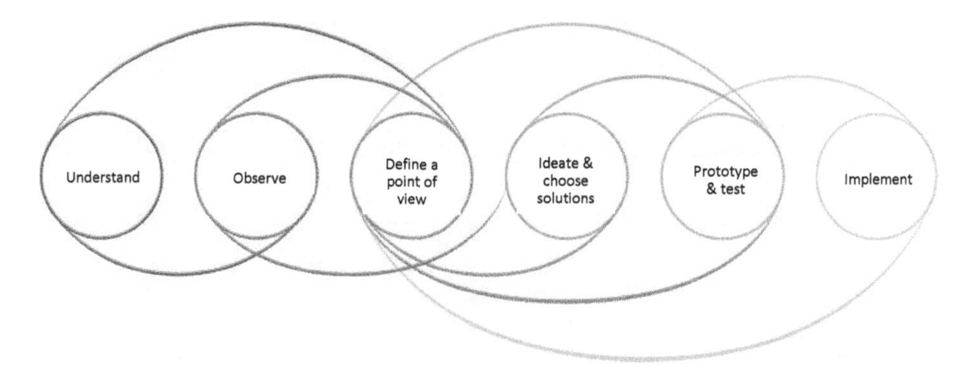

Fig. 6.4: Design-thinking process.[5]

In the first three phases, a comprehensive understanding of the problem is developed. Here primarily "human-centered" methods such as customer journey mapping, interviews, and observations are used. In the following three phases, solutions for the previously defined problems are developed. The use of creative methods for generating ideas, such as brainstorming, change of perspective, provocation, or visual suggestion, as well as the early development of prototypes and testing with the involvement of users, are characteristic.

The concepts of agility are important for project initiating and project managing. Elements of agile working forms can be used to design the processes "Project initiating"

4 Lockwood, T., 2010, p. xi.
5 See Hasso Plattner Institut, 2017.

and "Project managing", even if, for example, Scrum is not used. For this, some of the rigid rules of Scrum have to be adapted. In particular, the implementation of flexible iterations instead of "timeboxed" sprints must be enabled, the role of the product owner must be seen as part of the project organization, and flexible part-time teams instead of full-time Scrum teams are to be used in projects.

The values of agility, resilience, and sustainability are considered in the values of the RGC management paradigm (see Chapter 2 and Figure 2.3 on page 22). The values of the RGC management paradigm are interpreted for project initiating and project managing as follows (see Fig. 6.5). This forms the basis for the systemic project management approach presented here.

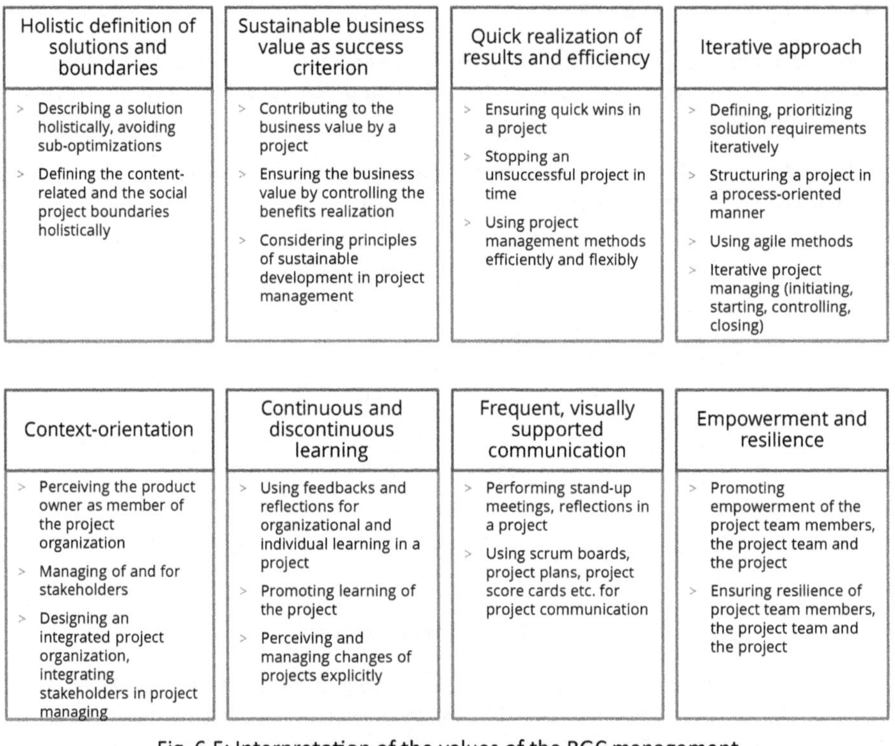

Holistic definition of solutions and boundaries
Describing a solution holistically, avoiding sub-optimizations
Defining the content-related and the social project boundaries holistically
Sustainable business value as success criterion
Contributing to the business value by a project
Ensuring the business value by controlling the benefits realization
Considering principles of sustainable development in project management
Quick realization of results and efficiency
Ensuring quick wins in a project
Stopping an unsuccessful project in time
Using project management methods efficiently and flexibly
Iterative approach
Defining, prioritizing solution requirements iteratively
Structuring a project in a process-oriented manner
Using agile methods
Iterative project managing (initiating, starting, controlling, closing)
Context-orientation
Perceiving the product owner as member of the project organization
Managing of and for stakeholders
Designing an integrated project organization, integrating stakeholders in project managing
Continuous and discontinuous learning
Using feedbacks and reflections for organizational and individual learning in a project
Promoting learning of the project
Perceiving and managing changes of projects explicitly
Frequent, visually supported communication
Performing stand-up meetings, reflections in a project
Using scrum boards, project plans, project score cards etc. for project communication
Empowerment and resilience
Promoting empowerment of the project team members, the project team and the project
Ensuring resilience of project team members, the project team and the project

Fig. 6.5: Interpretation of the values of the RGC management paradigm for "Project initiating" and "Project managing".

Holistic Definition of Solutions and Boundaries

The development of a holistic solution in a project requires a holistic definition of project boundaries. Taking a holistic view of a project means that all related consequences of a project are considered when planning the project objectives and project structures. For example, a purely technical definition of the objectives of an IT project leads to suboptimal results. Usually IT projects must also consider the consequences for the services of the organization, its business processes, its organizational structures and personnel structures, and possibly also its stakeholder relations.

A holistic approach to project management means that all project dimensions—that is, solution requirements, project objectives, project strategies, project scope, project organization, project contexts, etc., and all sub-processes of project management—are considered.

Sustainable Business Value as a Success Criterion

Performing the planned project scope and staying within budget and schedule are not the only relevant factors for project success. Rather, what is striven for is the securing of sustainable business value for the organization by performing a project. The economic, ecological, and social consequences of an investment implemented by a project, but also its short-, medium-, and long-term consequences, as well as its local, regional, and global consequences are considered in this respect. Because projects are temporary by definition, they cannot themselves be sustainable. Therefore, projects are to be perceived in the contexts of investments and medium-term organizational objectives. The project organization is responsible for contributing to securing sustainable investment benefits—for example, by controlling benefits realization.

The principles of sustainable development also influence the business process "project managing" and its methods. So, for example, the objectives of project managing are differentiated into economic, ecological, and social objectives, and stakeholders can be involved in project managing. Sustainability principles can be considered in the application of project management methods—for example, in project objectives planning or project risk analysis.

Quick Realization of Results and Efficiency

Managers want quick project results. Quick realization of results means that stakeholder expectations must be met not only by the final solution achieved at the end of the project, but also by quickly realizable interim results. In project structuring, the quick wins realizable in a project can thus be defined in an optimization step. Another advantage of defining interim results and of project controlling by "minimum viable products" is the possibility of terminating a possibly unsuccessful project at an early stage.

A rapid realization of project managing results can be achieved through the use of appropriate communication formats and project management methods. So, for example, for small projects, the planning of project milestones may suffice in place of more sophisticated scheduling methods such as applying bar charts or CPM (Critical Path Method) networks.

In projects, however, not only should project managing be efficient and lean, but the business processes for developing solutions should be performed efficiently. In repetitive projects, standardized processes can often significantly improve efficiency. The wheel should not be reinvented in the project every time . . .

Iterative Approach

One can use an iterative approach in projects, on the one hand for developing the solution, and on the other hand for project managing. Usually not all solution requirements can be defined in detail before the project begins. An iterative approach makes it possible to specify requirements during the performance of the project. The prioritization of requirements and the definition of new requirements as well as the finding of new ways to develop solutions all increase the flexibility of projects and reduce the risks of projects. The quality of the solution can be increased by continuous improvements.

If appropriate, Scrum can be used for individual phases of a project. However, an iterative approach is also useful even when agile methods are not used. Flexible iterations can be used in projects instead of formal sprints. Also, it is possible to use only a few iterations instead of the at least eight to ten sprints required by Scrum. Iterations can also have different durations, which is an adaptation of fixed "timeboxed" sprints.

An iterative approach is also used in project management: Initial project planning is performed in project initiating, detailed project plans are developed in project starting, and the project plans are adapted and further developed in project controlling. This iterative approach enables realistic project planning.

Context Orientation

Context orientation in a project means that, to ensure the success of the project, consideration is given to the pre-project and post-project phases, to the planning of relations with other projects implemented at the same time, to project stakeholders, to the realization of the objectives of the investment implemented by the project, and to the objectives of the organization carrying out the project.

Stakeholder orientation in a project means that project stakeholders are identified and relations with them are explicitly managed. Representatives of project stakeholders can be involved in project managing. "Management for stakeholders" is encouraged—that is, the interests of the stakeholders are considered as far as possible, even if these stakeholders cannot directly influence the success of the project.

The interests of the "product owner" regarding the fulfillment of business requirements and solution requirements are realized by the project. The role of the product owner is defined and integrated into the project organization. Product-owner interests can be realized on the one hand by the project owner, and on the other by project team members and project personnel. In this way, a sustainable solution in line with market requirements should be achieved.

Continuous and Discontinuous Learning

The process of project implementation is dynamic. The complexity and dynamics of projects necessitate continuous learning, but sometimes also discontinuous learning. Learning by members of the project organization as well as by the project as a temporary organization is to be encouraged in order to enable innovation and to assure competitive advantages. Including feedback and reflections as elements of the project culture promotes individual and organizational learning.

For the management of formal changes of projects—namely, for the further development, transformation, or repositioning of projects—change management methods can be used.

Frequent, Visually Supported Communication

The complexity and dynamics of a project necessitate frequent interactions between members of the project organization as well as communications with representatives of project stakeholders. Leadership in projects is characterized by a high need for communication.

In order to define and prioritize requirements for the intended solutions, to process feedback, and to make decisions, different communication formats are to be used in projects. In addition to the usual formats such as project workshops or project team meetings, stand-up meetings, and formal project reflections can also be conducted. Project boards, project plans, project scorecards, etc. can be used as visualization tools. An appropriate ICT and office infrastructure for projects supports project communication.

Empowerment and Resilience

Empowerment can take place at the level of the project, the project team, or the subteam, but also at the level of the individual project team member. Empowerment means that members of the project organization are given responsibility and assume responsibility for themselves. This decentralization of project responsibility requires ongoing development of the project personnel, as well as trust and clear rules in projects. Team orientation is called for in projects, instead of hierarchical structures.

Project resilience requires project cultures which are ready to learn, fault tolerant, but also prepared for confrontation. The resilience of projects can be promoted by agile and flexible structures, as well as by redundant structures. Redundancy results, for example, from the use of members of the project organization with overlapping qualifications, or from the use of different cooperation partners for similar tasks. The robustness and resilience of projects can also be promoted by participation in formal and informal networks such as networks of projects.

Values determine the objectives pursued in project initiating and project managing, the tasks to be performed and the methods to be used in these processes, as well as the roles to be fulfilled.

6.6 Managing Different Project Types

Project types can be distinguished by industry, location, content, phase in the investment life cycle, degree of repetition, duration, and relation to business processes.

The differentiation of projects into different project types (see Table 6.3) makes it possible to analyze specific challenges and opportunities for managing each type of project.

Table 6.3: Differentiation of Project Types

Differentiation criterion	Project type
Industry	Construction, engineering, IT, pharmaceutical, NPO, etc.
Location	National, international
Content	Customer relation, products and markets, infrastructure, personnel, organization
Phase in the investment life cycle	Study, conception, realization, relaunch or maintenance
Degree of repetition	Unique, repetitive
Customer	Internal customer, external customer
Project duration	Short, medium and long term
Performed business process	Primary, secondary and tertiary process

In the case of differentiation of projects by industry, one can differentiate between construction projects, engineering projects, IT projects, pharmaceutical projects, etc. Project team members need specific technological and market knowledge for projects of different industries. It is possible to develop industry-specific project management roles (e.g., IT project manager).

Regarding the location of a project, a distinction can be made between domestic and foreign projects. Specific personnel and organizational requirements must be fulfilled for foreign projects. On the one hand, the mobility of project team members and their knowledge of foreign languages must be ensured. On the other hand, foreign-language project documentation must be produced, and different national cultures in the project team and possibly also different time zones must be considered.

In the case of differentiation according to content, projects can be differentiated into contracting projects, product development and marketing projects, infrastructure projects, personnel projects, and organizational projects. For projects with different content, in addition to different project team member competencies, different project cultures can be required. Thus, for example, contracting projects are characterized by a higher commitment to objectives than organizational development projects. Furthermore, specific project phases can be defined for each project type, and standard project plans can be developed.

With regard to the phase in the life cycle of the investment for which a project is carried out, it is possible to differentiate between conceptual and implementation projects. In order to reduce the risk of unsuccessful investments and to optimize the quality of investment decisions, it is advisable—for example, before implementing a new IT system, establishing a new training program, or building a new industrial facility—to develop a concept. The process of developing a concept can itself be so complex and strategically significant that it is advisable to carry it out in the form of a project. Conceptual projects are characterized by a high need for creativity, project marketing, and openness with regard to content. If the decision makers decide in favor of implementation at the end of the conceptual project, a project chain is created by the conceptual project and the implementation project.

Regarding the repetition rate, one can distinguish between real unique projects and repetitive projects. Obtaining a quality certificate for the first time, for example, is a unique project for every company. The performing of a contracting project by a construction company represents a repetitive project, since the same processes (engineering, procuring, setting up the site, constructing, etc.) are always fulfilled for all contracting projects. Nevertheless, it is advisable to fulfill (socially) complex contracts (with new customers, partly new suppliers, or partners, etc.) in the form of a project.

For repetitive projects, in contrast to unique projects, some project management methods (e.g., work breakdown structure, milestone plan, work package specifications) can be standardized. For repetitive projects, less creativity is required than for unique projects. The use of creativity techniques and the use of multidisciplinary teams for innovative problem solving are primarily necessary for unique projects. In repetitive projects, however, there is a risk that the project performance will obtain a routine character, that standard project plans will not be questioned, and that no project autonomy will be enabled.

Pilot projects are a special form of unique projects. On the basis of the assumption that repetitive projects with similar objectives will be implemented after the performance of a pilot project, organizational and individual learning are explicit objectives of pilot projects. The organization of learning about the technology used; the market conditions, etc., will therefore be part of the project content.

Differentiation into internal and external projects is on the basis of different customers. In external projects (contracting projects), a customer orders an organization to provide a service for a fee. The objective of an internal project, on the other hand, is to solve an internal problem for an internal customer.

Only complex contracts are to be implemented by projects. For less complex contracts of smaller scope (e.g., deliveries of machines, provision of personnel, technical planning, etc.), the use of project management is not necessary (see Fig. 6.6).

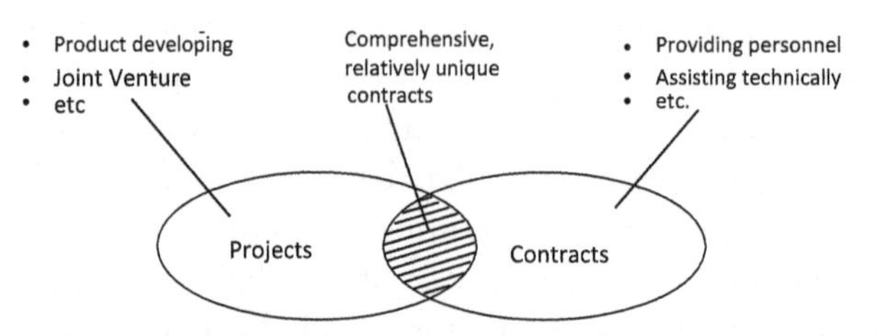

Fig. 6.6: Relation between contracts and projects.

All projects are internal projects except the processing of contracts of large scope. For external projects, there is usually a greater degree of clarity about project objectives than for internal projects, as a result of the extensive preparatory work performed as part of the tendering processes. However, this difference should become less significant in the future through more frequent implementation of conceptual projects. As regards the setting of priorities, preference is generally given to external projects. This is despite the fact that internal projects usually have more strategic importance than external projects.

As regards duration, it is possible to differentiate between short-term and medium-term projects. Since no project should last less than three months or longer than twelve months, projects with a duration of three to six months are defined as short-term, and those with a duration of seven to twelve months are defined as medium-term. Exceptions to this rule are projects for the implementation of infrastructure investments, such as buildings, factories, etc., which last longer than twelve months. The project starting and project closing of short-term and medium-term projects are to be performed as quickly as possible, and the frequency of project controlling is to be kept low. The possibility of changes in the project team has to be taken into account in long-term infrastructure projects, and provisions must be made for changes in the project in the case of possible technological advances and new legal provisions.

With regard to business processes performed by an organization, projects can be differentiated into projects for performing primary, secondary, and tertiary processes. Projects for performing primary processes are tendering projects and contracting projects. Projects for performing secondary processes are, for example, product development projects or advertising campaign projects. Projects for performing tertiary processes are, for example, reorganization projects or IT projects.

In various industries (e.g., construction, engineering, IT), projects were in the past primarily defined for performing primary processes. It is only in recent years that a more general project orientation and the use of projects for performing secondary and tertiary processes can be observed.

6.7 Project Managing: Benefits

The benefits of project managing are basically to enable projects to be performed efficiently and to ensure the quality of the project results. Without appropriate management, projects cannot be carried out at all or only very inefficiently because of their complexity and dynamics. The results of projects are not only the solutions developed, but also their contributions to a sustainable business value of the project-oriented organization.

In detail, the following benefits of project managing can be seen:

> ensuring competitive advantages through the design of appropriate project organizations and professional project planning,
> ensuring short project durations and low project costs—for example, by avoiding potential penalty payments and by optimizing interest income and interest costs,
> ensuring individual and organizational learning through reflections in the project,
> creating transparency through realistic project plans and consistent reports,
> providing clarity about the project status and providing a basis for management decisions,
> fulfilling stakeholder expectations and ensuring the acceptance of project results through appropriate stakeholder communications and involvement of stakeholders,
> contributing to the optimization of the benefit-cost difference of the investment implemented by a project, and
> providing the necessary information for professional project portfolio management.

The benefits of professional project managing also include the avoidance of fundamental errors, such as unclear definitions of project objectives, lack of information about risks, unclear understanding of roles, etc.

The dangers of an inappropriate use of projects and project management lie in an inflationary and undifferentiated use of the term "project" and unrealistic expectations regarding the integration function of project managing. If the term "project" is used for everything that is relatively unique and time limited, and if no clear distinction is made between a non-project, a project, and a program, then there will be "projects" for which project management is not useful. Moreover, attempts may be made to manage programs (not identified as such) as projects. The integrative functions required for a program cannot be fulfilled by project management. Both of these situations are dysfunctional and impede the acceptance of project management.

Sometimes it is also expected that functions of the permanent organization, such as ongoing technological development or continuous development of personnel, will also be fulfilled by the performed projects. These expectations usually are unrealistic. Project managing is no substitute for weak management of the permanent organization.

Literature

Duncan, W.R.: *A Guide to the Project Management Body of Knowledge* (*PMBOK® Guide*), Project Management Institute (PMI), Newton Square, PA, 2000.

Hanisch, R.: *Das Ende des Projektmanagements: Wie die Digital Natives die Führung übernehmen und Unternehmen verändern,* Linde, Vienna, 2013.

Hasso Plattner Institute (HPI): *Was ist Design Thinking?,* accessed at https://hpi-academy.de/design-thinking/was-ist-design-thinking.html (16.01.2017).

Lockwood, T.: Forward, in: Lockwood, T. (ed.), *Design Thinking: Integrating Innovation, Customer Experience and Brand Value,* pp. vii–xvii, Allworth Press, New York, 2010.

Radatz, S.: *Das Ende allen Projektmanagements. Erfolg in hybriden Zeiten—mit der projektfreien Relationalen Organisation, Relationales Management,* Vienna, 2013.

7 Project Organization and Project Culture

The design of project organizations acquires particular importance as a result of perceiving projects as temporary organizations. Specific project organizations have to be designed for the specific business processes to be performed. This should secure competitive advantages for the project-oriented organization. Project organizations are designed during project starting, adapted as necessary during project controlling, and dissolved during project closing. Organizational tasks must therefore be performed in all sub-processes of project managing (see overview below).

The objects of consideration in designing project organizations are the structural project organization and the process organization of projects. The elements of the structural organization are project roles and relations between these roles. These can be represented in project organization charts. Traditional and systemic organizational models for projects are presented. Methods for the application of these models are described in Chapter 9.

The process organization of projects is not discussed in this chapter. The elements of the process organization are the business processes and the methods and communication formats used in projects. It is possible to distinguish between content-related business processes to be performed in projects and the business process "Project managing". Since content-related business processes are different for each type of project, they cannot be dealt with here. The business process "Project managing", which is common to all types of projects, is dealt with in Chapter 6, while the sub-processes of project managing and specific methods for their performance are described in Chapters 9 to 12.

As a temporary organization, a project has a project-specific culture. The project culture can be observed in the behavior of the members of the project organization as well as by looking at the methods and artifacts used in the project. Symbolic project management supports the development of a project-specific culture. The establishment of project-specific values and rules in projects can be supported by symbolic management.

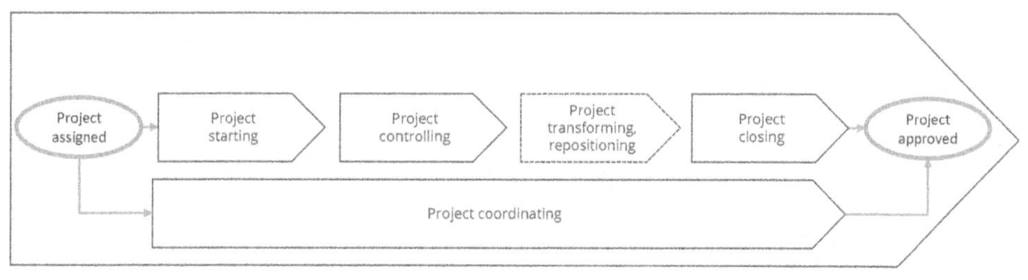

Overview: Designing the project organization and developing a project-specific culture as tasks in the business process "Project managing".

7 Project Organization and Project Culture

7.1 Project Roles

In project-oriented organizations, in addition to the roles of the permanent organization—for example, managing director, head of department, or expert pool manager—there are also project roles. As in the permanent organization, it is possible to differentiate between individual roles performed by individuals and team roles performed by teams. Project-specific individual roles are project owner, project manager, project team member, and project contributor. Project-related team roles are project owner team, project team, and sub-team.

Sociologically, the term "role" is defined as a set of expectations, values, and behaviors. Expectations of roles exist before "social actors" are assigned to these roles. Roles are thus independent of people, but people can influence roles by their specific interpretation of a role.

A project role can be described by specifying its objectives, its organizational position, its tasks, and its decision-making authority. Project roles must be described relationally—that is, taking into account the relations between roles. This creates clarity about cooperation in the project and the social boundaries of the project.

Project roles can generally be documented in role descriptions and have to be adapted project specifically, if required. Basic descriptions of project roles are given below.

7.1.1 Individual Roles in Projects

Role: Project Owner

The success of a project is extremely dependent on the professional fulfillment of the role "Project owner".

The project owner assigns the project to the project manager and the project team in the business process "Project initiating" and relieves them in project closing. The project owner provides contextual information, makes strategic project decisions, and gives feedback to the project manager and the project team. One of the main tasks of the project owner is to lead the project manager. The project manager has a right to be led.

The project owner is responsible for the application of the governance structures of the project-oriented company for project managing. He or she thus bears the responsibility for ensuring management quality in the project. The project owner also contributes significantly to project marketing. The project owner can communicate the objectives and the strategic significance of a project to the project stakeholders in a particularly authentic way.

Appropriate competencies are necessary to perform the tasks of a project owner. In addition to competencies regarding the organization performing the project and special competencies with regard to project content, a strategic orientation, social competence, and project management competence are also necessary. Although project owners do not have to be able to develop project plans themselves, they should be able to assess their completeness and formal correctness. Most of all, they should be able to use project plans as communication tools. The role project owner is to be performed actively. Depending on the complexity of the project, performing the tasks of the project owner may take up half a day per week.

The integration of a project into the project-oriented organization and the management attention that a project receives depends primarily on the person assigned to the project owner role. If the role is not appropriately filled, he or she will not be able to give the project enough support or will show little interest in the project.

The filling of the role is dependent on the project scope. The larger the scope, the more organizational units are involved in the project. The person chosen as a project owner should be hierarchically positioned in such a way that he or she covers several, but not necessarily all, of these organizational units.

In principle, project owners should be recruited from the "lowest" possible hierarchical level. This makes it possible to expand the pool of possible project owners in the project-oriented organization and assures the availability of the person performing the project owner role for the project.

Key characteristics of the project owner role are shown in Table 7.1. A general description of the role is given in Table 7.2.

Table 7.1: Characteristics of the Role "Project Owner"

Characteristics of the role: Project Owner	
Name	> Project owner, project steering committee, project sponsor, project supervisor, etc.
Relevance for the project success	> Very high > Mostly not perceived in practice accordingly
Number of persons	> 1 in small projects > 2 to maximum 3 in projects > Of the same or of different hierarchial levels
Competences	> Knowledge of the business, the industry > Project management competence > Strategical-orientation > Social competence
Recruitment	> Managers from the business units, most interested in the project results

Table 7.2: Description of the Role "Project Owner"

Role: Project Owner	
Objectives	> Interests of the company represented in the project > Project manager and project team assigned > Project manager directed > Project team supported > Application of project management governance rules assured
Non-objectives	> Operative project management performed
Organizational position	> Part of the project organization > Project manager reports the project owner
Tasks in project starting	> Selecting a project manager and essential project team members > Assuring allocation of resources > Agreeing on the project objectives with the project team > Contributing to the construction of the project context > Contributing to the initial project marketing
Tasks in project coordinating	> Communicating with representatives of project stakeholders > Performing project marketing
Tasks in project controlling	> Taking strategic decisions > Performing strategic project controlling > Informing about the project context > Assuring allocation of resources > Performing project owner meetings > Contributing to project marketing > Promoting the learning of the project
Tasks in project transforming and project repositioning	> Assigning the project transforming or project repositioning (if required) > Taking decisions about strategies and measures > Contributing to project transforming and project repositioning > Contributing to the implementation of measures and to monitoring > Closing project transforming or project repositioning
Tasks in project closing	> Assuring appropriate structures for the post-project phase > Assessing the project results, the performance of the project manager and of the project team > Assuring know-how transfer into the permanent organization > Approving the project formally
Formal decision-making authority	> Selecting the project manager and important project team members > Changing project objectives > Taking purchasing decisions over € ... > Assigning a required project transforming or project repositioning > Stopping a project > Approving a project

Role: Project Manager

The project manager is the central integrating role in a project. The project manager is the contact person for members of the project organization and representatives of project stakeholders. The project manager "drives" the project forward and ensures progress and the achievement of the project objectives.

A project must be started, continuously coordinated, controlled, and closed. A project may also have to be transformed or repositioned. It is the task of the project manager to design these sub-processes of project managing. Appropriate communication formats must be selected, project management methods must be used, and appropriate ICT must be used. The project manager is responsible for the professional management of a project.

In project managing, the project manager cooperates with members of the project team and with the project owner. The project manager is therefore not solely responsible for the project success, but shares this responsibility with the project owner and the project team. The fulfillment of the tasks of project managing by the project manager is a service provided to the project. The role is therefore not to be perceived as a position of power.

The core competency for fulfilling the project manager role is project management. A project manager also needs social competence, competencies with regard to the organization performing the project and the project content, and—in international projects—language skills and inter-cultural skills.

The role "Project manager" can be fulfilled in addition to other roles in the permanent organization. If project managers are considered in an organization as its own profession, project managers can belong to an expert pool of project managers from which they can be recruited (see Chapter 15).

The key characteristics of the project manager role are shown in Table 7.3. A general description of the role is given in Table 7.4.

Table 7.3: Characteristics of the Role "Project Manager"

Characteristics of the role: Project Manager	
Name	> Project manager, project leader, project coordinator, etc.
Relevance for the project success	> Very high > In practice often left alone
Number of persons	> Always just one person > In practice sometimes two persons (e.g. technical and commercial project manager)
Competences	> Project management competence > Social competence > Knowledge of the project-oriented company, of the business and the relevant products
Recruitment	> From the permanent organization > From a project management expert pool > From the external personnel market

Table 7.4: Description of the Role "Project Manager"

Role: Project Manager	
Objectives	> Project interests represented > Realization of the project objectives assured > Project team and the project contributors directed > Project represented to relevant project stakeholders
Non-objectives	> Contents work performed
Organizational position	> Reports to the project owner > Is part of the project team
Tasks in project starting	> Designing the project starting with project team members > Transferring know-how from the pre-project phase into the project along with the project team members and the project owner > Agreeing project objectives with project team members > Developing project plans with project team members > Performing project team building > Performing initial project marketing with project team members > Developing the project management documentation "Project starting"
Tasks in project coordinating	> Allocating project resources for work packages with project team members > Accepting work packages > Attending to sup-team meetings > Communicating with representatives of project stakeholders > Performing project marketing constantly
Tasks in project controlling	> Performing project controlling with project team members > Determining the project status with project team members > Agreeing and implementing directive measures with project team members > Further developing project plans with project team members > Redefining project objectives with project team members > Developing project status reports with project team members > Performing project marketing measures with project team members
Tasks in project transforming and project repositioning	> Suggesting project transforming or project repositioning to the project owner (if required) > Designing project transforming or project repositioning with project owner > Contributing to project transforming or project repositioning
Tasks in project closing	> Designing project closing with project team members > Planning the post-project phase > Transferring know-how into the permanent organization along with project team members and representatives of the permanent organization > Developing the project closing report > Performing a final project marketing along with project team members > Performing an emotional closing of the project with project team members
Formal decision-making authority	> Organizing project owner meetings and project team meetings > Taking purchasing decisions up to € . . . > Selecting project team members

Roles: Project Team Member and Project Contributor

Differentiating between a project team member and a project contributor makes it possible to differentiate their responsibility for the project success and their participation in project managing. Project team members are responsible for fulfilling individual work packages, but they also assume responsibility for project success through membership in the project team, contributing to project managing, and being responsible for project marketing. Project team members participate in project meetings and communicate the project objectives and their results to project stakeholders.

Project contributors, on the other hand, primarily focus on the fulfillment of content-related work. Appropriate expert skills are required for both roles. A project team member also requires project management skills and social competence. Project team member and project contributor are generic roles, which must be defined in detail project specifically by specifying the content-related tasks to be fulfilled. The roles also require project-specific names. Examples are the role names "Project team member: Developer" and "Project contributor: Tester".

Project team members and project contributors are mainly recruited from within the organization performing a project. In the case of integrated project organizations, representatives of partners, suppliers, etc. also are considered as project team members or project contributors.

Key characteristics of the roles project team member and project contributor are shown in Table 7.5. General descriptions of the roles are given in Tables 7.6 (page 130) and 7.7 (page 131).

Further project roles must be defined depending on the approach used for managing requirements. Possible project roles for managing requirements are described in the Excursus: Project Roles for Managing Requirements.

Excursus: Project Roles for Managing Requirements

A requirements manager[1] can be appointed as a specific project team member for managing requirements in a project (see Chapter 4). The tasks of the requirements manager depend on the approach for requirements managing (sequential or iterative) used in the project. In a sequential approach, the requirements manager handles any desired changes to the solution requirements—so-called "change requests", supports future users in testing solutions, and possibly supports the project manager in the acceptance of work packages.

[1] The role "requirements manager" is also referred to in practice as "business analyst" and "requirements engineer".

Table 7.5: Characteristics of the Roles "Project Team Member" and "Project Contributor"

Characteristics of the role: Project Team Member	
Name	> Project team manager, project expert, etc.
Relevance for the project success	> High, because there is no good quality of results without content experts
Number of persons	> One
Competences	> Competences for the content work, project management and social competence
Recruitment	> Organization-internally

Characteristics of the role: Project Contributor	
Name	> Project contributor; expert, etc.
Relevance for the project success	> High, because there is no good quality of results without content experts
Number of persons	> One
Competences	> Competences for the content work > Some project management competence
Recruitment	> Organization-internally

In an iterative approach, the product owner performs the demand analysis, the initial identification, definition and prioritization of requirements, and the approval of Minimum Viable Products (MVP). The multidisciplinary development team agrees on the solution requirements with the product owner.

Project role canvases are an alternative to role descriptions in list form. An example of this form of representation is given in Figure 7.1 (on page 131).

7.1.2 Team Roles in Projects

Role: Project Owner Team

The role project owner can be fulfilled not only as an individual role, but also as a team role. If several people are owners of a project, they form a project owner team. In practice, the terms "steering committee", "project sponsoring team", or "project supervisors" are also used.

Table 7.6: Description of the Role "Project Team Member"

Role: Project Team Member	
Objectives	> Project interests represented > Contributed to the realization of the project objectives > Possibly: Project contributors directed in sub-teams > Project company-internal and external represented > Qualitatively and quantitatively adequate results of work packages delivered > Contributed to project managing
Non-objectives	> Worked as content expert only
Organizational position	> Reports to the project manager > Is part of the project organization > Possibly: Reports to the line manager in the permanent organization > Possibly: Project contributor reports to the project team member
Tasks in project starting	> Performing project starting along with the project manager > Contributing to the know-how transfer from pre-project phase into the project > Agreeing project objectives along with the project manager > Developing project plans along with the project manager > Contributing to project team building > Contributing to the initial project marketing > Contributing to the development of the project management documentation "Project starting"
Tasks in project coordinating	> Contributing to the allocating of project resources for work packages > Contributing to acceptance of work packages > Attending sub-team meetings > Communicating with representatives of project stakeholders in accordance with the project manager > Contributing to project marketing
Tasks in project controlling	> Performing project controlling along with the project manager > Determining the project status along with the project manager > Agreeing and implementing directive measures along with the project manager > Contributing to the further development of project plans > Redefining project objectives along with the project manager > Contributing to the development of project status reports > Contributing to project marketing
Tasks in project transforming and project repositioning	> Contributing to project transforming or project repositioning (if required)
Tasks in project closing	> Performing project closing along with the project manager > Transferring know-how into the permanent organization along with the project manager and representatives of the permanent organization > Contributing to the project closing report > Performing a final project marketing along with the project manager > Performing an emotional closing of the project along with the project manager
Formal decision-making authority	> Taking decisions regarding the approach of performing delegated work packages > Taking decisions to assure the quality of delegated work packages > Possibly: Assigning project contributors in a sub-team for performing work packages > Possibly: Coordinating project contributors in sub-teams

Table 7.7: Description of the Role "Project Contributor"

Role: Project Contributor	
Objectives	> To the realization of the project objectives contributed > Qualitatively and quantitatively adequate results of work packages delivered
Non-objectives	> Project team meetings attended > Project marketing performed
Organizational position	> Is assigned by the project manager and reports to the project manager or a coordinating project team member > Is part of the project organization
Tasks in project starting	> Agreeing on project objectives along with the project manager or the coordinating project team member
Tasks in project controlling	> Attending sub-team meetings, if required > Attending project team meetings as a guest (if required) > Reporting the work progress regularly to the project manager or the coordinating project team member
Tasks in project transforming and project repositioning	> Contributing to project transforming or project repositioning (if required)
Tasks in project closing	> Contributing to project closing > Contributing to the know-how transfer into the permanent organization
Formal decision-making authority	> Taking decisions regarding the approach of performing delegated work packages > Taking decisions to assure the quality of delegated work packages

Role Model Canvas		Role ____________		
Responsibilities and decisions	No / Indications	Objectives and mission	Tools	Primary tasks
Support		Information transfer		

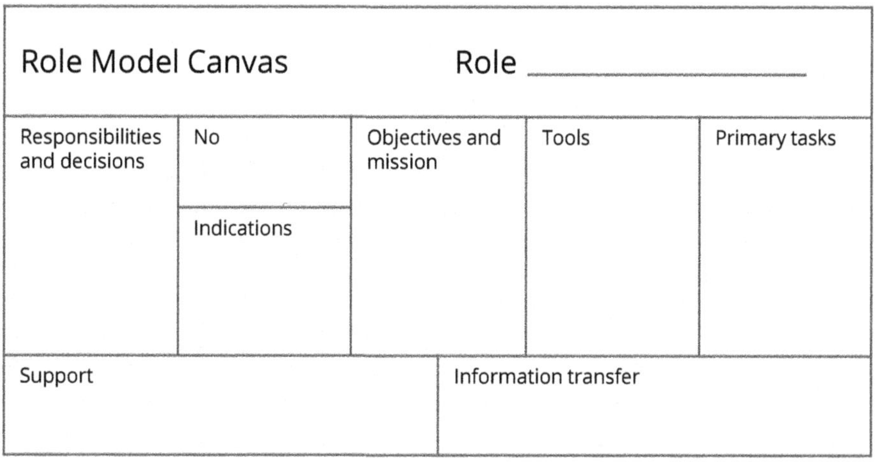

Fig. 7.1: Canvas for describing project roles[1]

The term "team" best expresses the understanding necessary for the fulfillment of the role—namely, bearing joint responsibility for the project, jointly representing the project both internally and externally, using the synergies of the owner team, etc. If, however, the members of a "steering committee" primarily represent the interests of

[1] See Botta, C., 2016.

their respective areas in the permanent line organization and do not assume joint project responsibility, this team role creates conflicts rather than benefits. The project owner role is a leadership role in the project-oriented organization. Project owners, either individually or as a team, must not represent their respective interests in a project, but the interests of the organization as a whole.

A project owner team should not include more than two or three people. In any case, a team speaker must be nominated as the primary contact person for the project manager in order to enable quick decisions. If there are larger project owner teams, possibly with members from higher hierarchical levels, the necessary intensity of communication is not assured, team members are often absent when strategic project decisions are made, and, above all, project responsibility is too broadly distributed. Nobody really feels responsible, and project commitment is lost.

Committees, however they are named, including managers from different organizational units who are mainly used to inform the employees in these units and who may support a project in an advisory capacity, can be beneficial for projects. They should not, however, be confused with the strategically important role of project owner.

The role description of the project owner team is the same as the description shown in Table 7.2 (page 125).

Roles: Project Team and Project Sub-Team

The project team and a project sub-team are also project roles. The expectations of the project team differ fundamentally from those of a project team member, and the expectations of a project sub-team also differ from those of a sub-team member. Only by differentiating between team and team member can the significance of teamwork, as well as the need to create competencies for teamwork, be understood.

The benefit of the project team is the creation of added value through teamwork. Likewise, the benefit of a project sub-team is in the creation of integrated work package results through teamwork. Sub-teams are needed when the scope of work packages is so large that a single project team member can no longer cope with them and the cooperation of several experts is necessary. Sub-teams are not to be seen as relatively separate "sub-projects" but represent integrated parts of the project.

The project team has to fulfill project managing tasks. The responsibility of the project team is to ensure professional project managing and high-quality project results. The project team does not fulfill any content-related work but integrates the results. The tasks of a sub-team are to carry out project work and to coordinate this work in the sub-team.

The project team consists of the project manager and the project team members. Project teams can consist of a minimum of three and a maximum of 10–15 people. A sub-team of a project is composed of several project contributors and a project team member who coordinates the sub-team and represents it in the project team.

Not only the individual project team members and sub-team members, but also the respective overall teams, require competencies for teamwork. The development of team competencies is a project managing task.

Key characteristics of the roles project team and project sub-team are shown in Table 7.8. General descriptions of the roles are given in Tables 7.9 (page 134) and 7.10 (page 135).

Table 7.8: Characteristics of the Roles "Project Team" and "Project Sub-Team"

Characteristics of the role: Project Manager	
Name	> Project manager, project leader, project coordinator, etc.
Relevance for the project success	> Very high > In practice often left alone
Number of persons	> Always just one person > In practice sometimes two persons (e.g. technical and commercial project manager)
Competences	> Project management competence > Social competence > Knowledge of the project-oriented company, of the business and the relevant products
Recruitment	> From the permanent organization > From a project management expert pool > From the external personnel market

7.1.3 Role Conflicts in Projects

The strong organizational differentiation in projects can lead to role conflicts. A conflict within a role, a so-called "intra-role conflict", occurs when two or more holders of expectations have inconsistent expectations of a role. So, for example, the customer's expectation that the project manager will further the customer's interests may be in conflict with the project owner's expectation that the project manager will further the organization's interests.

Figure 7.2 (page 136) shows an example of an intra-role conflict of the project owner. In this example, the results-related expectations are the same for all holders of expectations, but the process-related expectations differ significantly. Here the members of the project owner team expect not to have to do too much work in the project, while retaining the ability to intervene continuously and to alter project decisions.

Intra-role conflicts are common and are often structurally determined. One speaks of the "sandwich situation" of the manager. It is the responsibility of the role holder to deal with the different expectations of a role.

Table 7.9: Description of the Role "Project Team"

Role: Project Team	
Objectives	> Project managing tasks performed > Common "Big Project Picture" developed > Synergies assured > Committment created > Conflicts solved > Learning of the project organized
Non-objectives	> Content work performed (this is done in the sub-teams and by the individual project team members and project contributors)
Organizational position	> Is part of the project organization > Members are the project manager and the project team members > Is assigned by the project owner
Tasks in project starting	> Informing each other in the project team > Taking common decisions regarding the design of the project organization and regarding project planning > Taking common decisions regarding the design of the project context relations > Agreeing to common project rules
Tasks in project controlling	> Determining project status together > Redefining and redeveloping project objectives, schedules, costs, etc. together > Adapting the project organization and project context relations together > Performing project marketing measures together
Tasks in project transforming and project repositioning	> Contributing to a proposal for project transforming or project repositioning > Designing the process for coping with a project discontinuity together > Defining ad-hoc measures together > Developing a cause-analysis together > Developing alternative strategies together > Performing directive measures and monitoring the results together > Closing project discontinuities together with the project owner team
Tasks in project closing	> Performing the project closing process together > Planning the post-project phase together > Transferring know-how into the permanent organization together > Performing a communal final project marketing > Performing an emotional closing of the project
Formal decision-making authority	> Taking decisions regarding the approach in the project team

If multiple roles are held simultaneously by one person (see Fig. 7.3, page 137), inter-role conflicts can occur. These result from an incompatibility between expectations of simultaneously performed roles—for example, from high resource demands.

The simultaneous performance of multiple roles can also lead to opportunities for integration. It is therefore common, especially in small and medium-sized projects, for a person to hold the project manager role as well as the project team member role. A combination of the project manager role with other roles in a project is possible,

Table 7.10: Description of the Role "Project Sub-Team"

Role: Sub-Team	
Objectives	> Synergies in the sub-team assured > Committment in the sub-team created > Conflicts in the sub-team solved > Learning of the sub-team organized
Non-objectives	> Project managing performed > Project marketing performed
Organizational position	> Is part of the project organization > The members are the sub-team members > Coordinator of the sub-team (project team member) reports to the project manager
Tasks	> Fulfilling work packages > Coordinating the work of the sub-team
Formal decision-making authority	> Taking decisions regarding the approach in the sub-team

but requires the person to have the corresponding competencies, and requires a good organizational understanding of all parties involved.

Rules for the performing of multiple roles by a single person in a project are shown in Table 7.11.

In order to ensure appropriate attention for the individual roles, the number of different project roles that a person can hold in a project-oriented organization is limited. For example, no one should manage more than two to three projects at the same time. Table 7.12 summarizes ratios in this regard. These are dependent on project size and are therefore to be understood as maximum values.

7.2 Project Organization Charts

A project organization chart is a model for the representation of organizational project structures. It serves to present project roles, relations between these roles, and hierarchical structures.

Project organization charts are illustrations of the organizational structure of a project at a definite point in time. In addition to the project roles and the relations between these roles, the names of the holders of project roles can also be made visible in the organization chart. This information can also be provided in an additional list of holders of project roles. A clear distinction is to be made between the roles of the project organization and those of the permanent organization. For example, the name "Head of construction department" is not appropriate for a project team member. The

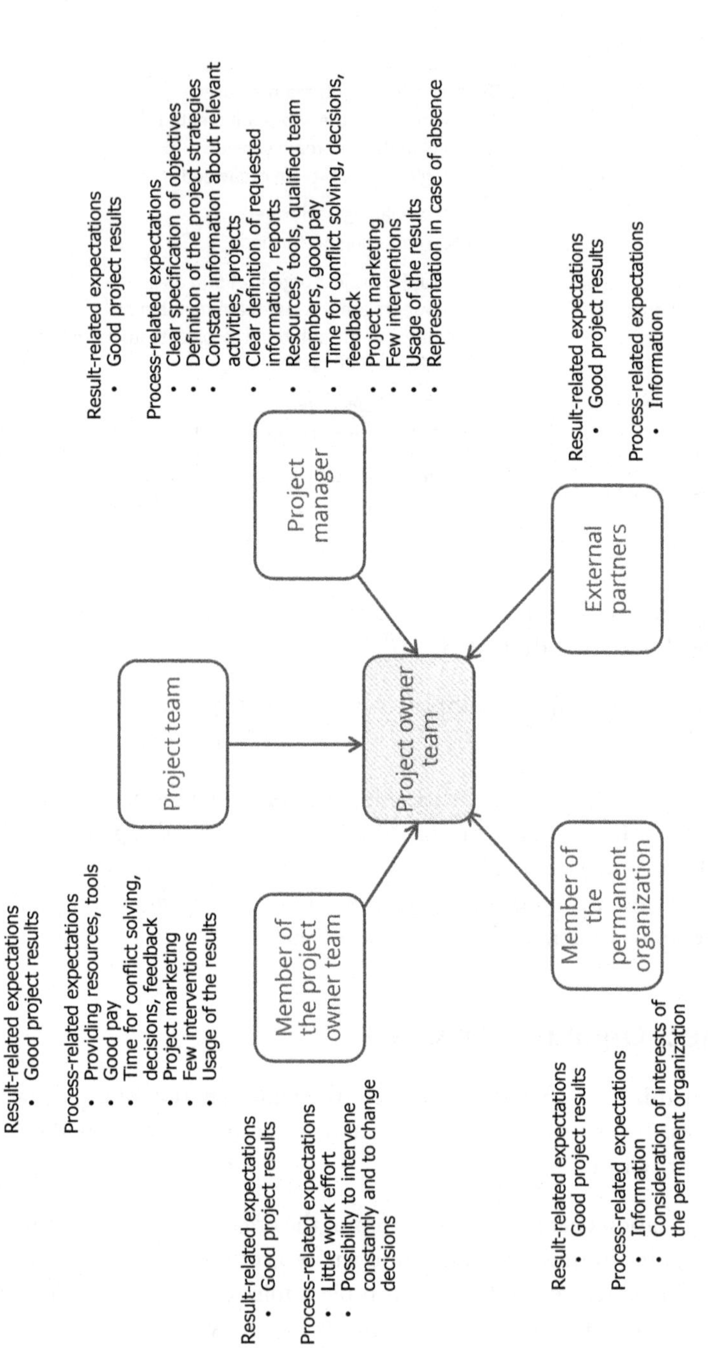
Result-related expectations
• Good project results
Process-related expectations
• Providing resources, tools
• Good pay
• Time for conflict solving, decisions, feedback
• Project marketing
• Few interventions
• Usage of the results
Project team
Result-related expectations
• Good project results
Process-related expectations
• Clear specification of objectives
• Definition of the project strategies
• Constant information about relevant activities, projects
• Clear definition of requested information, reports
• Resources, tools, qualified team members, good pay
• Time for conflict solving, decisions, feedback
• Project marketing
• Few interventions
• Usage of the results
• Representation in case of absence
Project manager
Result-related expectations
• Good project results
Process-related expectations
• Little work effort
• Possibility to intervene constantly and to change decisions
Member of the project owner team
Project owner team
External partners
Result-related expectations
• Good project results
Process-related expectations
• Information
Member of the permanent organization
Result-related expectations
• Good project results
Process-related expectations
• Information
• Consideration of interests of the permanent organization

Fig. 7.2: Intra-Role Conflict of the Project Owner (Example).

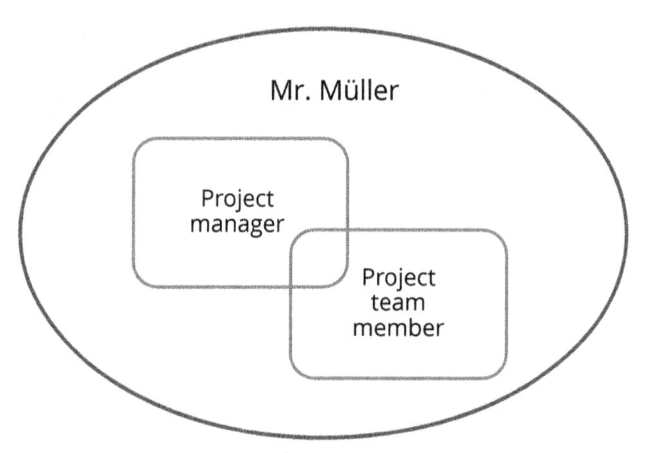

Fig. 7.3: Mr. Müller as a player of multiple roles in a project.

Table 7.11: Rules for the Performing of Multiple Roles in a Project

Role Combinations in a Project	Rules
Project owner and project manager	*NO*
Project owner and project team member	*NO*
Project owner and project contributor	*YES*
Project manager and project team member	*YES, but...*
Project manager and project contributor	*YES*

Table 7.12: Number of Project Roles per Person

Role	Number of Project Roles per Person
Project owner	4 – 6
Project manager	2 – 3
Project team member	2 – 4
Project contributor	5 – 8

name should be "Project team member: Construction", since the holder of this role does not perform any head-of-department functions in the project.

For the design of a project organization chart, different symbols, such as boxes, circles, ellipses, arrows, and dashes, can be used, but they can also be of different colors and sizes, and different line thicknesses can be used. The chosen symbols contribute to forming a picture of the project; the project organization chart is an artifact of the project culture. As examples, the project organization charts for a contracting project for an Austrian engineering company and for an airline's reorganization project are shown in Figures 7.4 and 7.5.

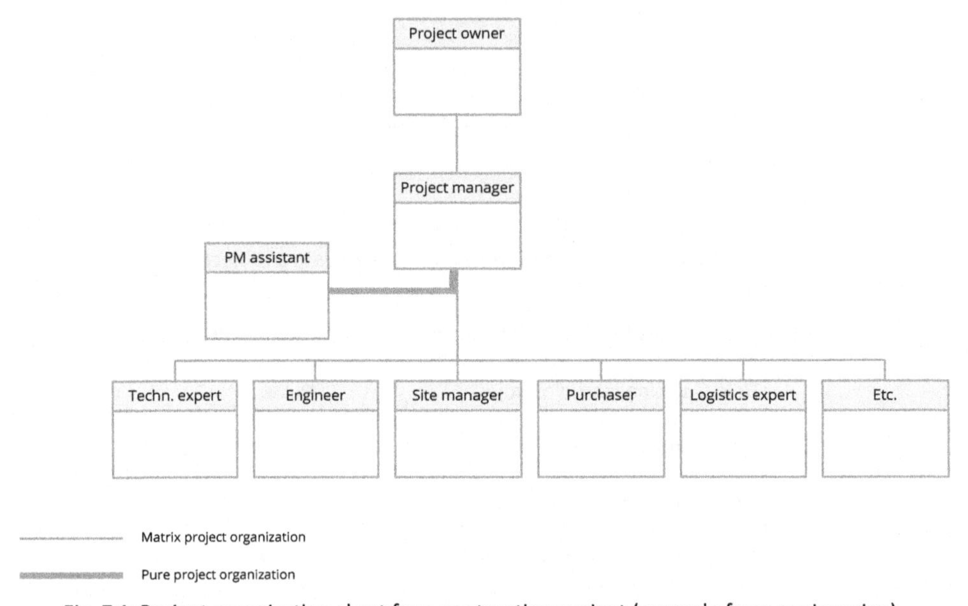

Fig. 7.4: Project organization chart for a contracting project (example from engineering).

The project cultures expressed in the two project organization charts have different attractions for potential members of these project organizations. Different representations provide different information. It is therefore important to pay attention to the symbols used to visualize the organizational structure of projects.

A project organization chart must be produced during project starting and adapted if necessary during "social" project controlling.

7.2.1 Traditional Project Organization Models

Leading a project team member or project contributor requires directions from the project manager and/or the line manager of the permanent organization. A collective responsibility requires coordination between the two managers. These relations are shown in Figure 7.6.

Traditional project organization models—namely, influence project organization, pure project organization, and matrix project organization—are differentiated by the different levels authority of the project manager and the line manager of the permanent

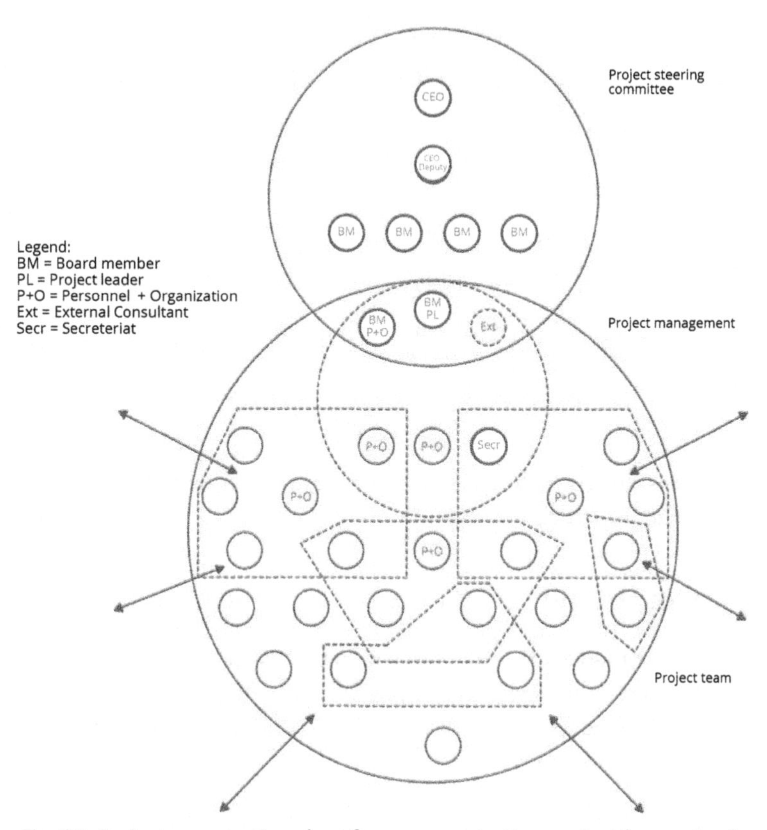

Fig. 7.5: Project organization chart for a reorganization project (example of an airline).

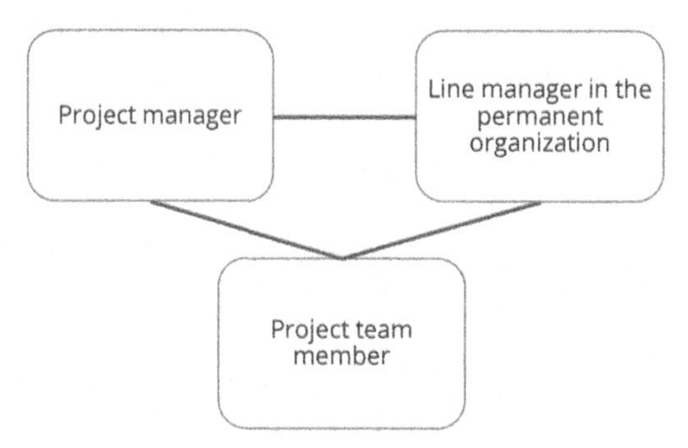

Fig. 7.6: Project roles involved in the performance of a work package.

organization over a project team member or project contributor. In the literature, these levels of authority are described with the help of "6 W questions" (see Table 7.13).[2]

The division of authority between the project manager and the line manager of a project team member can be regulated in different ways. This results in different project

[2] See, for example, Cleland, D. I., King, W. R., 1968, or Schulte-Zurhausen, M., 2013.

Table 7.13: Relevant Authorities in Project Organizations

Authority	Authority Description
What?	Authority to assign work packages and to control the progress of work packages
How well?	Authority to control the quality of work packages
Who?	Authority to select the person to perform a work package
How?	Authority to decide which methods, tools etc. to apply when performing a work package
How much?	Authority to agree on project resources & costs for the performance of a work package and to control resources and costs
When?	Authority to agree on the schedule for a work package and to control it

organization models—namely, influence project organization, pure project organization, and matrix project organization.

Influence Project Organization

In influence project organization, the project manager has a staff function without formal decision making and directive authority. All of the formal authority—that is, all "6 Ws"—lie with the leaders of the organizational units of the permanent organization that perform work packages.

Since the project owner expects the project manager to perform coordinating functions without formal authority, the project manager must use his or her informal competencies to ensure the success of the project. Informal competencies result from professional knowledge, experience, personal relationships and friendships, possession of information, proximity to power, and charisma. Informal competencies are more dependent on people than on roles.

Influence project organization can be represented in a project organization chart (see Fig. 7.7). It can be seen that there are no formal relations (to be shown by lines) between the project manager and the project team members. The advantages and disadvantages of influence project organization are summarized in Table 7.14.

Pure Project Organization

In pure project organization, the project manager has all formal authority over the project team members. From the graphical representation of the pure project organization model (see Fig. 7.8, page 142), it can be seen that the project creates a parallel organization to the permanent organization, in which all project team members are subordinated to the project manager (in project matters). This does not mean that

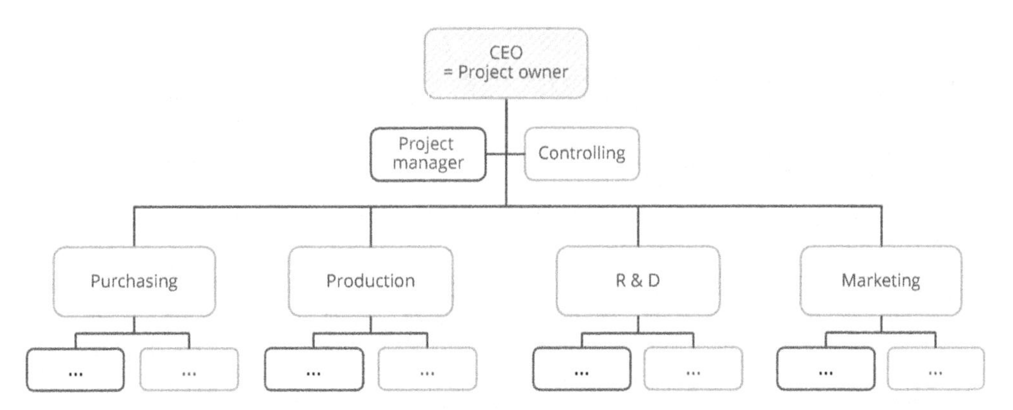

Fig. 7.7: Influence project organization model.

Table 7.14: Advantages and Disadvantages of the Influence Project Organization Model

Advantages of influence project organization	Disadvantages of influence project organization
> Low organizational effort > Project team members are formally subordinated only to their superiors in the permanent organization > Project team members work in a group of like-minded people, > No secondment and reintegration problems > Assurance of know-how in the departments of the permanent organization	> Little opportunity to integrate by the project manager > Dominance of departmental interests > Little concentration on the project > Slow decision-making

the project team members are working full time in the project. The advantages and disadvantages of pure project organization are summarized in Table 7.15.

Matrix Project Organization

Matrix project organization is characterized by a division of authority between the project manager (W-questions: What? When? How much?) and the line manager (W-questions: Who? How? How well?)

Matrix project organization can also be represented in a project organization chart (see Fig. 7.9 on page 143). The double subordination of the project team members to the project manager and the line manager can be seen in this representation. The advantages and disadvantages of matrix project organization are summarized in Table 7.16 (page 143).

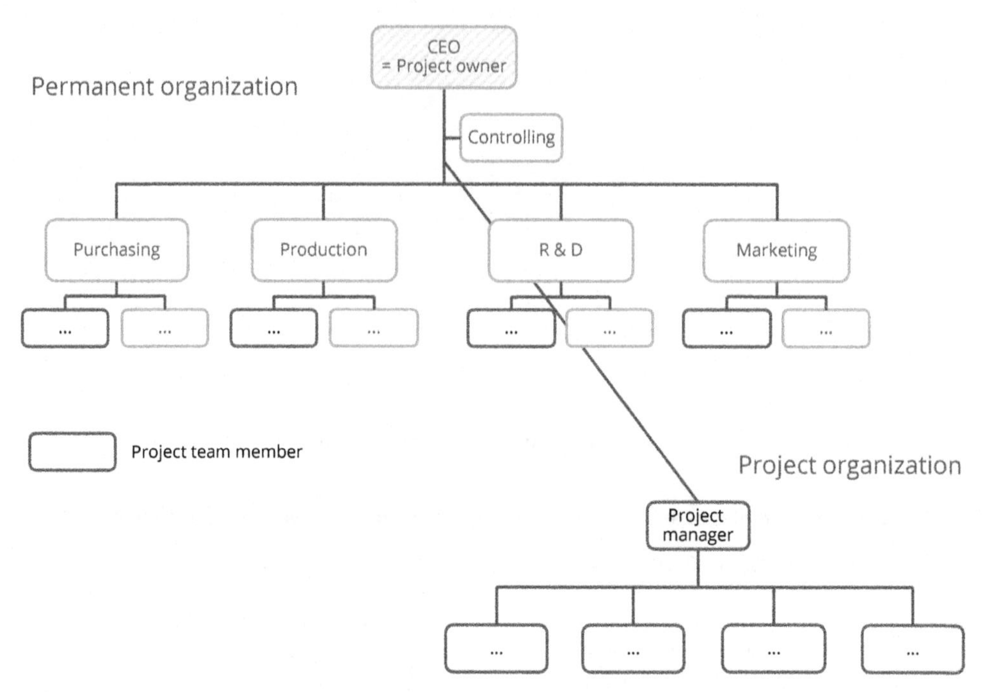

Fig. 7.8: Pure project organization model.

Table 7.15: Advantages and Disadvantages of the Pure Project Organization Model

Advantages of pure project organization	Disadvantages of pure project organization
> Full concentration of the project team members on the project > Rapid decision-making as a result of short channels of communication > Strong identification of members of the project organization with the project objectives	> Secondment and reintegration of project team members from/into departments of the permanent organization > Continuous utilization of members of the project organization > Assurance of know-how for the project-oriented organization

Through different distributions of authority between the project manager and the line manager, one can design matrix project organizations of "different strengths". For example, the project manager may also be given the authority to select project team members. In principle, direct communication between the project manager and the line manager is assumed in matrix project organization.

Use of Traditional Project Organization Models in Practice

In practice, the traditional project organization models described above are often used either consciously or unconsciously. Matrix project organization is particularly

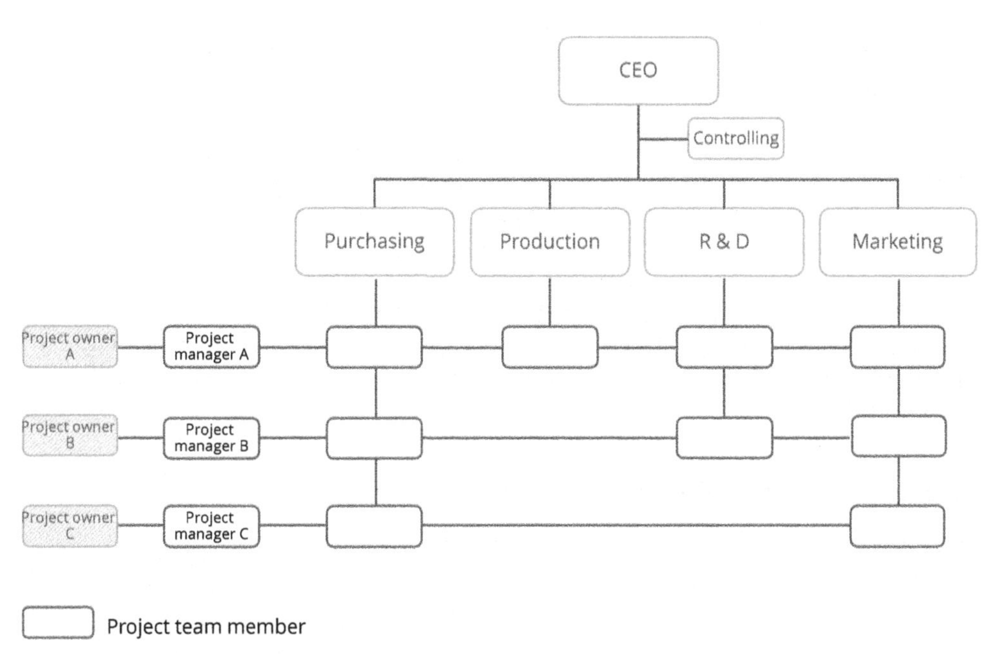

Fig. 7.9: Matrix project organization model.

Table 7.16: Advantages and Disadvantages of the Matrix Project Organization Model

Advantages of matrix project organization	Disadvantages of matrix project organization
> Formal integration of experts of departments in the project > Intentional, "constructive conflicts" resulting from dual supervision of the project team members	> "Dual supervision" places high demands on the organizational competence of the project team members > Need to resolve intentional "constructive conflicts" > High level of management effort for agreements

popular in companies that carry out repetitive contracting projects. The reasons for this are to secure an organizational home for the project team members in the departments of the permanent organization and to involve the managers of the permanent organization in project-related leadership tasks.

However, adapted applications of traditional project organization models are also observed. For example, the American large-scale construction company Bechtel uses an adapted matrix project organization for contracting projects. Project team members are employed "full time" and brought together in the same office room. The line managers of the permanent organization, however, retain their leadership functions, and "walk" from project to project to lead the project team members. So in contrast with traditional matrix organization, the project team members don't remain in their departments.

For small but strategically important projects, pure project organizations are used, instead of the expected matrix project organizations. The demand for full-time employment and spatial consolidation of the project team members is dropped, but all directive authority is transferred to the project managers to ensure the necessary integration.

Combinations of multiple project organization models are also used in practice, in order to benefit from the strengths of the different organizational models (see Fig. 7.4, page 138, in which matrix project organization and pure project organization are combined).

7.2.2 Systemic Project Organization Models

The traditional project organization models influence, matrix, and pure project organization date back to the 1960s and do not meet the requirements of agile concepts. Modern, systemic project organization models use new concepts and values, such as holistic setting of boundaries, integration of project stakeholders, organizational learning, visually supported and virtual communication, empowerment, and resilience, as well as an iterative approach (see Chapter 2) to design project organizations.

Empowerment can take place in the project at the level of the project, the project team, or the sub-team, but also at the level of the individual project team member. Empowerment means that members of the project organization are given responsibility and assume responsibility for themselves.

Self-organization is encouraged in projects instead of hierarchical structures. Agile and flexible, but also redundant, project organization structures ensure the resilience of projects. Redundant organization structures arise, for example, through the use of project organization members with similar qualifications.

The social project boundaries can be defined holistically—for example, by integrating into the project organization representatives of all organizational units that perform work packages in the project. Stakeholder orientation can also be achieved by including representatives of external stakeholders, such as suppliers or cooperation partners.

The development of project organization charts and the specification of project communication formats are performed iteratively. Initial project planning is performed in project initiating, detailed project plans are produced in project starting, and the project plans are adapted as necessary in project controlling. This iterative approach enables organizational learning and takes into account the evolution of the project organization over time. It can also be useful to develop different project organization charts for different project phases.

Project organization charts are cultural artifacts whose visual design supports communication in the project. In the project organization chart, the implementation of the described concepts and values can be made visible by means of symbols.

In the generic project organization chart used by RGC (see Fig. 7.10), individual roles are also displayed as team roles. There is no formal relation with managers of the permanent organization. This expresses the empowerment of the project team members. An ellipse around the project organization symbolizes the relative autonomy of the project as a social system. The demarcation of the project team and the subteams should convey their empowerment. Project roles represented with shading are fulfilled by representatives of project stakeholders. This promotes their integration. This visualization corresponds to the model of an "empowered and integrated project organization".

The "systemic" aspect of the generic project organization chart is the representation of the social project boundaries, the consideration of the relations between project roles, and the flat and non-hierarchical representation of the roles, focusing on the functions and not on the hierarchical position of the project roles.

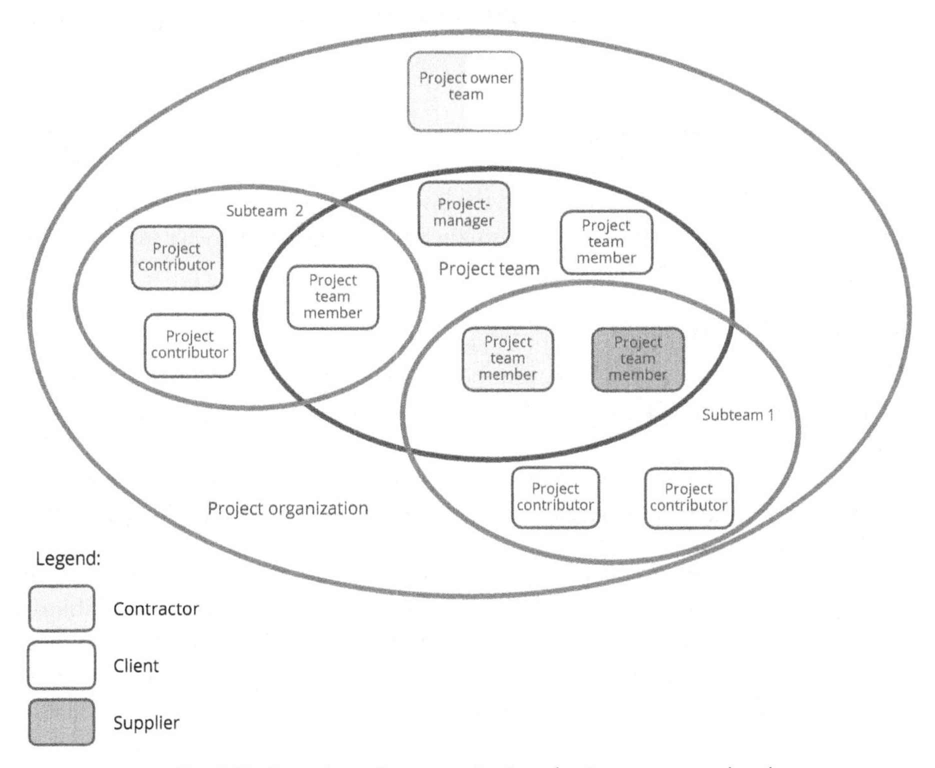

Fig. 7.10: Generic project organization chart—empowered and integrated project organization model.

In the following, empowerment in and of projects, integration of project stakeholders, partnering, and the creation of virtual structures are described as essential elements for the organizational design of projects. The objective of

> empowerment is to transfer decision-making powers and responsibilities to project team members, project teams, and projects,
> integration is to design holistic project organizations by integrating representatives of project stakeholders,
> partnering is to promote the integration of representatives of customers, suppliers, and subcontractors through the use of joint contracts and incentive systems, and
> virtuality is to use virtual communication structures for cooperation between members of the project organization and various partners who work in different locations.

The systemic project organization models presented here provide for a combination of these design elements. In concrete project situations, they usually involve an empowered and integrated project organization with virtual communication elements. The integration of partners through joint contracts and incentive systems is a further design option, but will only rarely be necessary for projects.

7.2.3 Empowerment as a Design Element for Project Organizations

Weaknesses of Matrix Project Organization

Matrix project organization uses a hierarchical organizational approach, which separates leadership functions from performing functions. Leadership functions are divided between the project manager and the line manager of the project team member. The project team member has the performance functions. The project team member receives directions from both managers.

Matrix project organization is . . .

> Not "lean", because of the large number of members of the project organization. Each project team member requires a manager from the permanent organization for leadership.
> Expensive and slow because of the need for coordination between the managers.
> Not customer oriented, because project team members lack decision-making powers.

"Lean management" requires the transfer of decision-making powers to those responsible for performance in order to speed up the processes, motivate the contributors, and ensure stronger customer orientation. In implementing the concept of empowerment in the design of project organizations, not only project team members, but also the project team and the project itself, can be "empowered".[3]

[3] The term "empowerment" means self-qualification and self-empowerment, strengthening of self-sufficiency, autonomy, and self-disposal.

Empowerment of Project Team Members

Empowerment of project team members is achieved by transferring the HOW? and HOW WELL? authority from the project team members' line managers to the project team members themselves (see Fig. 7.11). As a result, project team members are able to decide how assigned work packages are fulfilled within the governing guidelines of the organization. They take responsibility for the work package results. In project meetings and workshops, project team members can agree upon objectives regarding their work packages with the project manager and other project team members.

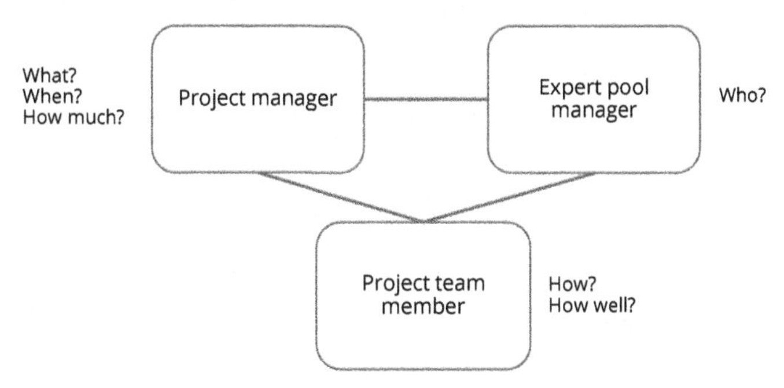

Fig. 7.11: "Empowerment" of the project team member.

To make clear the difference in the allocations of decision-making authority, the role name of the project team members' line managers in the permanent organization can be changed from "Head of department" to "Expert pool manager". An expert pool manager does not have a self-understanding as a "super-expert" who must make all decisions for the project team members, but is the leader of a pool of differently qualified experts. The expert pool manager still has the WHO? authority and, therefore, decides on the disposition of personnel.

If there is a need for a department head or expert pool manager to contribute to the content of a project, he or she can perform all project roles, from member of the project owner team to project team member, as required. If such a contribution is not necessary, the department head or expert pool manager should be considered as a stakeholder and must be kept informed accordingly to ensure the success of the project.

Empowerment of the Project Team

The project team is empowered by participation in project managing and by being given the WHAT? WHEN? and HOW MUCH? authority (see Fig. 7.12). The project team has the opportunity to self-organize. The project team assumes joint responsibility for project success. The project manager, as a special role in the project team, is responsible for the professionalism of project managing.

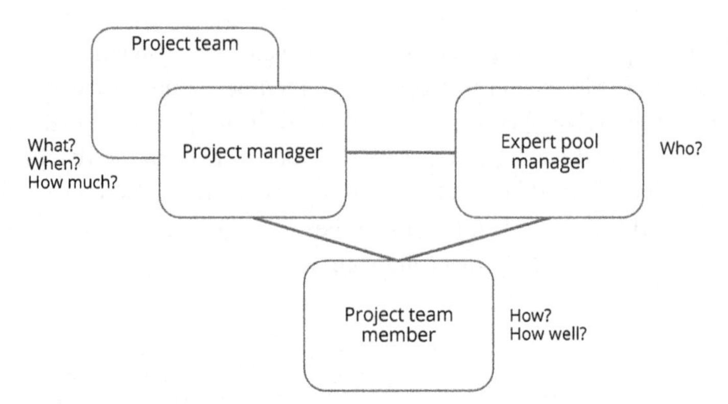

Fig. 7.12: "Empowerment" of the project team.

Empowerment of the Project

Empowerment of the project is achieved by setting boundaries between the project and its social context. This is to prevent frequent interventions by the permanent organization. A minimum level of project autonomy must be ensured in order to ensure efficient project performance. Symbolically, the empowerment of the project is represented in the project organization chart by the ellipse creating a project boundary (refer to Fig. 7.10 on page 145).

7.2.4 Integration as a Design Element for Project Organizations

Hierarchies of Project Organizations and Parallel Project Organizations

In practice, hierarchies of project organizations and/or several parallel project organizations are often used to perform a project.

For example, an investor defines a project to establish a new IT system. In the framework of this project, a general contractor is awarded the contract. This general contractor defines a project to fulfill this customer order. Within this project, the general contractor further awards one or more main contractors with contracts, and those again award various subcontractors with contracts. Because of the scope of the services of individual contributing organizations, it is appropriate from their point of view to define projects for their respective scope of work. The result of this approach is a hierarchy of project organizations and several parallel project organizations following the contractual structures (see Fig. 7.13).

Different project management terms and methods are likely to be used for all the projects in this hierarchical and parallel structure. Project organizations with different

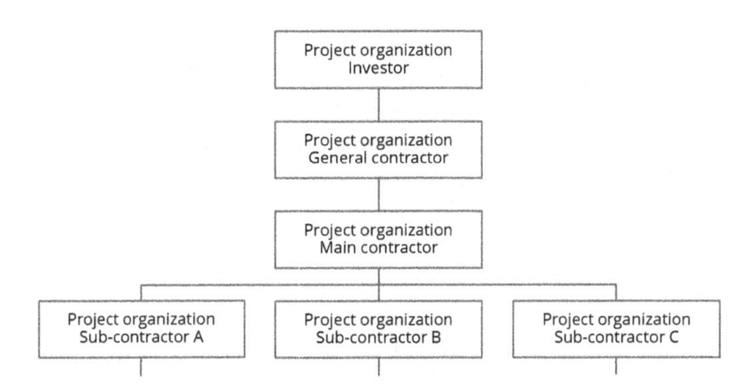

Fig. 7.13: Hierarchy of project organizations and parallel project organizations.

project managers are created, and different project plans are developed for the corresponding scope of services. Cultural and structural uncertainties arise. Multiple project managers work with different plans and tools to implement the investor's solution. The demands of the investor, which are ultimately realized by the subcontractors, are often misinterpreted as a result of "Chinese whispers", and this leads to conflicts and repeated work.

Design of an Integrated Project Organization

An "integrated project organization" is a project organization in which representatives of various organizations are members. It can basically be divided into horizontal and vertical collaborations in integrated project organizations. In vertical collaboration, representatives of an investor, suppliers, and subcontractors may be members of an integrated project organization. In horizontal collaboration, representatives of several partners of equal rank may be members.

The project for which an integrated project organization is designed is to be planned by the investor who demands a solution. The project organization has to be designed in such a way that the investor's business case can be optimized. The project boundaries are therefore defined by the investor's business case and not by any contractual structures.

Collaborations in integrated project organizations require a decoupling of organizational structures from contractual structures. The multi-stage customer-supplier relations are not considered when designing the project organization, but common project organizations are created which are as flat as possible, and in which the representatives of the different organizations fulfill different roles. Representatives of a supplier and of the investor can form a project owner team, a representative of the supplier can be the project manager, a representative of a subcontractor can be a project team member.

Even without a contractual agreement, this integrated organizational design results in of the use of a common project management approach, a single project manager, and the development of common and integrated project plans. All these measures require a high degree of openness and trust among the collaborating organizations. The collaborating organizations must recognize the benefits of a holistic project view and be willing to assume overall responsibility for the project. It also has to be accepted that not every collaborating organization has its own project, its own project manager, and its own project plans.

The objectives of an "integrated project organization" are:

> optimized processes and project results through a holistic project view, and a "big project picture" shared by all partners involved in the project,
> open project communication between all project partners,
> project objectives shared by all participants, creation of win–win situations, avoidance of sub-optimality,
> reduced project costs by using only one project manager, by producing project management documentation only once,
> efficient processes as a result of a common project language and project rules, and
> common project plans, possibly "open books".

Possible risks of integrated project organizations are:

> loss of know-how through collaboration with potential competitors,
> non-compatible cultures of collaborating organizations,
> complex collaboration processes, and
> unclear responsibilities and liabilities.

The case study "Integrated Project Organization" shows that it is difficult to implement the integrated project organization model.

Case Study: Integrated Project Organization

Company performing the project

> Austrian telecommunications company with about 200 employees
> Products and services: telephony, internet services, business solutions project

Project

> Building of a regional network as a pilot infrastructure to enable services for commercial users.
> Provision of a billing system for commercial handling of the services.
> Development of the organizational and personnel requirements for handling the services.

Contractual structures

- > Award of two main contracts to a technical contractor for the construction of the technical infrastructure and to a commercial contractor for the implementation of the billing system.
- > Development of the required organizational and personnel structures by the telecommunications company.
- > Planning of the project organization according to the contractual structures.
- > Parallel project organizations for the investor and the technical and commercial main contractors (see Fig. 7.14).
- > Three project managers for what is *de facto* a single project.
- > Informal communications expected between the role holders of the three projects.
- > Three sets of project plans based on different project management approaches.

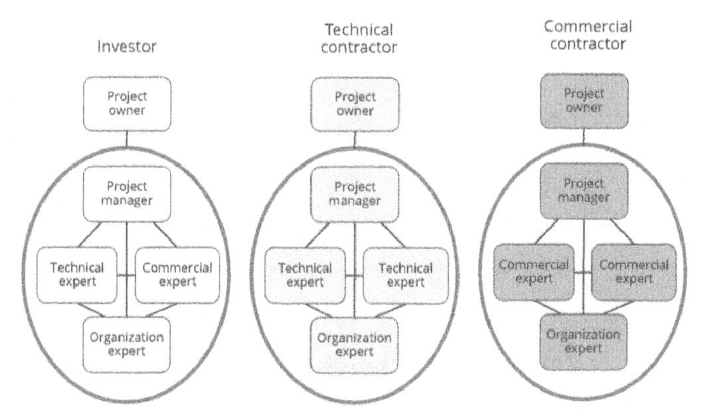

Fig. 7.14: Parallel project organizations.

Integrated Project Organization as an Alternative Design Proposal

- > Representatives of the investor and the two main contractors (possibly also the selected subcontractors) as members of a common project organization (see Fig. 7.15).
- > Use of only one project manager (nominated, for example, by the main technical contractor).
- > Common formal project communication structures.
- > Common project plans.
- > Decoupling of the organizational solution from the contractual structures.

Decision About the Organization

The investor opted for the traditional form of parallel project organization, because . . .

- > There were fears about unclear liability issues in the case of integrated project organization.
- > Informal communication structures and the power of the contract were trusted.
- > There was a fear of applying a new model.
- > Organizing was not recognized as a success factor.
- > Lawyers were dominant in the company and influenced the decision.
- > There was a tension between "execution of contracts" and "successful project performance".

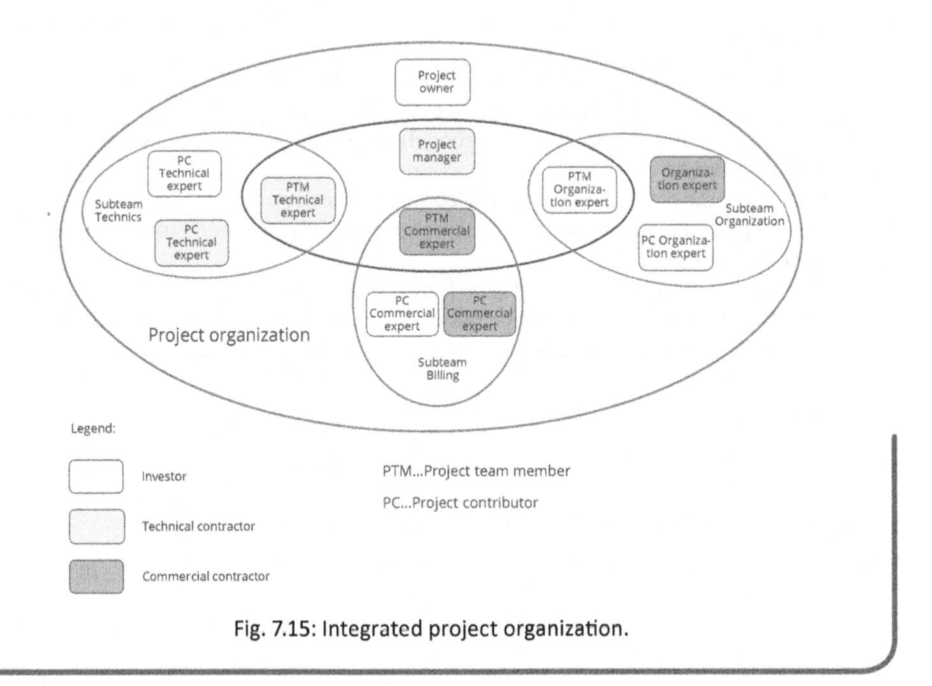

Fig. 7.15: Integrated project organization.

The concept of "partnering" goes beyond the definition of the objectives of the integrated project organization, in that these are supported by contractual framework conditions. Joint incentive systems for all partners involved in the project should formally promote their cooperation.

7.2.5 Virtuality as a Design Element for Project Organizations

Projects can be perceived as virtual organizations because they usually require cooperation between several partners in different locations. Virtual project organizations allow flexibility, fast responsiveness, low costs, and good resource utilization.

In the case of cooperation between several partners in projects, virtual structures are to be created in which team members can work together across space, organization, and time with the help of IT-supported communication formats.[4] Standardized project management and office software has to be used, and the necessary hardware has to be made available. Communication tools such as project management portals, collaboration software, telephone conferencing, and video conferencing can be applied.

Virtual working in projects may result in a lack of personal and informal contacts between the members of the project organization. The project team members mostly cannot be observed in their project work. This can make it difficult to identify

[4] See Henderson, L. S., 2008.

undesirable developments in the project. Communication, coordination, and the establishment of relationships between project team members, which previously took place by means of personal contact, is primarily based on the use of ICT.

Verburg et al. identified two categories of conditions that must be met for a virtual project organization to be effective: clear communication rules, openness, and trust within the team, as well as the provision of a technical infrastructure by the organization.[5] Because of the complexity of virtual project organizations, appropriate training is essential for the success of projects.[6]

Members of virtual project teams who have previously received training on topics such as teamwork, intercultural competence, or use of technology are more likely to be in a position to exploit the potential of virtual structures.

"Face-to-face" communications are also important in virtual project organizations. These are to be combined with virtual communication formats. A "big bang" communication in project starting with as many project team members as possible is necessary to achieve the necessary project management quality. Rules for project communication must be specified, and on-site visits to the project team members by the project manager must be performed.

7.2.6 Consequences of the Use of Scrum for Project Organizations

Scrum can be used for project phases in which tasks can be performed iteratively. Examples could be the project phases "planning a solution" and "developing a solution". If Scrum is used, the Scrum roles and the Scrum communication formats (see Chapter 5) must be taken into account in the design of the project organization.

The Scrum roles to be considered are the "product owner" or the "product owner team", the "Scrum master", and the "Development team". The product owner is a member of the project team, as his or her views and interests are to be coordinated with the other project team members. Above all, the solution requirements must be agreed upon jointly in the project team. It is advisable to define the product owner role explicitly. This promotes the perception of a holistic interest in the content. If several backlog parts are defined (e.g., for the technological solution, the organizational solution, and the marketing solution) this can be dealt with by a product owner team.

A combination of the roles of product owner and project manager is not advisable. In order to maintain the tension between the content-related interests of the product owner and the process-related interests of the project manager, it is advisable for them to be represented by different persons.

[5] See Verburg, R. M. et al., 2013.
[6] See Gilson, L. L. et al., 2015.

The development team is in most cases a sub-team in the project whose coordinator is the Scrum master. The Scrum master is therefore also a project team member. Figure 7.16 shows a generic project organization chart that takes the Scrum roles into account.

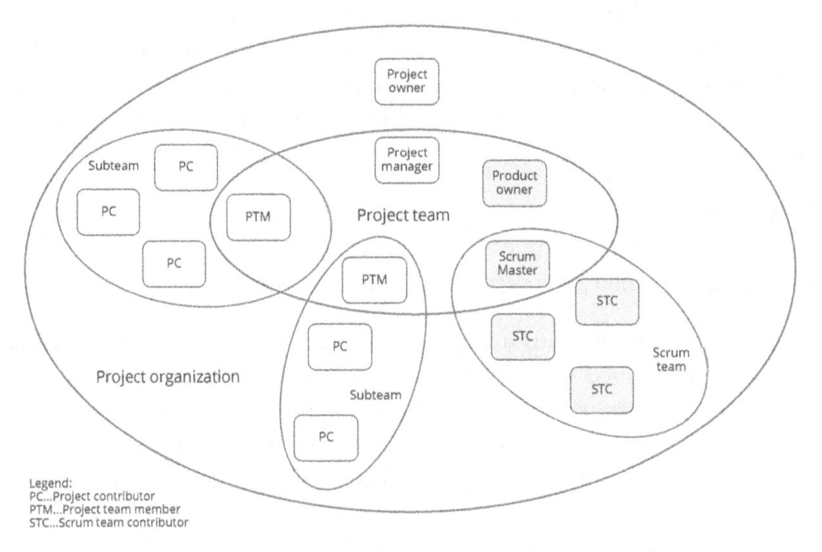

Fig. 7.16: Project organization chart when Scrum is applied.

The Scrum communication formats sprint planning, daily scrum, sprint review, and sprint retrospective can be used in the Scrum team. The specific Scrum roles and communication formats create a specific subculture in the Scrum team that differs from the cultures of the other sub-teams of a project. But this contributes to the success of the project. The necessary integration has to take place at the project team level.

7.3 Project Culture

Corporate culture can be defined "as a set of shared mental assumptions that guide interpretation and action in organizations by defining appropriate behavior for various situations".[7] The mostly unspoken assumptions reveal themselves in a web of formal and informal practices as well as in visual, verbal, and material artifacts.

Culture derives from the human capacity for socialization. Individuals bring the cultural standards that they have learned in primary socialization into an organization as basic assumptions. There, these basic assumptions are given an "organizational-cultural specification" by means of secondary socialization. This model can be applied to both permanent and temporary organizations.

[7] Ravasi, D., Schultz, M. 2006, p. 437.

7.3.1 Developing a Project-Specific Culture

The objectives of developing a project-specific culture are

> to encourage the identification of members of the project organization with the project,
> to provide orientation in the project,
> to reduce project complexity,
> to ensure the recognizability of the project, and
> to create a basis for project marketing.

The members of a project organization who come from different cultural backgrounds can develop a project-specific culture for temporary collaboration in a project. This is not based on tradition as is the culture of a permanent organization, but is to be created by appropriate use of resources and methods. The development of a project culture must be given time and space during project starting. Further development of the values and rules requires reflection in the project, which can primarily be achieved in project controlling.

The project name, a project logo, project values and possibly a project mission statement, project-specific rules, project slogans and anecdotes, the project language, project-related artifacts, a project office, and project-related events are elements of the project culture which are to be developed. The use of symbols supports the development of a project culture.

7.3.2 Elements of a Project Culture

Project Name and Project Logo

The project name should enable the project to be identified. The project name can provide information about the project type and should relate to the project objectives. Conceptual projects must be distinguished from implementation projects. For example, it is clear from the project names "Conceptualizing Values4Business Value" and "Developing Values4Business Value" that a chain of innovation projects is involved. Project names should be short and should use no abbreviations that are not understandable to project stakeholders.

The use of a project logo seems to require complex design work. However, it is rarely necessary to create a fancy graphical solution as a project logo. Often it is sufficient to use a typeface. So, for example, a project logo can be created by writing the project name in colored italics. A specific "project color" can be used for this purpose.

Project Values and Project Policy

Project values provide orientation for what is considered good and desirable in a project. These values consciously and unconsciously determine the behavior of members of the project organization. They are a management tool. In defining them, one can distinguish between results-related and process-related values. Project values can be summarized in a project mission statement.

Project-Specific Rules

Project-specific rules should provide orientation for behavior in the project. These rules can be established in addition to the general rules of the project-oriented organization. These may relate to signatory rights, decision-making authority, project documentation and filing, behavior in meetings, etc.

Project Slogans and Project-Related Anecdotes

Project slogans are intended to convey briefly and concisely what is important in a project or project phase. The image of a project can also be shaped by project-related anecdotes. An anecdote about a constructive and cooperative discussion between the project manager and a customer can make a significant contribution to the management of project-related customer relations.

Project-Related Artifacts, Project Language, and Project Office

Project-related artifacts are primarily the project plans. The project culture is designed and communicated through their content as well as their form. The language used in a project is shaped by the communication formats, project plans, and other project management documents. The use of uniform terms and names contributes to the reduction of project complexity.

A dedicated project office is an "organizational home" for a project. These project rooms do not always have to be available to the project. They nevertheless contribute to the project identity if the same rooms are always available for the project meetings. An appropriate design of the project rooms with selected project artifacts such as printouts of the work breakdown structure and the project organization chart, as well as some project photos, strengthens the perception of an "organizational home".

Project-Related Events

Various objectives can be pursued by means of project-related events. Outdoor events are used for teambuilding, project vernissages for providing information about the project to project stakeholders, and social events for networking between members of the project organization.

The organizing of project-related events depends on the size and the strategic importance of the project, as they incur costs. However, even in small projects, it is advisable, for example, to have "social" interaction at the end of the project start workshop.

Project-Related Symbols

A symbol is a behavior or an artifact which conveys meanings that go beyond the actual content of the behavior or thing. A symbol enables interpretations to be made in a larger context. It is "a sign, which denotes something much greater than itself, and which calls for the association of certain conscious or unconscious ideas, in order for it to be endowed with its full meaning and significance".[8]

One can distinguish between verbal, interactional, and objectified symbols (see Table 7.17). Verbal symbols such as anecdotes, slogans, and speeches use language as a tool. Interactional symbols, such as celebrations, meetings, and awards, are symbolic behavior. Objectivized symbols or symbolic objects, such as rooms design, logos, and organization charts, are artifacts of an organization. Many elements of the project culture contribute to its development. Their symbolic character is interpreted in Table 7.17.

Table 7.17: Project-Related Symbols and Their Interpretations

Verbal symbols	Possible interpretation
Project slogan	Key objectives and values of the project
Project-related anecdotes	Key values and standards of the project
Project language, jargon	Belonging to the project
Interactional symbols	
Provision of scarce resources	High strategic importance of the project
Seating arrangement for project meetings	Power in the project
Social events	Personal interest of the members of the project organization in one another
Milestone celebration	Start of a new project phase
Informal address	Belonging to the project, setting boundaries with other organizations
Burning of old project plans	Agreement of new project objectives, dates, etc.
Objectified symbols	
Graphical representation of the project organization chart	Power in the project, significance of individual roles, of relations between roles
Project logo, project name	Challenges and objectives of the project
Provision of a project room	High importance of project teamwork, home for the project
Size, furnishing of the project office	Status of the project in the organization

The functions of symbols can be divided into three categories: describing, energy controlling, and system maintaining.[9]

> Symbols have a descriptive effect: In temporary organizations, it is particularly important to describe the organization for the representatives of project stakeholders and the members of the project organization. Symbols can be used to pass on information in a complex form. A project can be described by facts and figures, but also by symbols. Symbols are, for example, the colors green, yellow, and red for assessing the project status in a project scorecard.

[8] Pondy, L. R. et al, 1983, p.4 f.

[9] See Dandridge, T. C. (Symbols) 1983, p. 71.

> Symbols have an energy-controlling effect: Symbols such as an event for project starting can be used to motivate members of the project organization and possibly to recruit personnel. Similar events might facilitate the "recalling" of good feelings and experiences.
> Symbols have a system-maintaining effect: Symbols are important for protecting, stabilizing, or supporting a project—for example, project plans as artefacts, which are continuously updated, assure continuity in a project. Project communication is to be supported by these project plans.

Symbols may be perceived in different ways. This requires the project manager to interpret the symbols used. Even in projects one does not have the choice about whether or not to use symbols. Management is always performed with symbols. The only question is whether one is conscious of it.

Literature

Botta, C.: Das Role Model Canvas—Rollen schnell und gemeinsam definieren, *Projekt Magazin,* 07, 2016.

Cleland, D. I., King, W. R.: *Systems Analysis and Project Management,* McGraw Hill, New York, NY, 1968.

Dandridge, T. C.: Symbols' Function and Use, in: Pondy, L. R., Frost, P., Morgan, G. (Eds.), *Organizational Symbolism,* pp. 69–79, Jai Press, Greenwich, CT, 1983.

Gilson, L. L., Maynard, M. T., Young, N. C. J., Vartiainen, M., Hakonen, M.: Virtual Teams Research: 10 Years, 10 Themes, and 10 Opportunities, *Journal of Management,* 41(5), 1313–1337, 2015.

Henderson, L. S.: The Impact of Project Managers' Communication Competencies: Validation and Extension of a Research Model for Virtuality, Satisfaction, and Productivity on Project Teams, *Project Management Journal,* 39(2), 48–59, 2008.

Pondy, L. R., Frost, P., Morgan, G. (Eds.): *Organizational Symbolism,* JAI Press, Greenwich, CT, 1983.

Ravasi, D., Schultz, M.: Responding to Organizational Identity Threats: Exploring the Role of Organizational Culture, *Academy of Management Journal,* 49(3), 433–458, 2006.

Schulte-Zurhhausen, M.: *Organisation,* 6th edition, Vahlen, Munich, 2013.

Verburg, R. M., Bosch-Sijtsema, P., Vartiainen, M.: Getting It Done: Critical Success Factors for Project Managers in Virtual Work Settings, *International Journal of Project Management,* 31(1), 68–79, 2013.

8 Teamwork and Leadership in Projects, Competencies for Projects

Projects require the use of interdisciplinary teams—namely, project owner teams, project teams, and project sub-teams. The teams in projects are temporary and follow a team life cycle, with the phases forming, storming, norming, performing, and adjourning, as covered in Section 8.1.2.[1] Team building is necessary for a team to work efficiently.

Leadership is a person or team-related intervention. Leadership in projects is a task to be fulfilled in all sub-processes of project managing. Leadership tasks are fulfilled by the project owner and the project manager, but also by project team members. Social competencies are necessary for managing social interactions in projects. Prerequisites for this are self-competencies and an appropriate self-understanding for fulfilling a project role.

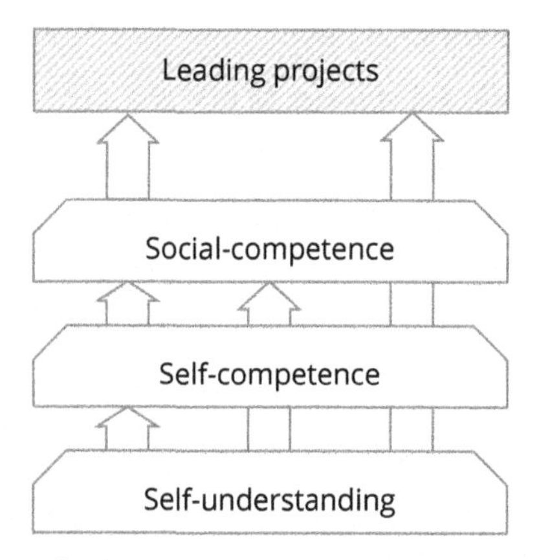

Overview: Leadership in projects based on social competencies, self-competencies, and self-understanding.

[1] Tuckman, B. W. et al., 1977.

8 Teamwork and Leadership in Projects, Competencies for Projects

8.1 Teams in Projects

Various teams—namely, project owner teams, project teams, and project sub-teams, can be required in projects. An essential feature of teams in projects is their temporary nature.

Teams in projects are often faced with strong time pressure. As a result, there is usually relatively little time available in projects for appropriate team building.

8.1.1 Team: Definition

A team is characterized by the involvement of several persons who interact over a period of time and build up a specific structure of roles and norms. Teams are established to produce results that could not be produced by individuals alone. Teams generate added value.

A team can be distinguished from a group on the basis of its objectives and the responsibility for achieving these objectives. A group is used to achieve the objectives of individuals. A team has a common objective, for which the team is collectively responsible. For instance, in a learning group, each group member pursues individual learning objectives. It tries to optimize the achievement of the objectives of the individual group members through group work. An example of a team is a football team that has a common objective, such as winning a championship. The team as a whole is responsible for achieving this objective.

Teams also differ from groups in that the skills of the team members are complimentary. The sense of belonging together is greater in teams than in groups. Working in teams requires a minimum level of mutual trust and the building of "team spirit".

Groups and teams can be perceived as social systems. In the typology of social systems developed by Luhmann, groups and teams are categorized as being between interactions and organizations (see Fig. 8.1).[2]

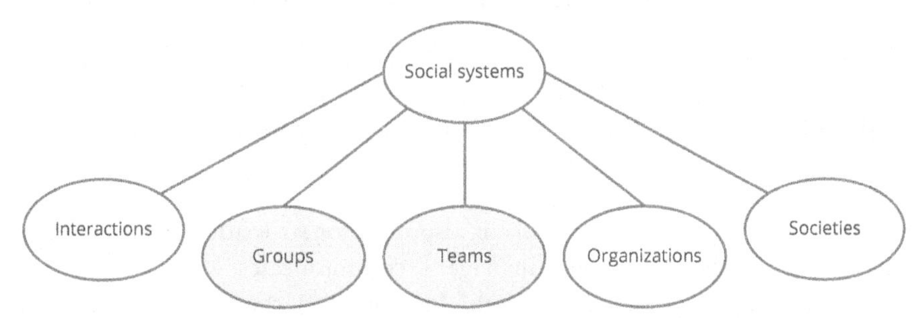

Fig. 8.1: Groups and teams in the typology of social systems.

[2] In practice, the term "project team" is sometimes used instead of the term "project". This is because projects are not perceived as standalone organizations, which can be distinguished from teams.

Teams are characterized by stronger task orientation and more pronounced formalism as compared to groups.

There are different types of teams. There are different leadership requirements for different teams (see Table 8.1).

Table 8.1: Types of Teams and Their Characteristics

Types of team	Characteristics
Permanent versus temporary team	> Permanent team: Oriented towards long-term existence, continuity is to be ensured > Temporary team: Establishing, developing and dissolving a team
Management versus working team	> Management Team: Focus on the performance of management processes > Work team: Focus on the performance of content-related processes
Large versus small team	> Large team: Recommended size up to 12 people > Small team: Up to 6 people. Management and coordination requirements decrease with the number of people involved.
Heterogeneous versus homogeneous teams	> Heterogeneous team: Different competences of team members > Homogeneous team: Similar competences of team members.
Locally concentrated versus virtual teams	> Local team: Working in a common place, personal interaction > Virtual team: Working spatially separated. Use of different project communication formats

Project owner teams and project teams are usually temporary, heterogeneous, and virtual management teams. Project sub-teams, on the other hand, are working teams that are small, temporary, relatively heterogeneous, and virtual. Homogeneous teams have less potential for conflict than heterogeneous teams. However, because of their diversity, heterogeneous teams have greater potential for dealing with new tasks.

Teams are needed when the scope and/or complexity of tasks cannot be dealt with by an individual. Teamwork should enable the exploitation of synergies and the creation of added value. This requires team members to communicate directly with one another.

In practice, however, the team concept is also used for structures in which only one person—for example, the project manager—communicates with the other team members. In this case of "false teamwork" (see Fig. 8.2), the above-described added value is lost as a result of selected information, and the potential for creativity is lost through the limitation of opportunities for interaction.

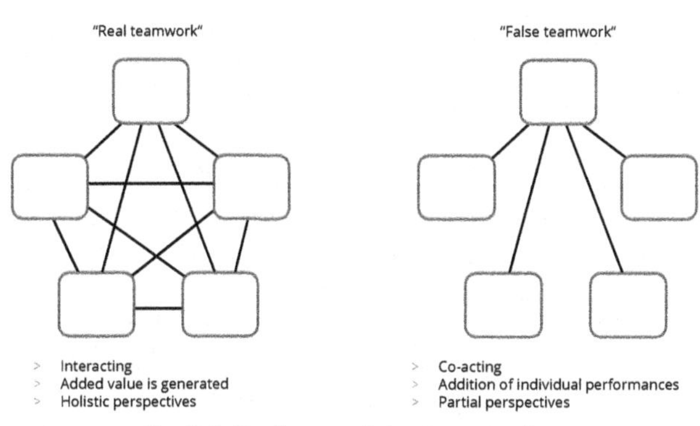

Fig. 8.2: Real versus false teamwork.

Project teams require social competence to accomplish their tasks and achieve their objectives. While team competence is based on the competencies of individual team members, it has to be developed specifically for a team. Team competence can be defined as the capability of a team

> to manage the team work,
> to learn as a team,
> to create commitment in the team,
> to construct common realities,
> to exploit synergies in the team, and
> to resolve conflicts in the team.

A capable team is characterized by team competence. The specific characteristics of a capable team are summarized in Table 8.2.

Table 8.2: Characteristics of a Capable Team

Objectives and tasks	• Objectives are clear to all team members and are accepted
	• Responsibilities for the tasks are defined
Collaboration	• Uncertainties are discussed and clarified
	• Agreements made are binding and are observed
	• Rules are established collectively
	• Management responsibility has been made clear and is accepted
	• Creativity is encouraged, synergies are exploited
	• Collective learning takes place in the team
Communication	• Communication in the team is open and spontaneous
	• All team members communicate with one another
	• Team members listen to one another, can involve themselves appropriately
	• Meetings are planned, prepared, conducted and followed up

(continues on next page)

Table 8.2: Characteristics of a Capable Team *(cont.)*

Dealing with conflicts	• Differences and different opinions are accepted
	• Conflicts are picked up on and dealt with; they are understood as an opportunity for further development
Decision making	• Decision making is transparent and comprehensible
	• Consensual decisions are sought
Working atmosphere	• Work is performed professionally
	• Team members have a good relationship with one another
	• Appropriate tools are available

8.1.2 Life Cycle of Teams in Projects

According to the Tuckman model, teams have a life cycle which includes the phases forming, storming, norming, performing, and adjourning (see Fig. 8.3).[3] Options for managing these phases are described in the following.

Fig. 8.3: Team life cycle according to Tuckman.

Forming: Selection of Team Members

The objective of "forming" is to select the team members and to perform team building. The basic composition of a team will already have been planned in project initiating. In project starting, the people who will hold the roles in the team are selected.

Expert competencies, project management competencies, and social competencies must be considered in selecting the members of the project owner team, the project team, and possible project sub-teams. These competencies should be developed differently for different project roles. The selection of team members in projects is based not only on the competencies of the team members, but also on their relationships with other team members and project stakeholders. This "relationship capital" can be a key success factor for projects.

The members of the project owner team are selected by the project portfolio group (see Chapter 16) or a similar committee of the project-oriented organization. The project manager is selected by the project owner, and the other project team members are selected by the project manager in accordance with the project owner and the expert pool managers. Sub-team members are selected by the corresponding project team member in consultation with the project manager and the expert pool managers.

[3] See Tuckman, B. W. et al., 1977.

Forming: Team Building

Team building contributes to orientation and should create a basis for efficient cooperation between team members. The team members get to know each other, and an understanding of the team roles is created. There is no "we-feeling" yet. Reference points are primarily the common objectives and tasks.

In addition to the formal structures, however, informal structures such as informal channels of communication, coalitions of team members, and informal roles are already forming in the team in this early phase (see Table 8.3). The existence of informal structures is also important for the success of teamwork.

Table 8.3: Possible Informal Roles in a Team

Informal roles	Focus in the team
Enforcer	Enforces the own opinion
Analyst	Analyzes situations and develops solutions
Integrator	Forms relationships in the team
Controller	Controls time and progress
Follower	Avoids conflicts
Worker	Performs the work

Team building in projects takes place mainly during project starting. However, additional sub-teams for subsequent project phases may become necessary during project performance. In this case, the selection of personnel and team building for these sub-teams must be carried out at a later time. Because the composition of project teams can also change over the course of time, special attention must be paid to the integration of new team members. "Disruptions have priority!" is a key rule for teamwork.

Storming

The storming phase is characterized by conflicts and confrontations within the team. These are caused by the "me" orientation of individual team members and their interest for self-promotion, but also by unclear agreements and missing rules. Rivalries regarding (informal) leadership positions can also arise. There is little progress in content-related work. All of this results in frustration. The need to create clarity and rules is recognized.

Norming

The objective of norming is to resolve conflicts and confrontations. This requires the clarification of roles, the forming of relationships between team members, and the establishment of project rules. A "we-feeling" is established. The established rules and structures are documented. A basis is created for working efficiently.

Performing

Once the relevant conditions have been clarified by "norming", the energy of the team focuses on the content and achievement of the objectives. Cohesion in the team provides motivation. The team organizes itself to a large extent. Team members are coordinated by the team leader. Leadership tasks are performed to enable the achievement of the team objectives.

Adjourning

Once the objectives have been achieved, the team is dissolved. The objectives of "adjourning" are documentation of the work results, reflection on the cooperation and "social" closing. Team members have an opportunity to thank each other, to give and to receive feedback.

Adjourning usually takes place during project closing. It is, however, possible that sub-teams are no longer needed and are therefore dissolved before the project end.

Optimizing the Life Cycle of Teams

The objective of team building is to establish a team as a social (sub-)system. Competencies have to be developed for team learning, for creating commitment, for exploiting synergies, and for resolving conflicts in the team.

The processes of "forming" and "norming" are usually given little attention in practice.[4] As a result, teams often do not work efficiently. Individual interests are in the foreground. "Storming" is part of everyday life, and the objectives achieved do not meet expectations.

It is therefore advisable to pay more attention to "forming" and to combine a first "norming" with "forming". This can reduce the frequency and intensity of "storming" and quickly lead to productive "performing".

Cohesion in the team, securing of shared views, clarification of the roles, creation of a common project language, etc. can be promoted by the use of project management methods in project starting. The collaborative development of project plans in project start workshops, the provision of collective project-related training, and "social" events support the social development of project teams.

In project controlling, the structures of the project owner team, the project team, and the sub-teams can be reflected upon, successful action and successful behavior can be reinforced by positive connotations, and new rules can be agreed upon. This results in a further process in the team's life cycle—namely, "reflecting and renorming".

[4] The process of team building is not performed explicitly in practice, especially for project owner teams.

8.2 Leading in Projects

8.2.1 Leadership Tasks and Leadership Styles

To lead means to intervene; it means to have a targeted influence on the behavior of persons or teams. Leadership "[. . .] aims to achieve objectives through communication processes".[5] Leadership secures the results orientation of those who are led, their concentration on what is essential, and their use of existing strengths.

"Leadership defines what the future should look like, aligns people with that vision, and inspires them to make it happen despite the obstacles".[6]

"Leadership involves establishing a clear vision, sharing that vision with others so that they will follow willingly, providing the information, knowledge and methods to realize that vision, and coordinating and balancing the conflicting interests of all members and stakeholders. A leader steps up in times of crisis, and is able to think and act creatively in difficult situations".[7]

Leadership Tasks

Basic leadership tasks include observing, constructing, informing, making agreements, allocating tasks, deciding, checking, and giving feedback.

By observing a person or team to be led, their strengths and weaknesses can be analyzed. Constructing common views of the leader and those who are led provides clarity with regard to a possible need for support. Informing those who are led about relevant contexts provides them with sense regarding their work. However, they must also be informed about the methods, available tools, etc. which are to be used. Agreements are primarily concerned with the objectives to be achieved and the tasks to be performed, but can also involve, for example, intended measures for personal development. Leadership decisions support the fulfillment of tasks. Checking that tasks have been fulfilled provides a basis for the leader to give feedback regularly.

Leadership Styles

A leadership style is to be understood as consistent leadership behavior of a given type. There are different types of leadership styles described in the literature. According to Tannenbaum and Schmidt, leadership styles can be distinguished according to the decision-making authority of the manager in comparison to authority of the group that is led (see Fig. 8.4).[8]

[5] See Weinert, A. B., 1989.

[6] Kotter, J.P., 2012, p. 25.

[7] See Business Dictionary: "Leadership" (www.businessdictionary.com/definition/leadership.html, retrieved on 24.2.2017).

[8] See Weibler, J., 2001, p. 300.

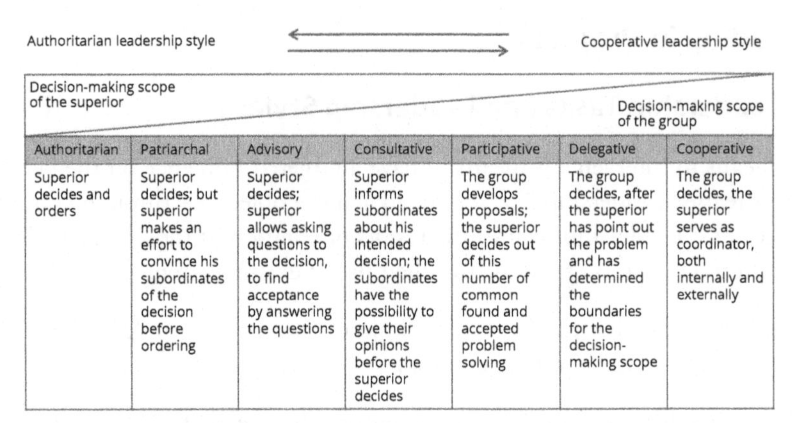

Authoritarian	Patriarchal	Advisory	Consultative	Participative	Delegative	Cooperative
Superior decides and orders	Superior decides; but superior makes an effort to convince his subordinates of the decision before ordering	Superior decides; superior allows asking questions to the decision, to find acceptance by answering the questions	Superior informs subordinates about his intended decision; the subordinates have the possibility to give their opinions before the superior decides	The group develops proposals; the superior decides out of this number of common found and accepted problem solving	The group decides, after the superior has point out the problem and has determined the boundaries for the decision-making scope	The group decides, the superior serves as coordinator, both internally and externally

Fig. 8.4: Leadership styles according to Tannenbaum/Schmidt.

Leadership styles can be adapted to a specific context. In his contingency model, Fiedler distinguishes between task-oriented and person-oriented leadership styles.[9] He derives the appropriate leadership style from the given situation.

Value-Oriented Leadership

The relatively new concept of "value-oriented leadership" perceives organizational values as the basis of leadership in organizations.[10] Values are relatively stable beliefs about what is important and desirable for an organization. In value-based leadership, values are seen as a basis for action, decision-making, and behavior. The values of an organization are visible, for example, in the priorities it sets.

In order to be able to provide value-oriented leadership, values must be defined by an organization and possibly operationalized by leadership principles derived from them. The values or management principles can then be referred to when performing leadership tasks. Since values can only be formulated in a relatively abstract manner, they have to be interpreted in each situation. They only acquire meaning for those who are led by being interpreted in a specific context.

The definition of values and their operationalization by means of leadership principles can be illustrated by the example of the RGC project management paradigm. In Figure 6.5 in Chapter 6 (page 112), the values underlying the project management approach are shown and briefly interpreted. These interpretations can be used as leadership principles. The leadership principles for the values "quick realization of results and efficiency" are:

> ensure quick wins in projects,
> discontinue unsuccessful projects at an early stage, and
> use project management methods efficiently and flexibly.

[9] See Steyrer, J. (Theorien der Führung), 2002, p. 202 ff.
[10] See Daxner, F. et al., 2005.

Project owners, project managers, and project team members can use these principles for orientation in fulfilling their leadership tasks. In leadership, the observations made, the information provided, the agreements and decisions made, etc. should be based on these values or should contribute to the fulfillment of these values.

A core competency of organizations for practicing value-oriented leadership is their ability to reflect. Regular reflection on the consistency of the values with actual behavior and action ensures the necessary authenticity and enables any necessary adaptations of the behavior and action or the values.

8.2.2 Leading in Projects as a Project Management Task

Leadership Roles in Projects

Leadership is an essential task of project managing. Leadership is to be distinguished from the other tasks of project managing, such as developing and updating project plans, creating minutes and project reports, and so on.

The perception of projects as social systems implies the forming of social relations between the members of the project organization through leadership. Leadership in projects can be understood as personal and team-related communication for achieving the project objectives.

In projects, leadership tasks must be carried out not only by project managers, but also by project owners and project team members. Leadership tasks in projects must be fulfilled both with regard to individuals and with regard to teams.

The leadership roles in projects and the corresponding "persons and teams being led" are shown in Table 8.4.

Table 8.4: Leadership Roles and Persons/Teams Being Led

Leadership roles in projects	Led roles in projects
Project owner	Project manager
Project manager	Project team member, project contributor, project team
Project team member (as sub-team manager)	Project contributor, sub-team

As explained in Chapter 7, project managers tend to have little formal directive authority. Project managers therefore use their informal authority—for example, authority based on knowledge and experience, personal relationships, and charisma—to carry out their leadership tasks.

Leadership in the Sub-Processes of Project Managing

Leadership tasks must be fulfilled in all sub-processes of project management. The context in which leadership tasks are to be fulfilled is different in the individual sub-processes. So, for example, decisions about transforming a project have to be made under greater time pressure than decisions in project controlling, and the provision of information in project starting is usually more important than in project closing.

Projects are complex and dynamic, which means that leadership in projects is also dynamic. Background conditions, assumptions, and information change. These changes need to be addressed: New situations are to be observed and analyzed, new realities are to be constructed, reflections are necessary. The dynamics and the energy in the project are to be managed.

The motivation of the members of the project organization and the productivity of the fulfillment of services can usually not be kept constant over the duration of the project. The "energy" of projects can be controlled by the definition of formal social interactions such as meetings, workshops, or presentations (see Fig. 8.5).

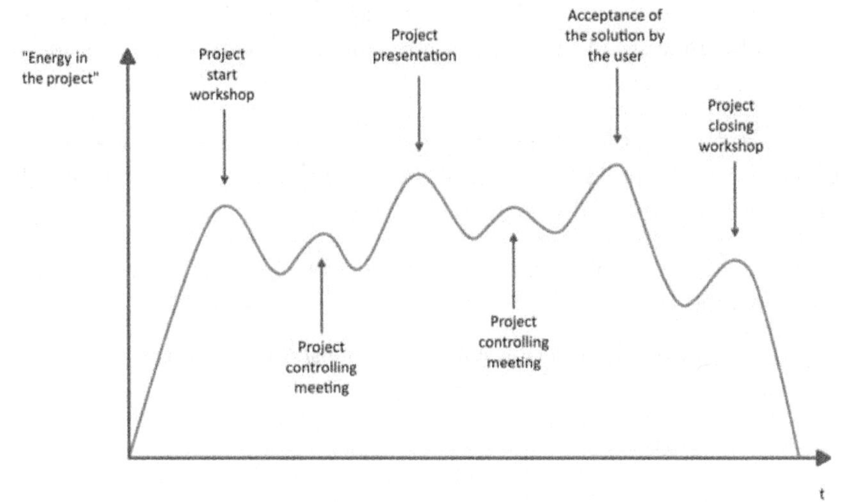

Fig. 8.5: Social interactions in projects as leadership tools.

One objective of periodic formal social interactions is to create pressure in the project by building external pressure. External pressure on a project arises, for example, from the expectations of participants in a project presentation to receive information about the status of the project and interim results. In order to provide projects with sufficient time for "regeneration" and for continuing work, care must be taken that these formal interactions do not take place too frequently.

In leading projects, emotions in the project should also be observed and controlled as required (see Excursus: Leading "By Emotions" in Projects).

Excursus: Leading "By Emotions" in Projects

In projects, there are emotions such as anger, fear, joy, sadness, and surprise. Emotions are intense feelings experienced by individuals or teams. The project team can, for example, be happy about a successful project presentation or angry about negative feedback from a stakeholder.

Emotions in projects can be caused by structural factors or by interventions. It is a leadership task in projects to identify structurally determined emotions and to plan and implement strategies and measures to deal with these emotions.

There are different structurally determined emotions in the different sub-processes of project managing. Typical positive emotions of individuals that can be expected in project starting are enjoying new, interesting tasks, getting to know new people, and cooperating in a new team. Typical negative emotions in project starting are fear of new things, fear of overload with project work, and fear of assuming responsibility in the project.

Being over-challenged or under-challenged both lead to low motivation of project team members (see Fig. 8.6). High motivation is to be ensured by assigning tasks in accordance with the competencies of the project team members.

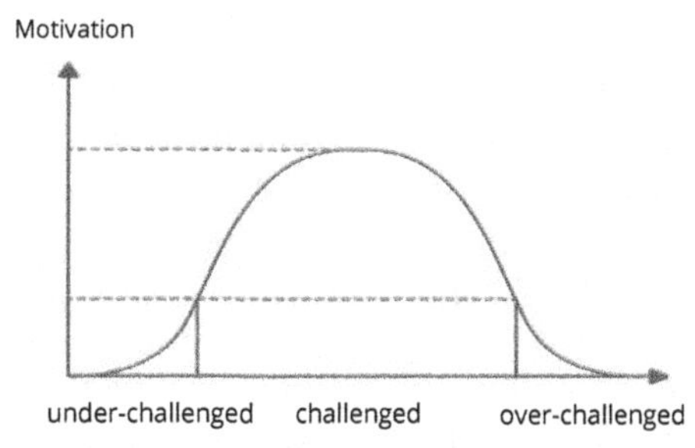

Fig. 8.6: Motivation of project team members in relation to challenge.

Measures for managing positive and negative emotions during project starting are, for example,

> communicating within the project team to shape shared expectations,[11]
> communicating the contribution of the project to achieving the strategic objectives of the organization,
> allocating tasks to individual project team members to ensure an appropriate level of challenge,
> clarifying the project roles to ensure acceptance of the responsibilities assigned to the project team members,
> team building for integrating team members with different cultural backgrounds and for establishing organizational rules,
> collective developing of the project plans to ensure their acceptance by the project team members.

[11] Communication can be supported by visualization. The project culture can also be expressed by the symbols and terms used.

Emotions not only are structurally determined, but can also be triggered by specific causes. Measures for "emotionalizing" in projects are, for example,

> holding project presentations to create periodic pressure in the project,
> addressing taboos in the project team to avoid fears or uncertainties,
> using associative methods such as images, parables, metaphors, etc. for reflecting in the project team to generate surprises,
> developing competitive situations in the project team to increase the quality of project work.

8.3 Competencies for Projects

Competency: Definition

"Competencies are self-organizational dispositions of thought and action".[12] The difference between competency and qualification consists in the fact that qualifications are not revealed by self-organized action but by isolated examination settings which do not determine whether a person can actually apply this knowledge in a self-organized manner.[13]

One can distinguish between the competencies of individuals, teams, and organizations. The competencies discussed in the following—namely, project management competencies, social competencies, and self-competencies—are individual competencies. Team competency has already been defined above. In connection with organization, one often speaks of core competencies, defined as "the ability to do something better than others and thereby generate a strategic advantage".[14]

Competencies as self-directed skills or self-organizational dispositions are context-related—that is, they enable a specific situation to be managed or a specific role in a specific social context to be fulfilled.

Competency is potential based on knowledge and experience. If the necessary competency is not available, this has a negative effect on the results to be achieved. The competency of a person can be assessed with respect to a defined standard and can be further developed. The knowledge and experience of a person are not independent of one another, but influence one another. Knowledge is the foundation on which experience can be gathered.

Project managers need not only project management competencies, but also social competencies and self-competencies, organizational competence, content-related solution competence, inter-cultural competence, and foreign language competence for international projects.

[12] Erpenbeck, J., Von Rosenstiel, L., 2007, p. xi.
[13] *Ibid*, p. xix.
[14] See Prahalad C.K., Hamel G., 1990.

Organizational competence refers to knowledge and experience with the organization performing a project. Project managers do not need specific competencies with regard to the content-related solutions to be developed. However, a basic understanding of the technologies and products used in projects supports the acceptance of the project manager by the members of the project organization, especially in smaller projects.

The competencies of the project manager are based on his or her self-understanding as a project manager.

8.3.1 Project Management Competencies

The project management competence of a project manager can be defined as the ability to fulfill the role of "project manager", based on project management knowledge and project management experience.

In order to ensure that the project managers of a project-oriented organization have the appropriate competencies, minimum requirements for project management knowledge and the project management experience can be defined for the various project management roles. This contributes to clarifying the stages of a project management career path (see Chapter 16). As an example, the requirement profiles for the career stages "project manager" and "senior project manager" are defined in Table 8.5. The knowledge and experience elements of project managing used to describe the requirements are mainly structured according to project management sub-processes.

It can be seen from Table 8.5 that, in addition to knowledge and experience of project managing, project managers also need (to a lesser extent) knowledge and experience of program managing and managing a project-oriented organization.

8.3.2 Social Competencies for Projects

Social competence can be defined as a potential for communicative and cooperative behavior that serves the realization of objectives in social interactions. Social interactions in projects and programs are, for example, meetings, workshops, events, presentations, and negotiations. Social competence is knowledge and experience in managing these situations using appropriate communication methods. Relevant methods are:

- managing others' emotions, networking, virtually communicating,
- negotiating, managing a conflict, facilitating a large group,
- problem solving, working in a team, giving and receiving feedback, and
- presenting and facilitating.

Figure 8.7 (page 175) represents these communication methods and techniques.[15]

[15] A "method" is a regulated process in which techniques for obtaining knowledge and results are applied. A "technique" is understood as a concrete tool or means for carrying something out.

Table 8.5: Minimum Requirements of Project Management Competencies for Project Managers and Senior Project Managers

	Project management knowledge					Project management experience				
Elements	5	4	3	2	1	1	2	3	4	5
Project definition, project management process	SPM	PM						PM	SPM	
Project starting methods: Project contexts	SPM	PM						PM	SPM	
Project starting methods: Project organization	SPM	PM						PM	SPM	
Project starting methods: Project planning	SPM	PM						PM	SPM	
Project coordinating methods	SPM	PM						PM	SPM	
Project controlling methods	SPM	PM						PM	SPM	
Project closing methods	SPM	PM						PM	SPM	
Project transforming and project repositioning methods	SPM	PM						PM	SPM	
Program definition and program management		SPM	PM				PM	SPM		
Managing the project-oriented organization		SPM	PM				PM	SPM		

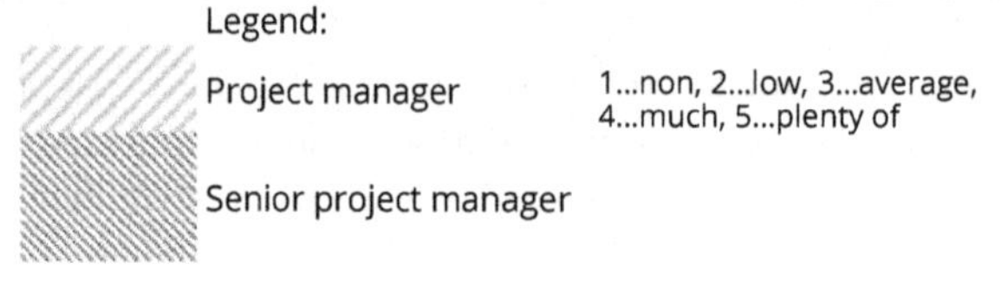

The methods and techniques represented can be used for leading members of the project organization and managing relationships with project stakeholders.[16] Social competencies are required to perform these leadership tasks.

8.3.3 Self-Competencies for Projects

Self-competence can be defined as a potential for self-managing, for managing one's own resources such as time, motivation, energy, and expertise. Self-management serves the realization of personal tasks and objectives. A high level of self-competence is a requirement for the professional fulfillment of the project manager role. Individuals

[16] A description of the methods and techniques depicted in Figure 8.7 is omitted, since there is an extensive literature on this topic, which is also relevant for project managing (see de Janasz et al., 2015 or Polzin, B., Weigl, H., 2014).

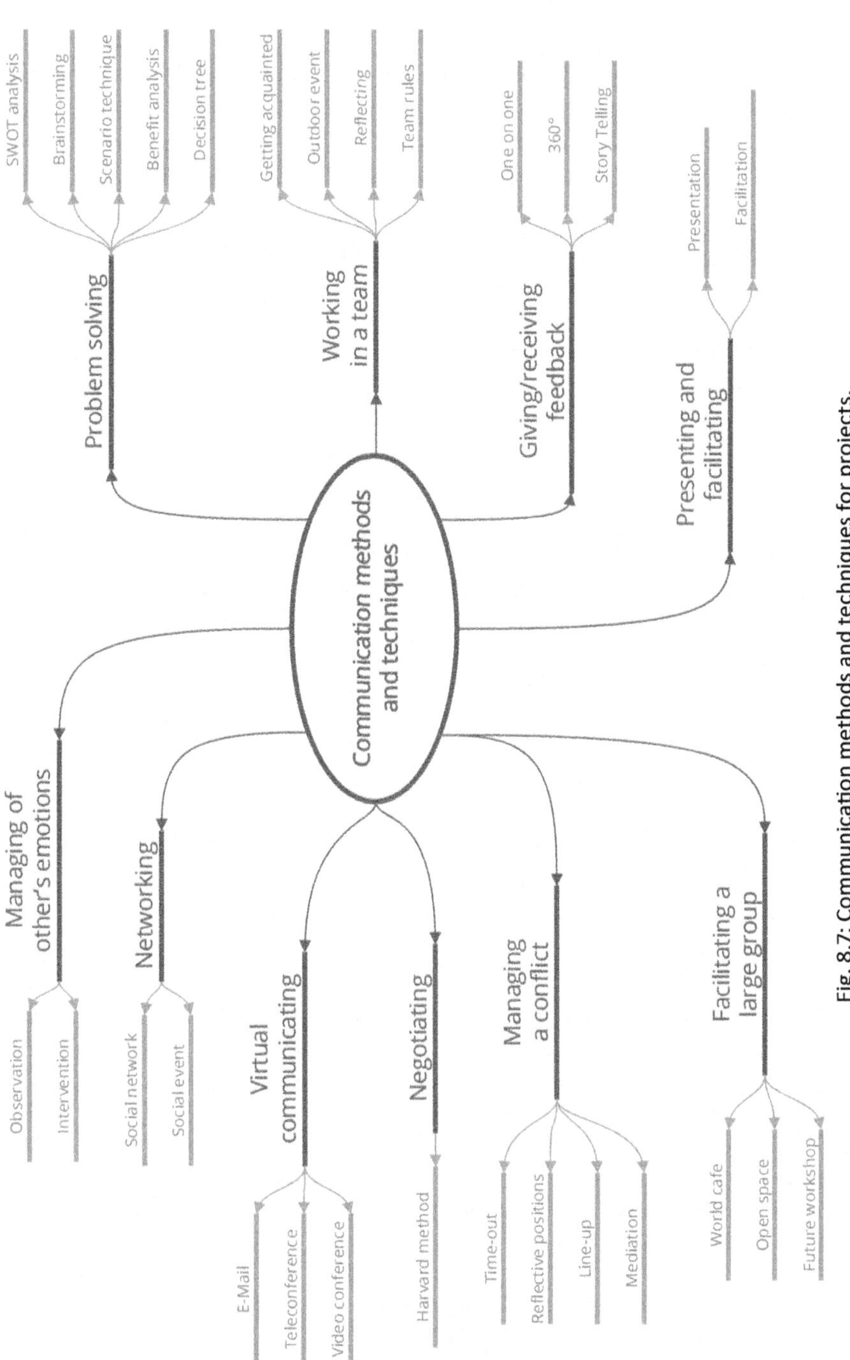

Fig. 8.7: Communication methods and techniques for projects.

who cannot adequately manage themselves are generally not qualified to manage organizations.

Self-competence is knowledge and experience in managing oneself using appropriate methods, such as stress managing, self-motivating, coping with one's own emotions, self-reflecting, problem solving, time managing, and managing the work-life balance (see Fig. 8.8). Techniques for self-management are also shown in Figure 8.8.

8.3.4 Self-Understanding as Project Manager

Self-understanding can be understood as how a person interprets and explains him or herself. A self-understanding is developed by individuals, not only in their private environment. Professional self-understanding can also be developed through reflection on professional requirements. It represents a perspective on how an individual perceives himself/herself in a professional role in a specific context.

"The expression 'self-understanding' means the establishment of subjectively binding norms, which are formed by balancing expectations placed on our roles, and in which individual action finds its orientation".[17] Self-understanding is strongly connected with individual values, identity, character, and with constructions of reality.

Fundamental elements of self-understanding are interpretations ...

> for dealing with oneself (e.g., frustration tolerance, sensitivity to personal emotions, self-control),
> for defining one's responsibility (towards others, towards society), and
> for dealing with others (e.g., transparency, fairness, openness, ability to cooperate).

These elements of self-understanding can be adapted and interpreted for different professional roles.

An individual's self-understanding regarding the fulfillment of a professional role is learned; it develops through interaction with stakeholders. This self-understanding may change but has a relatively stable character. It is the basis for the self-competencies and social competencies deemed necessary for fulfilling the professional role.

The self-understanding of a person as project manager is characterized by the project management approach used by the person. How one deals with oneself, one's responsibility towards others and society, and how one deals with others all differ according to whether a mechanistic or a systemic project management approach (see Chapter 2) is used.

An example of the self-understanding of project managers who use a systemic project management approach is shown in Figure 8.9.

[17] Sachs-Hombach, K., 2000, p. 189.

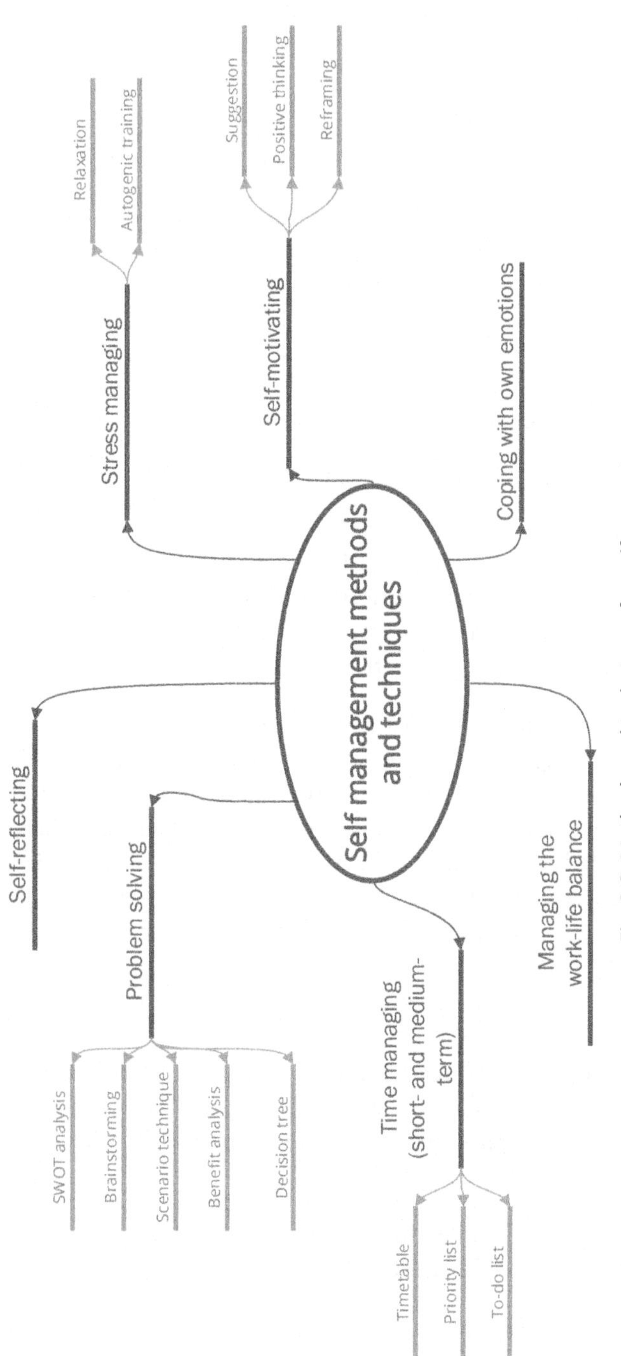

Fig. 8.8: Methods and techniques for self-managing.

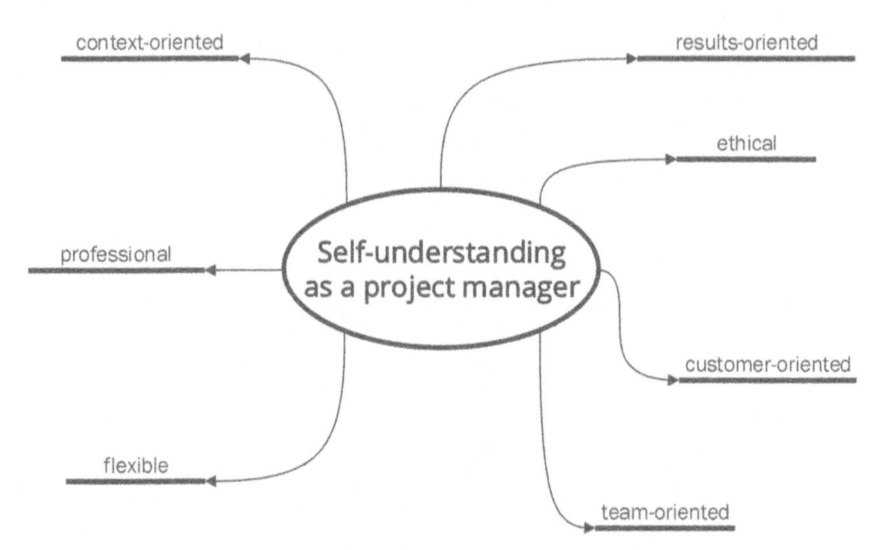

Fig. 8.9: Dimensions of self-understanding as a project manager.

The values shown need to be interpreted in the specific context. For example, "results-oriented" can mean securing a holistic and sustainable solution. "Customer-oriented" can mean ensuring customer benefits, and "team-oriented" can mean empowering teams in projects and making collective decisions.

An example of the formulation of the self-understanding of project managers is the PMI Project Management Institute standard provided by the "Code of Ethics and Professional Conduct".[18] This applies to all PMI members, as well as non-members who have a PMI certificate or work as volunteers for PMI. The values which are relevant for this global project management community are "responsibility", "respect", "fairness", and "honesty". In the description of the values, a distinction is made between "mandatory standards" and "aspirational standards".

Literature

Daxner, F., Gruber, T., Riesinger, D.: Werteorientierte Unternehmensführung: Das Konzept, in: Auinger, F., Böhnisch, W. R., Stummer, H. (Eds.), *Unternehmensführung durch Werte. Konzepte – Methoden – Anwendungen,* p. 3–34, Deutscher Universitäts-Verlag, Wiesbaden, 2005.

De Janasz, S., Dowd, K. O., Schneider, B. Z.: *Interpersonal Skills in Organizations,* 5th edition, McGraw-Hill Education, New York, NY, 2015.

Erpenbeck, J., Von Rosenstiel, L. (Eds.): *Handbuch Kompetenzmessung: Erkennen, verstehen und bewerten von Kompetenzen in der betrieblichen, pädagogischen und psychologischen Praxis,* 2. Auflage, Schäffer-Poeschel, Stuttgart, 2007.

[18] See Project Management Institute.

Gester, P.: Warum der Rattenfänger von Hameln kein Systemiker war? Systemische Gesprächs- und Interviewgestaltung, in: Schmitz C., Gester P., Heitger B. (Eds.), *Managerie – Systemisches Denken und Handeln im Management,* 136–164, 1. Jahrbuch, Carl Auer, Heidelberg, 1992.

Kotter, J. P.: Leading Change, *Harvard Business Review Press,* Boston, MA, 2012.

Polzin, B., Weigl, H.: *Führung, Kommunikation und Teamentwicklung im Bauwesen,* 2nd edition, Springer, Wiesbaden, 2014.

Prahalad, C. K., Hamel, G.: The Core Competencies of the Corporation, *Harvard Business Review,* 68(3), 79–91, 1990.

Project Management Institute (PMI): *Code of Ethics and Professional Conduct,* downloaded at http://www.pmi.org/-/media/pmi/documents/public/pdf/ethics/pmi-code-of-ethics.pdf?sc_lang_temp=en (21 Feb. 2017).

Sachs-Hombach, K.: Selbstbild und Selbstverständnis, in: Newen, A., Vogeley, K. (Eds.), *Selbst und Gehirn. Human Self-Awareness and Its Neurobiological Foundations,* p. 189–200, Mentis, Paderborn, 2000.

Steyrer, J.: Theorie der Führung, in: Kasper, H., Mayrhofer, W. (Eds.), *Personalmanagement, Führung, Organisation,* p. 25–94, Linde, Wien, 2002.

Tuckman, B. W., Jensen, M. A. C.: Stages of Small-Group Development Revisited, *Group & Organization Studies,* 2(4), 417–427, 1977.

Weibler J.: *Personalführung,* Franz Vahlen, München, 2001.

Weinert, A. B.: Führung und soziale Steuerung, in: Roth, E. (ed.), *Organisationspsychologie (Enzyklopädie der Psychologie,* Vol. 3), 552–577, Hogrefe, Göttingen, 1989.

9 Sub-Process: Project Starting

Because of the time pressure of projects, there is a great temptation to begin work on a project immediately upon receiving the project assignment, without starting the project appropriately. This can lead to unrealistic project objectives, unclear project role definitions, and unclear project stakeholder relations. A professional project start can prevent these problems and ensure adequate project management quality.

Adequate communication formats and project management methods should be used during the project start. The design of the sub-process "project starting" and the methods relevant to the project start are described below. Examples of project planning for the project "Developing Values4Business Value" are presented as a case study.

In the following overview, the sub-process "project starting" is shown in the context of the business process "project managing".

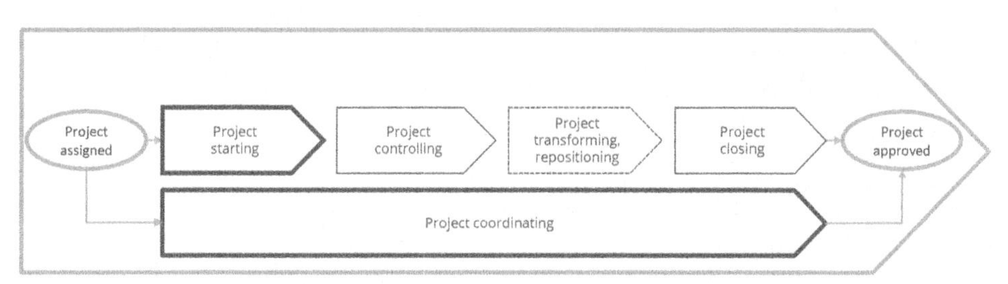

Overview: The sub-process "project starting " in the context of the business process "project managing".

9 Sub-Process: Project Starting

9.1 Project Starting: Objectives and Process

9.2 Project Starting: Organization and Quality

9.3 Methods: Planning Objects of Consideration, Planning Project Objectives, Planning Project Strategies, Developing a Work Breakdown Structure, Specifying a Work Package

- 9.3.1 Planning Objects of Consideration
- 9.3.2 Planning Project Objectives
- 9.3.3 Planning Project Strategies
- 9.3.4 Developing a Work Breakdown Structure
- 9.3.5 Specifying a Work Package

9.4 Methods: Planning Project Schedules, Planning a Project Budget, Planning Project Resources

- 9.4.1 Planning Project Schedules
- 9.4.2 Planning a Project Budget
- 9.4.3 Planning Project Resources

9.5 Methods: Analyzing Project Context Relations

- 9.5.1 Analyzing Relations to Strategic Objectives of the Organization
- 9.5.2 Analyzing and Planning Relations to other Projects
- 9.5.3 Analyzing the Pre-Project Phase and Planning the Post-Project Phase
- 9.5.4 Analyzing and Planning Relations to Project Stakeholders

9.6 Methods: Designing a Project Organization, Developing a Project Culture, Managing the Project Personnel

- 9.6.1 Designing a Project Organization
- 9.6.2 Developing a Project Culture
- 9.6.3 Managing the Project Personnel

9.7 Methods: Managing Project Risks

9.1 Project Starting: Objectives and Process

Project Starting: Objectives

The sub-process "project starting" begins when the project is assigned and ends when the project start documents are filed. Project starts can last for two to four weeks, depending on the scope and complexity of the project. The "project initiating" process takes place before project starting. As a result of the project initiation, initial project plans, which are to be further detailed and elaborated by the nominated project team, are available for starting a project.

The processes "project coordinating", "working out a solution", "project administrating", and "change managing" run in parallel. The processes "project controlling", "project closing", and "controlling the benefits realization" take place later.

The objectives of the sub-process "project starting", differentiated into economic, ecological, and social objectives, are described in Table 9.1.

Table 9.1: Objectives of the Sub-Process "Project Starting"

Objectives of the sub-process: Project starting
Economic objectives > Information transfer from the pre-project phase into the project performed > Appropriate project plans developed > Project organization designed > Project context relations analyzed and planned > Project established as a social system > First project marketing performed > Structures of the sub-processes "project coordinating", "project controlling" and "project closing" defined > Sub-process "project starting" performed efficiently
Ecological objectives > Ecological consequences of the project starting optimized
Social objectives > Project personnel recruited and allocated > Incentive systems developed and agreed on > Development of project personnel planned > Stakeholders in the project starting involved

Project Starting: Process

The sub-process "project starting" and the other project managing sub-processes can be divided into the following phases: planning, preparing communicating, performing communications, and following up. This structure of the sub-process "project starting" is shown in the responsibility chart of Table 9.2.

Table 9.2: Sub-Process "Project Starting"—Responsibility Chart

Tool/document: 1 ... List of project management methods to be used 2 ... Invitation for the project start workshop for participants 3 ... Documents for project start workshop 4 ... Documentation of project starting Legend P .... Performing C...Contributing I... Being informed Co...Coordinating Process tasks		Roles						
		Project owner	Project manager	Project team	Project team members	Expert pool managers	Project stakeholders	Tool/Document
1	Planning the project starting process							
1.1	Analyzing the situation	C	P		C		C	
1.2	Selecting the communication formats for starting		P					
1.3	Recruiting, allocating members of the project organization		A			P		
1.4	Performing one-on-one meetings		P		C		C	
1.5	Defining the project management methods to be applied		P					
1.6	Agreeing with the project owner	C	P					1
2	Preparing the project start communications							
2.1	Preparing a kick-off meeting, the start workshop, the project owner meeting, etc.		P					

(continues on next page)

The planning of project starting involves an analysis of the initial project plans and of essential decisions and documents from the pre-project phase. The results of this analysis are used to plan the application of further project management methods, to define necessary project communication formats, to plan the use of information and communication technology, and, if necessary, the involvement of a project management consultant. The members of the project organization must be recruited either from the permanent organization units of the project-oriented organization or from the external personnel market. These persons are allocated to the project roles.

The participants for the different communication formats for starting the project are to be selected and invited. For performing project start communications, a combination of individual meetings, kick-off meetings, project start workshops, and project owner meetings are required. In large projects, several project start workshops and

Table 9.2: Sub-Process "Project Starting"—Responsibility Chart *(cont.)*

2.2	Inviting participants		P			C		2
2.3	Documenting results from the pre-project phase		P		C		C	
2.4	Developing drafts of the detailed project plans		P		C		C	
2.5	Preparing the documents for the start communication		P		C		C	3
3	Performing the project start communications							
3.1	Distributing information to the participants		P					
3.2	Performing the project kick-off		P	C				
3.3	Performing the project start workshop	C	Co	P			C	
3.4	Developing a draft of the documentation of project starting		P					
3.5	Performing the project owner meeting	P	C					
4	Following up the project start communications							
4.1	Finalizing the documentation of project starting		P					4
4.2	Performing a first project marketing	C	P		C		C	
4.3	Distributing and filing the documentation of project starting		P		C	I	I	

also "social events" may be necessary. To follow-up, the project start documentation is to be prepared and distributed to the members of the project organization, and initial project marketing can take place.

The communication necessary to establish a project as a social system is promoted and structured by applying project management methods. This and not the creation of control instruments is the benefit of applying project management methods. The methods to be used when starting the project can be differentiated according to the project dimensions—namely, the objects of consideration, project objectives, project scope, project schedule, etc. The results of the application of these methods are project plans such as the objects of consideration plan, the project objectives plan, the work breakdown structure, and the project schedules.

An overview of the project management methods to be used was given in Table 6.2 (on page 106), differentiated according to the project managing subprocesses. As described in Chapter 6, the listed methods are also to be used in programs. The descriptions of the methods are thus relevant for both project managing and program managing, although for simplicity we shall, in what follows, speak only of "project management methods".

Examples of project plans created while starting a project are presented on page 188 for the project "Developing Values4Business Value" as a case study.

9.2 Project Starting: Organization and Quality

Project Starting: Organization

The project owner, the project manager, and the project team are responsible for starting a project. In socially complex situations, a project management consultant can be involved.

The planning of project starting and the preparing of the project start communications can be carried out by the project manager alone for small projects and by the project manager in cooperation with project team members for larger projects. When starting repetitive projects, standard project plans can be used, which considerably reduces the planning effort. Unique projects, however, require novel structures and solutions, which necessitates the use of specific working forms and of creativity techniques.

At the end of a project start workshop, the results achieved can be presented by the project team to the project owner and, possibly, to representatives of project stakeholders. There, open issues can be discussed. Usually more attention to agreements is required in unique projects than in repetitive projects. The sub-process "project starting" should be documented by the minutes of meetings and workshops.

The resources required for starting a project (personnel, materials, information and communication technology, consulting) can be estimated and assigned to the "project starting" work package of the work breakdown structure.

Project Starting: Quality

The quality of project starting can be evaluated on the basis of the project management results achieved and on the process quality—that is, the duration, costs, involvement of stakeholders, etc., of the subprocess. In addition to the project plans developed, for example, the structure and culture of the established social system "project", the development of stakeholder relations, and the implementation of the project marketing are results of project starting. The project plans can be evaluated with regard to their existence and the quality of their content (completeness, level of detail, fulfillment of formal criteria, etc.)

9.3 Methods: Planning Objects of Consideration, Planning Project Objectives, Planning Project Strategies, Developing a Work Breakdown Structure, Specifying a Work Package

9.3.1 Planning Objects of Consideration

Planning Objects of Consideration: Definition and Example

An objects of consideration plan is a breakdown of the objects to be considered in a project. Work packages are to be performed in a project for objects. They are the subject matter for work packages. On the one hand, objects of consideration are the dimensions of an organization to be changed. For example, services or products, markets, organizational structures, personnel, etc. can be changed by a project. On the other hand, the plans, documents, contracts, etc. necessary for performing a project can be objects of consideration.

Therefore, the objects of consideration plan must be distinguished from the work breakdown structure. The objects of consideration plan is structured in an object-oriented manner. The work breakdown structure is structured in a process-oriented manner. The connection between the two plans is that work packages that relate to objects are presented in the work breakdown structure.

The objects of consideration of a project and their relationships to one another can be represented either graphically or as a list. As an example, Figure 9.1 shows the objects of consideration plan for the project "Developing Values4Business Value".

Planning Objects of Consideration: Objectives and Process

The objective of developing an objects of consideration plan is to provide information about the objects to be considered in a project. It should be ensured that the members of the project organization and the representatives of project stakeholders develop a shared point of view in this regard. The labeling of the objects of consideration contributes to the development of a shared project language. The objects of consideration not only are the basis for work breakdown structuring but also provide a basis for the planning of the project objectives and the designing of the project organization (see Fig. 9.2 on page 189).

When starting a project, the objects of consideration can be defined analytically as a first working step. An alternative is to derive the objects of consideration from the project objectives plan and the work breakdown structure after these have been created.

The solution requirements developed in conceptualizing an investment provide a basis for defining the objects of consideration. For example, a "specification document" or an "initial backlog" can be the basis for defining objects of consideration.

Case Study: "Developing Values4Business Value"—Objects of Consideration Plan

The objects of consideration plan for the project "Developing Values4Business Value" is presented as a mind map with two levels. Not all objects of consideration are subdivided in this representation. The second level is sometimes "switched off" in order to provide appropriate information in project communication.

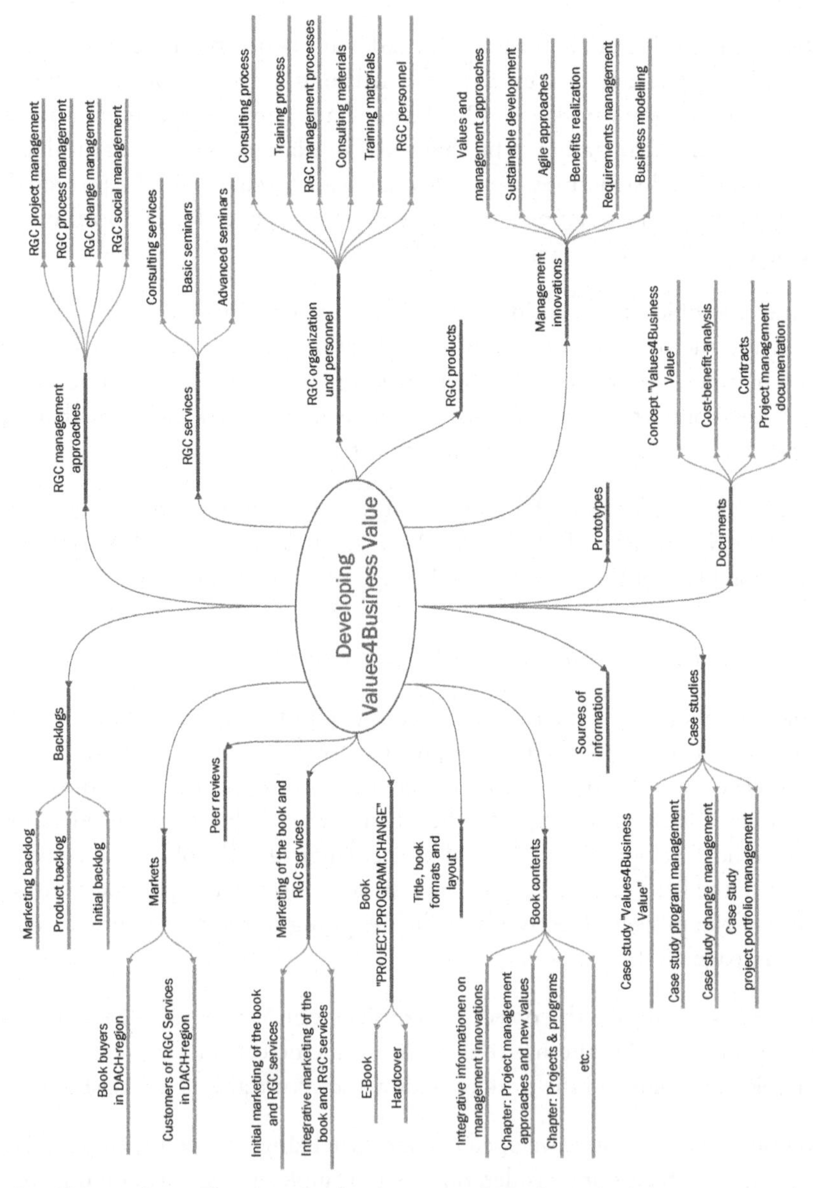

Fig. 9.1: Objects of consideration plan for the project "Developing Values4Business Value".

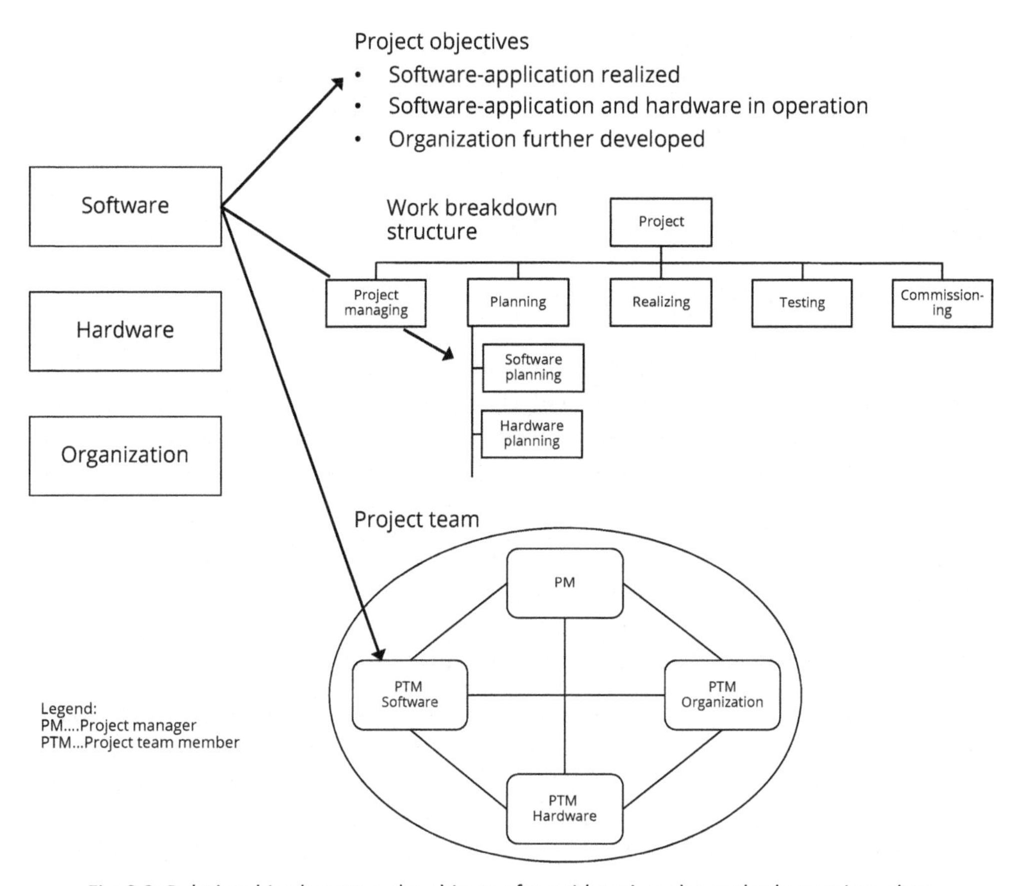

Fig. 9.2: Relationships between the objects of consideration plan and other project plans.

The dimensions of organizations—that is, their services, products, organization, personnel, etc.—have already been mentioned as relevant criteria for structuring the objects of consideration. Table 9.3 gives some tips and trick for planning objects of consideration.

Table 9.3 Planning Objects of Consideration: Tips and Tricks

Planning objects of consideration: Tips and Tricks
> Ensuring a holistic project view in the objects of consideration plan
> Clustering the objects to assure an overview
> Depicting objects of consideration in a unified manner
> When creating further project plans, optimizing the objects of consideration plan iteratively
> Assuring consistency of the objects of consideration plan with the project objectives plan and work breakdown structure
> Not presenting tasks but objects

For repetitive projects, standard objects of consideration plans can be used. The objects of consideration must be coded in order to ensure orientation. Since the objects of consideration are objects and not activities, object-oriented labeling (with nouns) must be ensured.

9.3.2 Planning Project Objectives

Planning Project Objectives: Definition and Example

Projects can be seen as goal-oriented temporary organizations. Objectives related to the content, the schedule, and the budget are to be achieved in projects, whereby the content-related objectives can be divided into economic, ecological, and social objectives. It is also possible to differentiate between short-term, medium-term, and long-term as well as local, regional, and global objectives.

In a narrow sense, project objectives are defined here as content-related results which should be obtained at the end of a project. These content-related project objectives are described in the project objectives plan. In order to plan other project objectives, additional project management methods, such as schedules and budgets, are used.

There is a connection between the content-related project objectives and the solution requirements defined in conceptualizing an investment (see Chapters 3 and 4). When planning project objectives, orientation can be provided by the initial solution requirements. When applying an iterative approach, the solution requirements are iteratively defined during the development of a solution. Not all requirements and project objectives can be defined in detail before starting a project. The requirements and objectives will only become concrete during the course of the project. At the end of the project, the results obtained should ideally correspond to the expectations of the project stakeholders at the end of the project. These expectations might deviate from the requirements defined before the start of the project.

The solution requirements are a project management dimension to be kept in mind, although the identification and the description of the requirements are not tasks of project managing but of the business process "managing requirements" (see Chapter 4).

The content-related objectives defined in the project objectives plan must be distinguished from the business case objectives or the benefits considered in a cost–benefit analysis. From the point of view of managing requirements, the business case objectives or the benefits of an investment are considered as business requirements. The connection between project objectives and business case objectives is that the fulfillment of the project objectives creates the basis for realizing the business case objectives.

As an example, in Table 9.4 an excerpt of the project objectives plan of the project "Developing Values4Business Value" is presented.

Case Study: "Developing Values4Business Value"—Project Objectives Plan

Table 9.4: Project Objectives Plan for the Project "Developing Values4Business Value"

Project Objectives Plan Developing Values4Business Value
Services-related and market-related project objectives
RGC management approaches further developed by innovations (values and management approaches, sustainable development, agile approaches, benefits realization management, requirements management, business modelling)
RGC management approaches documented in the book PROJECT.PROGRAM.CHANGE (hardcover and e-book)
Practical relevance of the developed RGC management approaches and readability of the book ensured by peer review groups and layout optimizations
Services (consulting, training, events, lectures) and products (sPROJECT, books) further developed by optimizing the processes and tools
Optimized services and book PROJECT.PROGRAM.CHANGE marketed in Austria, Germany and Switzerland; additional marketing planned
Organization-related and personnel-related project objectives
RGC management further developed
RGC personnel further developed
Infrastructure-related and budget-related project objectives
Basis for an increase in sales of RGC services and RGC products established
Stakeholder-related project objectives
Cooperation with publisher Manz continued
Publishing partners C.H. Beck (Germany) and Stämpfli (Switzerland) involved
Clients bound by the optimized management approaches
Peer review group involved
Additional objectives
Basis for certification of the project manager established
Case study for RGC trainings developed
Non-objectives
English version of the book developed
Advanced RGC seminars further developed

The project objectives plan for the project "Developing Values4Business Value" is broken down according to the structure dimensions and the context dimensions of an organization. A differentiation was made between service-related and market-related objectives; organizational and personal objectives; and infrastructure-related, financial objectives, and stakeholder-related objectives. The breadth of the changes pursued by RGC can be seen from the project objectives.

The defined additional objectives were to be achieved in the project, although this would not have been necessary for the fulfillment of the main objectives. The definition of the non-objectives contributed to the clarification of the project boundaries.

Planning Project Objectives: Objectives and Process

The objective of planning the project objectives is to define the content-related objectives to be achieved by the project. This requires the construction of a shared view amongst the members of the project organization and the representatives of project stakeholders, as well as shared agreements. The construction of a "holistic" view of the project is striven for. A holistic solution is to be enabled by taking into account all closely linked objectives.

The project objectives are to be operationalized by appropriate differentiation and quantification. Differentiation can be supported by using the objects of consideration and the defined solution requirements. Later on, this enables operational controlling of the project objectives and evaluation of the project success.

The project objectives can be divided into main, additional, and non-objectives. The main objectives describe the solution requirements that should be fulfilled at the end of the project to meet a need of the organization. Additional objectives may be the development of a supplier relationship or the further development of the project personnel involved. The additional objectives do not relate to the solution requirements. They must not be confused with "can-objectives"—that is, they must also be achieved and resources and budgets have to be made available for this. By defining non-objectives, by explicitly excluding possible solution requirements, the project objectives are concretized.

9.3.3 Planning Project Strategies

Planning Project Strategies: Definition and Example

Project strategies can be understood as the basic approaches for realizing project objectives.

The following project strategy types can be distinguished:

- project strategies regarding the procurement of services and products,
- project strategies regarding the use of technologies,
- project strategies regarding the application of methods,
- project strategies regarding project personnel, and
- project strategies regarding the managing of project stakeholder relations.

Project strategies are to be defined when there are options with respect to basic approaches. Options in procurement are, for example, in-house production or third-party production. Options regarding the application of methods are, for example, the use of a waterfall model or an iterative model for performing a project. As an example, Table 9.5 shows the project strategies plan for the project "Developing Values4Business Value".

Planning Project Strategies: Objectives and Process

The objective of planning project strategies is to define and to agree on basic approaches for realizing the project objectives. The decisions made in project strategies planning influence further project planning. For example, the decision to proceed iteratively

Case Study: Developing Values4Business Value—Project Strategies Plan

Table 9.5: Project Strategies Plan for the Project "Developing Values4Business Value"

Project Strategies Plan Developing Values4Business Value
Project strategies regarding procuring
Peer Review Group: Cooperation with management experts to ensure the practical orientation of the RGC management approaches
Publishing partners: Involving publishers in Germany and Switzerland, in addition to the Austrian publisher, to ensure the marketing of the book in the DACH-region
Layout expert: Employment of a layout expert to revise the consulting and training materials, to document the RGC management approaches and to ensure an appealing layout of the book
ProcPurchasing a Scrum-board
Project strategies regarding the applied technologies
E-book: Designing the book to enable the realization of an e-book
Graphics: implementing the RGC CI in graphical solutions to contribute to the RGC image
Print: Ensuring the instant usability of RGC's graphical solutions by the printing company
Project management documentation: Use of sPROJECT
Project strategies regarding the use of methods
Change managing: Perceiving the further development of the RGC management approaches as a change
Project managing: Iterative approach in the project "Developing Values4Business Values"
Researching: Literature analysis, case studies, interviews, reflection workshops, ...
Visualizing: Use of a Scrum-board and all project plans and change plans for the communication
Project strategies regarding the project personnel
Project team members: Ensuring the personnel development, learning as incentive
Peer review group: Promoting the networking between the group members
Project strategies for managing the project stakeholder relations
RGC clients: Information on the further developed RGC management approaches and invitation to the book presentation
RGC key accounts: Involvement of representatives in the peer review group, information about the benefits of the management innovations
RGC consultants and network partners: Involving in the development process

The concept of "procuring" was understood in a broad sense in the project and was thus used not only for procuring suppliers, but also for cooperation partners. "Technological solutions" needed to be specified for the book production and also became relevant for the application of project and change management methods.

The personnel-related project strategies should ensure the motivation and the further development of the project personnel.

and not sequentially defines the structure of the work breakdown structure. Or, for example, the decision to implement quality assurance measures through external experts affects the project organization and the project budget.

The planning of the project strategies must be based on the project objectives plan. Strategic options for realizing the project objectives are to be identified and evaluated. The optimal approach in each case must be defined as a project strategy.

The definition of project strategies for procurement should not be restricted to procurement strategies and the treatment of suppliers but should also take into account potential cooperation partners who provide essential resources to ensure the success of the project. The technologies to be used depend on the project content. For strategies concerning the application of methods, a distinction can be made between methods for the content of project work and methods for managing the project and the change. The definition of the project strategies for managing project stakeholder relations can be a result of the project stakeholder analysis. The consideration of project strategies for project personnel reflects the importance of the project personnel for the success of the project.

The defined project strategies must be communicated to the members of the project organization and to selected representatives of project stakeholders. This should give them orientation regarding behavior in the project.

9.3.4 Developing a Work Breakdown Structure

Developing a Work Breakdown Structure: Definition and Example

The work to be performed within the framework of a project can be planned in a work breakdown structure (WBS) and concretized by work package specifications. The objects of consideration plan, the project objectives plan, and the project strategies plan form the basis for work breakdown structuring. A WBS can be represented graphically as a tree structure and/or as a list.

A WBS is a model of a project. It breaks down the work to be performed into plannable and controllable work packages. A WBS is not a schedule, a cost plan, or a resource plan. It is, however, the common structural basis for scheduling, cost planning, and resource planning. The WBS can also be the basis for a project-related filing system. A WBS is a relatively stable project plan because it is not influenced by changes in schedules, costs, or resources.

As examples, in Figure 9.3 (page 195 through page 197) the work breakdown structure of the project "Developing Values4Business Value" (breakdown to the 3rd level) and in Figure 9.4 (page 198) the breakdown of a phase of this project (breakdown to the 4th level) are represented as tree structures.

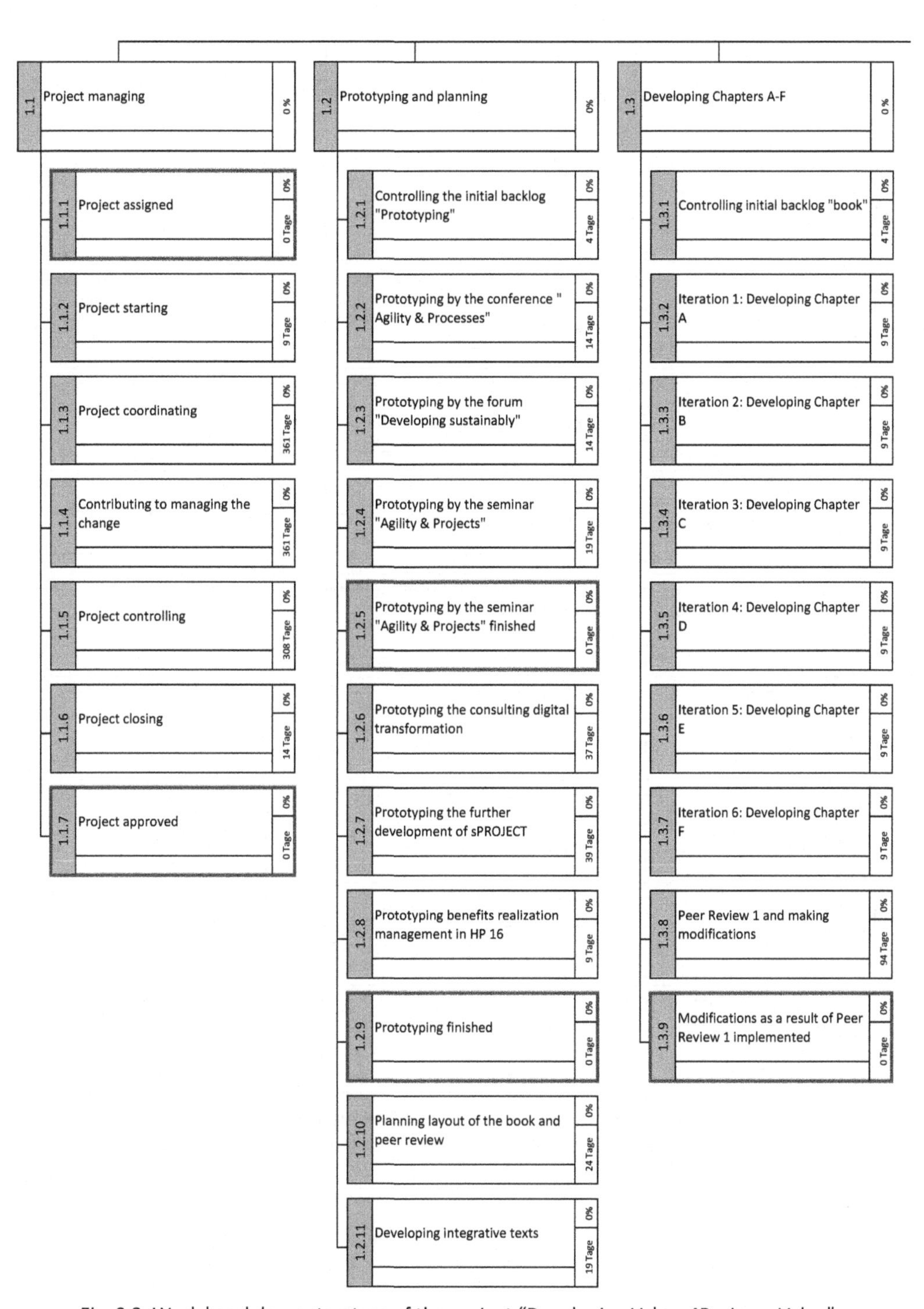
1.1 Project managing 0 %
1.1.1 Project assigned 0% 0 Tage
1.1.2 Project starting 0% 9 Tage
1.1.3 Project coordinating 0% 361 Tage
1.1.4 Contributing to managing the change 0% 361 Tage
1.1.5 Project controlling 0% 308 Tage
1.1.6 Project closing 0% 14 Tage
1.1.7 Project approved 0% 0 Tage
1.2 Prototyping and planning 0%
1.2.1 Controlling the initial backlog "Prototyping" 0% 4 Tage
1.2.2 Prototyping by the conference " Agility & Processes" 0% 14 Tage
1.2.3 Prototyping by the forum "Developing sustainably" 0% 14 Tage
1.2.4 Prototyping by the seminar "Agility & Projects" 0% 19 Tage
1.2.5 Prototyping by the seminar "Agility & Projects" finished 0% 0 Tage
1.2.6 Prototyping the consulting digital transformation 0% 37 Tage
1.2.7 Prototyping the further development of sPROJECT 0% 39 Tage
1.2.8 Prototyping benefits realization management in HP 16 0% 9 Tage
1.2.9 Prototyping finished 0% 0 Tage
1.2.10 Planning layout of the book and peer review 0% 24 Tage
1.2.11 Developing integrative texts 0% 19 Tage
1.3 Developing Chapters A-F 0 %
1.3.1 Controlling initial backlog "book" 0% 4 Tage
1.3.2 Iteration 1: Developing Chapter A 0% 9 Tage
1.3.3 Iteration 2: Developing Chapter B 0% 9 Tage
1.3.4 Iteration 3: Developing Chapter C 0% 9 Tage
1.3.5 Iteration 4: Developing Chapter D 0% 9 Tage
1.3.6 Iteration 5: Developing Chapter E 0% 9 Tage
1.3.7 Iteration 6: Developing Chapter F 0% 9 Tage
1.3.8 Peer Review 1 and making modifications 0% 94 Tage
1.3.9 Modifications as a result of Peer Review 1 implemented 0% 0 Tage

Fig. 9.3: Work breakdown structure of the project "Developing Values4Business Value" (breakdown to the 3rd level). *(continues on next two pages)*

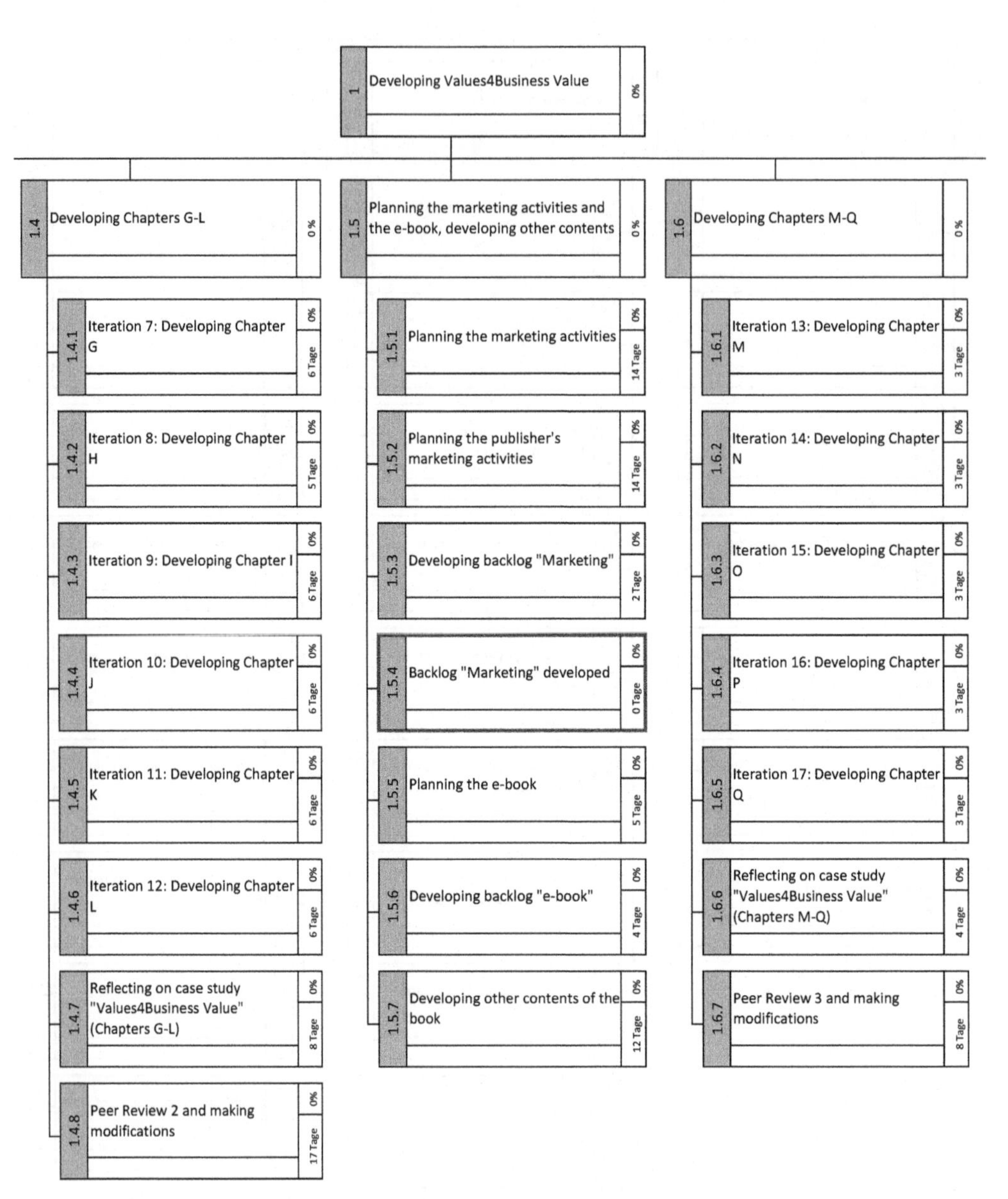

Fig. 9.3: Work breakdown structure of the project "Developing Values4Business Value" (breakdown to the 3rd level). *(continues on next page)*

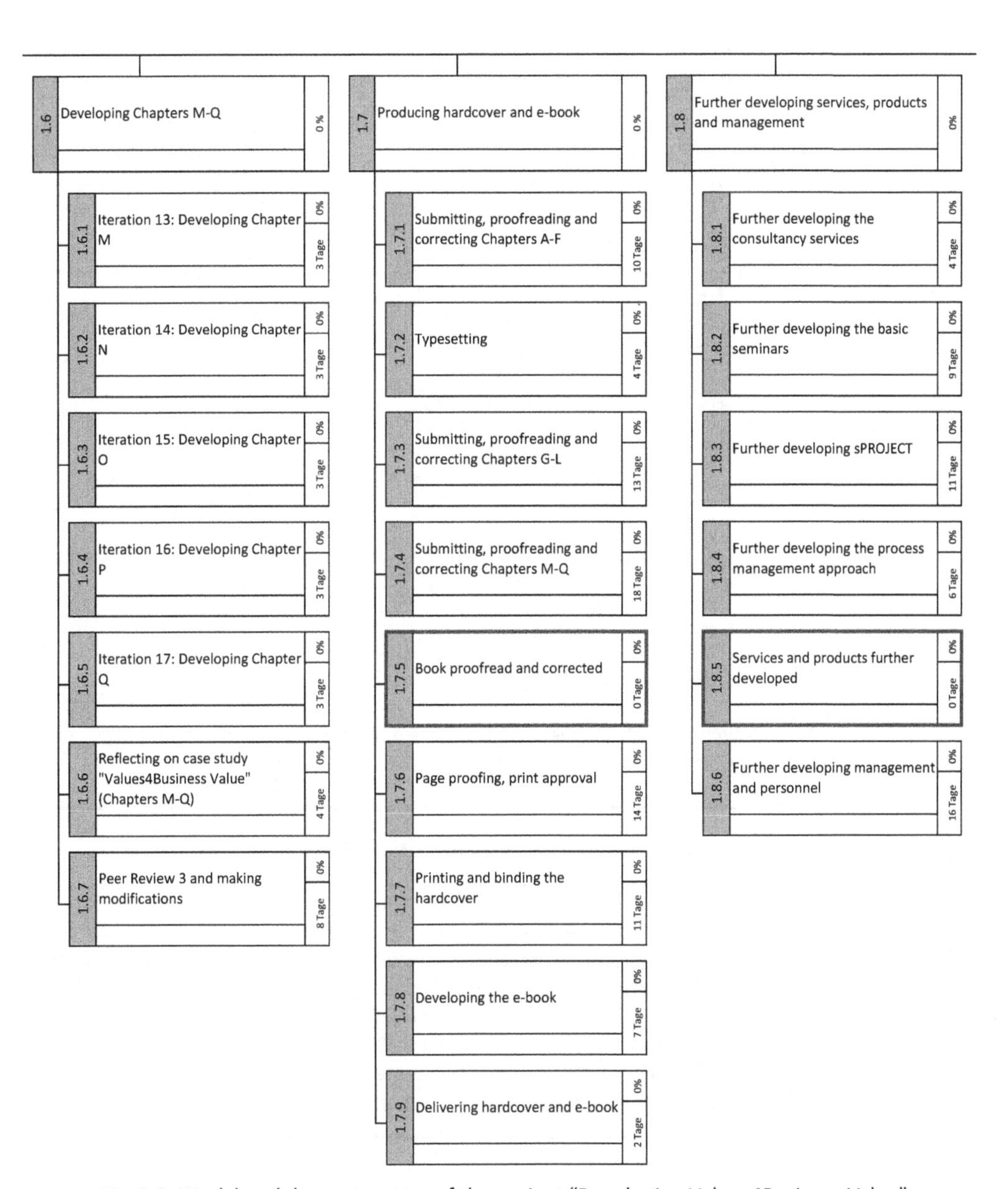

Fig. 9.3: Work breakdown structure of the project "Developing Values4Business Value" (breakdown to the 3rd level). *(end)*

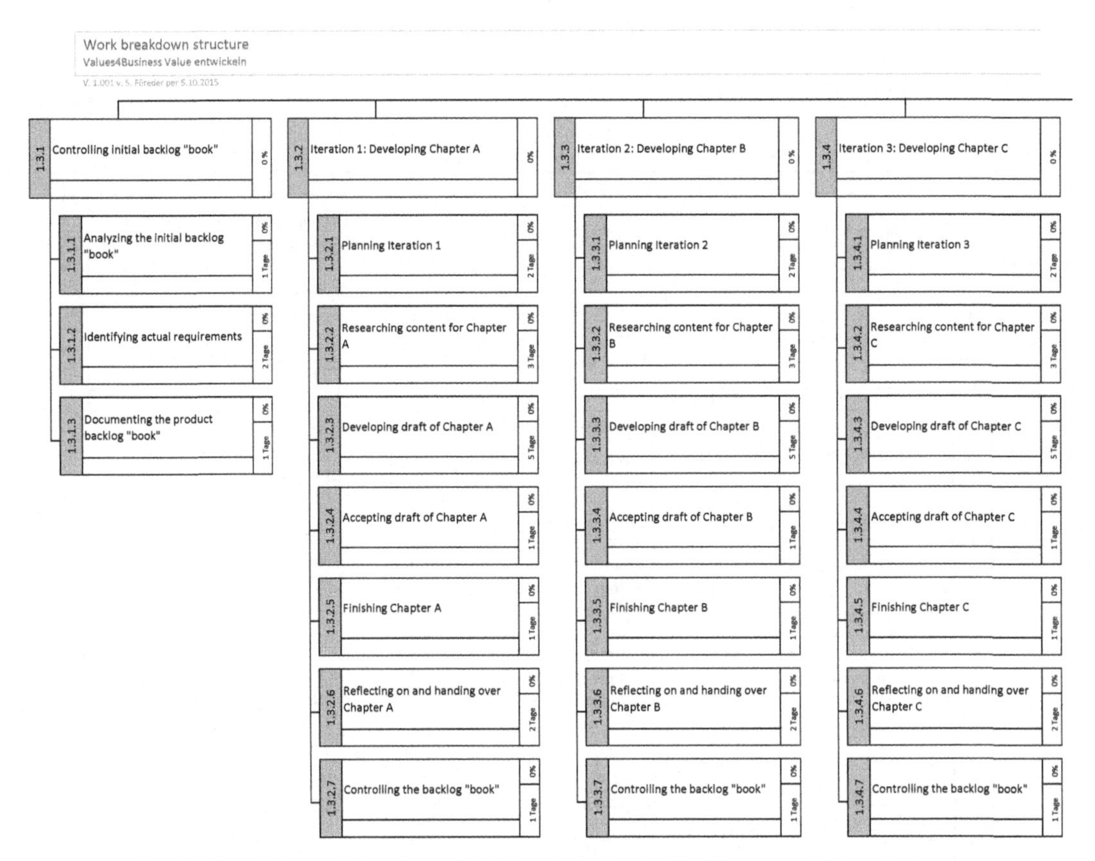

Fig. 9.4: Breakdown of the phase "Develop Chapters A–F" of the project "Developing Values4Business Value" (4th level of the WBS). *(continues on next page)*

Case Study: Developing Values4Business Value—Work Breakdown Structure

Most phases of the project "Developing Values4Business Value" were broken down to the 4th level. The phase "Developing Chapters A–F" is an example of this. The need to subdivide phases and a group of work packages into work packages is a result of the varied nature of the work to be performed and its relatively long duration. This can be seen, for example, in the group of work package 1.2.5 "Prototyping the seminar 'Agility & Projects'". The following work packages were to be completed within this group of work packages: "Planning seminar content", "Establishing the cooperation with a co-trainer", "Conducting content-related research", "Developing seminar content", "Conducting the seminar", and "Reflecting on and optimizing the seminar".

Developing a Work Breakdown Structure: Objectives and Process

The objective of developing a work breakdown structure is to break down the project into plannable and controllable work packages. For this the project scope needs to be defined as holistically as possible. The WBS should ensure a shared view of the project amongst the members of the project organization and representatives of project stakeholders. The WBS is a central communication tool in project managing.

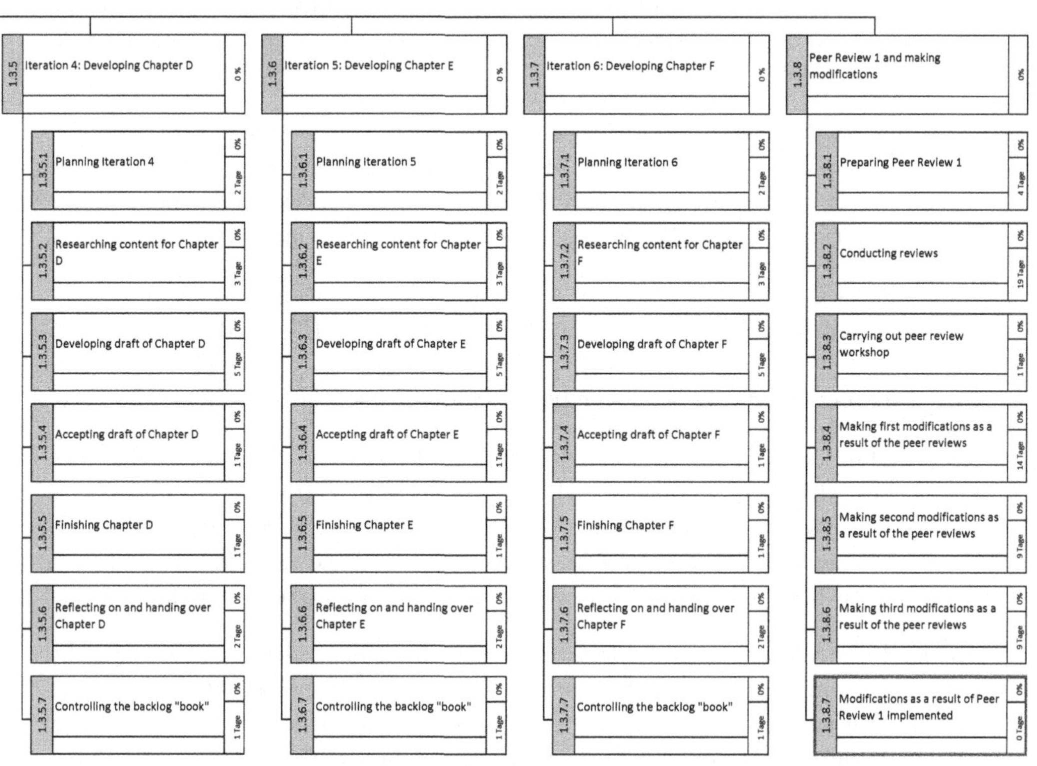

It contributes to the unification of the project language and to the development of commitment to the project and enables a clear assignment of work packages to project team members and project contributors. A process-oriented breakdown of the project creates an appropriate basis for controlling the progress of the project.

A WBS is created by a horizontal and vertical breakdown of the project scope. The breakdown is based on business processes and on the objects of consideration. In order to arrive at detailed work packages, it is necessary to combine several structuring criteria. In structuring the WBS it is advisable to proceed in a process-oriented manner at the 2nd level (phase level) as well as within the individual project phases.

In a process-oriented structure, the focus is not on a strict chronological sequence but rather on a general process logic. Each breakdown criterion, and thus also the phase schema, can be disrupted or extended. So, for example, the phases of planning, preparing, executing, and following up an event can be extended to include the marketing and the project managing, which must be carried out in parallel with these project phases.

The positioning of a task as a phase at the 2nd WBS level is to be ascribed to the importance of this process. Project managing is always seen as a business process at the 2nd level of the WBS and is broken down into the subprocesses project starting, project coordinating, project controlling, and project closing.

The further structuring of the WBS is carried out for each phase or group of work packages by their division according to either processes or objects of consideration. In the case of breakdown by objects, the appropriate level of the objects of consideration plan is to be referred to.

When using agile methods for carrying out specific project phases, these are to be broken down into iterations such as "time-boxed" sprints. It is possible to use different sprint or iteration structures for different project phases. This can be seen from the above example (Fig. 9.3 starting on page 195), from a project which is performed iteratively.

In the case of new, unique projects, it is recommended to collect ideas for work packages to be performed. This can be done by brainstorming. The listed work packages can be aggregated into groups of work packages or phases. In repetitive projects, standard work breakdown structures can be used for creating project-specific WBSs.

The detailing of the WBS is to be continued until plannable and controllable work packages for which responsibilities can be assigned have been arrived at. Work packages form the basis for the agreement of quantitative and qualitative results between the project manager and project team members, suppliers, partners, and customers. Project team members may further subdivide their work packages for the carrying out of the work.

The level of detail of the work breakdown structure is dependent on the scope, the complexity, and the dynamics of the project. During project starting, the early project phases are usually planned in more detail than later phases. It can be assumed as a matter of principle that a WBS is to be broken down to the 4th level and, if necessary, to a 5th level. In the case of a project with seven to nine phases at a 3rd level, this may result in some 30 to 50 work packages. A further breakdown to the 4th level produces some 200 to 300 work packages.

The names of the work packages are "labels" which should be understandable for all members of the project organization and representatives of stakeholders. Each work package represents a task to be performed in the project which relates to an object of consideration. Work packages must therefore be task-oriented and named with reference to the corresponding object of consideration (e.g., "procuring hardware"). The coding of work packages allows them to be unambiguously identified and also enables the sorting and selecting of work packages. The number of digits in the code indicates the WBS level on which the work package is located. The work package "1.3.2.1 Plan iteration 1" of the project "Developing Values4Business Value", for example, is located at the 4th level of the WBS.

The WBS, just like the other project plans, is to be developed by the project team and agreed upon with the project owner. Through teamwork, the necessary creativity can be assured and the acceptance of the jointly achieved solution can be guaranteed.

9.3.5 Specifying a Work Package

Specifying a Work Package: Definition and Example

Work packages are to be specified at the lowest level of a work breakdown structure. A work package specification includes a quantitative and qualitative description of the tasks to be performed in a work package and the results to be achieved. A work package specification can specify how the progress in the performance of the work package is to be measured. The scheduling and the budgeting of work packages are not part of specifying work packages. This information can be found in the project schedule and in the project cost plan.

As an example, Table 9.6 shows the specification of the work package "1.3.2.3 Developing a draft of Chapter A".

Case Study: Developing Values4Business Value—Work Package Specification

Table 9.6: Specification of the Work Package "1.3.2.3 Developing the Draft of Chapter A"

Work Package Specification
Developing Values4Business Value

WBS Code:	1.3.2.3	Work package:	Developing a draft of Chapter A	Progress measurement
Content of the work package				
Structuring Chapter A and documenting available information				10%
Developing text modules and relating to the chapter structure				50%
Drawing up a draft of the texts and figures				70%
Finishing a draft of the texts and figures				100%
Result of the work package				
A draft of Chapter A drawn up; scope approx. 25 A4-pages; about 5 figures in Powerpoint				

Specifying a Work Package: Objectives and Process

The objective of specifying a work package is to determine which tasks are to be performed within the framework of a work package, what the work package results are, and possibly how the progress of the work package is measured.

The description of the results of a work package should also specify the form of the results. Work package specifications are the basis for agreements between the project manager and the project team members and project contributors.

Through specification, work packages are differentiated from each other, and interfaces between work packages can be identified. Work package specifications save further detailing of the work breakdown structure.

Work package specifications are to be prepared by the project team members responsible for the corresponding work packages. Agreements regarding the specifications are made by the project manager and the responsible project team member or project contributor. This also makes it possible to deal with interfaces between work packages.

In the case of unique projects, it is advisable to specify a relatively large number of work packages. In repetitive projects, one can fall back on standard specifications.

9.4 Methods: Developing Project Schedules, Planning a Project Budget, Planning Project Resources

9.4.1 Developing Project Schedules

Developing Project Schedules: Overview

Before planning the logic and the dates for a project, the object to be scheduled, the level of detail of planning, and the planning methods to be used must be defined.

Objects for which scheduling can be performed are the project as a whole and individual project phases. Different scheduling methods can be used for different objects to be scheduled. So, for example, a bar chart can be used for the entire project and network planning for a project phase.

The work breakdown structure is the basis for the planning of the project logic and for scheduling. The position in time of (groups of) work packages can be planned. A distinction can be made between rough and detailed scheduling. If the WBS is not sufficiently subdivided to enable detailed scheduling, further subdivisions of work packages must be carried out. So-called "activities" are created as elements of the detailed scheduling.

The planning of the project logic and the scheduling can be carried out using the methods of listing project dates, developing a project bar chart, and/or developing a project network. These methods complement each other. Of these methods, network planning is the most sophisticated, and the listing of dates is the simplest. Milestone planning is a specific form of listing dates. Different information is required for the application of different methods for scheduling (see Table 9.7).

Planning Project Milestones: Definition and Example

A project milestones plan lists the dates of important project events, so-called project milestones. Project milestones are usually associated with symbolic events of a

Table 9.7: Information Required for the Different Scheduling Methods

Method	Required information
Planning project milestones	> List of work packages > Dates of milestones
Listing project dates	> List of work packages > Start and/or end dates of work packages
Developing a project bar chart	> List of work packages > Duration of each work package > Position of each work package on a time scale
Developing a project network plan	> List of work packages > Duration of each work package > Dependences between work packages due to technological and/or resource requirements

project, such as the ground-breaking ceremony, the topping-out ceremony, and the house-warming of a project "house construction". As an example, Table 9.8 shows the project milestones plan for the project "Developing Values4Business Value".

Case Study: Developing Values4Business Value—Project Milestones Plan

Table 9.8: Project Milestones Plan for the Project "Developing Values4Business Value"

Project Milestone Plan

Developing Values4Business Value

WBS Code	Milestone	Basic Plan
1.1.1	Project assigned	05.10.2015
1.2.6	Prototyping by the seminar "Agility & Projects" finished	18.03.2016
1.2.9	Prototyping finished	29.04.2016
1.5.4	Backlog "Marketing" developed	18.11.2016
1.3.9	Modifications as a result of Peer Review 1 implemented	23.11.2016
1.7.5	Book proofread and corrected	13.01.2017
1.8.5	Services further developed	24.02.2017
1.9.8	Marketing activities performed	10.03.2017
1.1.7	Project approved	31.03.2017

V. 1.001 v. S. Füreder per 5.10.2015

The project milestones up to November 2016—that is, for the early phases—are more spread out from each other than in the later project phases. The number of milestones is relatively high, but this makes it possible to provide an overview of major events of the project.

Planning Project Milestones: Objectives and Process

The objective of project milestones planning is to define milestones and plan the dates of these major project events. Milestones provide orientation in the project. The object scheduled in milestones planning is the project as a whole. The milestones plan is a rough scheduling method.

A project milestones plan should not have more than seven to nine milestones. Structurally, milestones relate to initial and final events of work packages. "Project assigned" and "project approved" are obligatory milestones. Milestones are to be formulated in an event-oriented manner. Dates must be planned for each milestone. These can be adjusted during periodic project controlling.

Listing Project Dates: Definition and Example

A list of project dates consists of the start and/or end dates of the work packages of a project. As an example, Table 9.9 shows an excerpt of the list of dates for the project "Developing Values4Business Value".

Case Study: Developing Values4Business Value—List of Project Dates

Table 9.9: Excerpt of the List of Dates for the Project "Developing Values4Business Value"

List of Dates

Developing Values4Business Value

WBS Code	Phase / Work package	Start	End
1.1	Project managing	05.10.2015	31.03.2017
...			
1.2	Prototyping and planning	19.10.2015	27.05.2016
1.2.1	Controlling the initial backlog "Prototyping"	19.10.2015	23.10.2015
1.2.2	Prototyping by the conference " Agility & Processes"	11.01.2016	29.01.2016
1.2.3	Prototyping by the forum "Sustainable development"	08.02.2016	26.02.2016
1.2.4	Prototyping by the seminar "Agility & Projects"	22.02.2016	18.03.2016
1.2.5	*Prototyping by the seminar "Agility & Projects" performed*	18.03.2016	18.03.2016
1.2.6	Prototyping the consulting of a digital transformation	07.03.2016	27.04.2016
1.2.7	Prototyping the further developing of sPROJECT	07.03.2016	29.04.2016
1.2.8	Prototyping benefits realization management in HP 16	18.04.2016	29.04.2016
1.2.9	*Prototyping finished*	29.04.2016	29.04.2016
1.2.10	Planning layout of the book and peer review	18.04.2016	20.05.2016
1.2.11	Developing integrative texts	02.05.2016	27.05.2016
1.3	Developing Chapters A-F	09.05.2016	18.11.2016
...			

V. 1.001 v. S. Füreder per 5.10.2015

The excerpt of the list of dates shows the start and end dates of the work packages of the project phase "Prototyping and planning". This phase lasted from 19 October 2015 to 27 May 2016. The long lead time is explained by the handling of the innovative content and the great importance of prototyping for further development. Two milestones are also included in the list of dates—namely, "Agility & Projects seminar prototyping performed" and "Prototypes completed".

Listing Project Dates: Objectives and Process

The objective of listing project dates is to plan the dates of the work packages. Either only the start dates, only the end dates of the work packages, or both the start and end dates can be listed. The duration of the work packages and the logical relationships between them are not explicitly planned and documented in the list of dates.

A list of dates for work packages is a detailed scheduling method. In a list of dates, the object to be scheduled can be either the project as a whole or one or several project phases. The work packages to be scheduled are to be listed on the basis of the WBS. Dates must be planned for each work package.

Developing a Project Bar Chart: Definition and Example

A project bar chart is a graphical representation of a project or a project phase that provides a visualization of the timing and of the duration of the phases and work packages of a project. The work packages are represented as time-proportional bars.

Knowledge of the timing of work packages is the basis for developing project bar charts. However, there is no explicit planning of the technological or resource-related dependencies between the work packages. As an example, Figure 9.5 shows a rough bar chart for the project "Developing Values4Business Value".

Case Study: Developing Values4Business Value—Project Bar Chart

In the bar chart for the project "Developing Values4Business Value", all project phases apart from the phase "1.3 Developing Chapters A–F" are represented by only one bar. This represents the 2nd level of the WBS. Phase 1.3 is shown at the 3rd WBS level. More time was planned for the development of Chapters A–F in comparison with the following chapters, since these dealt with particularly innovative contents. Cooperation with the peer review group also took place for the first time for this group of chapters.

The process-oriented structure of the project is clearly visible in the bar chart from the "flow" of the phases from the top left to the bottom right.

Project bar chart

Project name

V. 1.001 v. S. Füreder per 5.10.2015

WBS Code	Phase / Work package / Milestones	Date Start	End
1	**Developing Values4Business Value**	05.10.2015	31.03.2017
1.1	**Project managing**	05.10.2015	31.03.2017
1.2	**Prototyping and planning**	19.10.2015	27.05.2016
1.3	**Developing Chapters A-F**	09.05.2016	18.11.2016
1.3.1	Controlling initial backlog "book"	09.05.2016	13.05.2016
1.3.2	**Iteration 1: Developing Chapter A**	09.05.2016	20.05.2016
1.3.3	Iteration 2: Developing Chapter B	23.05.2016	03.06.2016
1.3.4	Iteration 3: Developing Chapter C	06.06.2016	17.06.2016
1.3.5	Iteration 4: Developing Chapter D	20.06.2016	01.07.2016
1.3.6	Iteration 5: Developing Chapter E	04.07.2016	15.07.2016
1.3.7	Iteration 6: Developing Chapter F	18.07.2016	29.07.2016
1.3.8	**Peer Review 1 and making modifications**	11.07.2016	18.11.2016
1.4	**Developing Chapters G-L**	12.09.2016	05.12.2016
1.5	**Planning the marketing activities and the e-book, developing other contents**	31.10.2016	20.12.2016
1.6	**Developing Chapters M-Q**	07.11.2016	20.12.2016
1.7	**Producing the hardcover and e-book**	21.11.2016	24.02.2017
1.8	**Further developing the services, products and management**	30.01.2017	28.02.2017
1.9	**Initial marketing of the book and services**	09.01.2017	10.03.2017

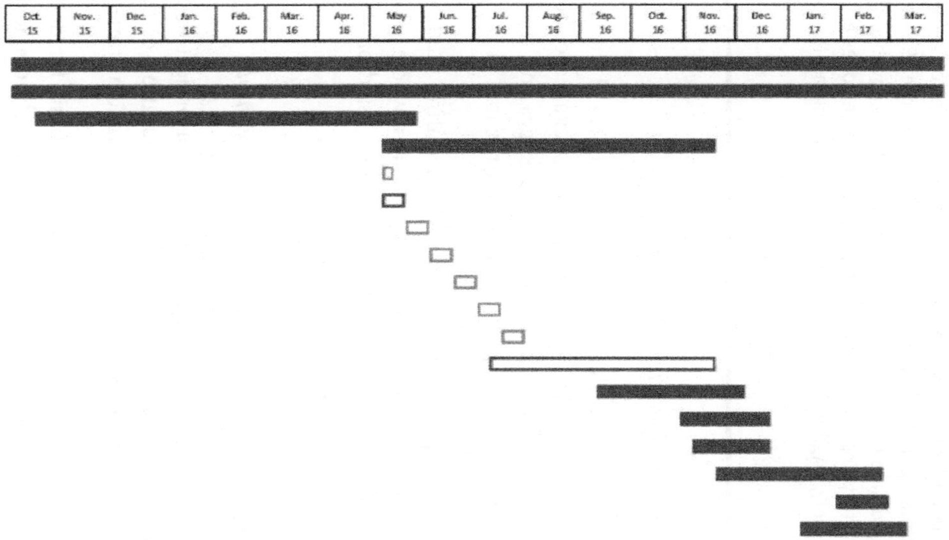

Fig. 9.5: Excerpt from the project bar chart for the project "Developing Values4Business Value".

Developing a Project Bar Chart: Objectives and Process

The objective of developing a project bar chart is to visualize the timing and durations of the phases and work packages of a project. The bar chart is an important communication tool because it provides an easy to read visualization of the project schedule.

The object to be scheduled in a bar chart can be the project as a whole and/or one or several project phases. A bar chart can be used as a rough or a detailed scheduling method. For example, a rough bar chart can be created for the project as a whole, and detailed bar charts can be created for individual project phases.

Knowledge of the duration of work packages is a prerequisite for developing a project bar chart. The duration can be estimated either intuitively or analytically. In any case, the estimate is based on experience.

An example of an analytical estimate of work package duration for the project "Developing Values4Business Value" is shown in Table 9.10. Tips for estimating work package durations are given in Table 9.11.

Case Study: Developing Values4Business Value—Analytical Estimation of the Duration of a Work Package

Table 9.10: Analytical Estimation of the Duration of the Work Package "1.8.2 Further Developing the Basic Seminars"

Analytic Estimation of the Duration of the Work Package "1.8.2 Further developing the basic seminars"

Developing Values4Business Value

Scope of work
Updating the contents and the designs of 3 basic seminars
For each seminar the scope of work is similar
Productivity of possible ressource combinations
Team consisting of 2 experts: Duration for 1 seminar = 3 days
Team consisting of 3 experts: Duration for 1 seminar = 2 days
Estimation of work package duration
Decision: Employment of a team of 2 experts
Calculation: 3 seminars x 3 days = 9 days

V. 1.001 v. S. Füreder per 5.10.2015

Table 9.11: Tips for Estimating Work Package Durations

Tips for estimating work package durations
> All durations must be estimated in the same time units. A "day" will usually be used as the time unit. For projects with long durations a "week" can be used as a time unit.
> The duration of work packages represent throughput times and not working times (in person days).
> Probable durations are to be estimated, without considering time reserves.
> The basis for estimates of duration are assumptions about the typical use of resources for carrying out each work package. The resulting "normal durations" give rise to minimal work package costs.
> Estimates of work package durations are to be made by those responsible for carrying out the work packages.

Developing a Project Network Plan (CPM Schedule): Definition and Example

A project network plan (e.g., a schedule based on the CPM, the critical path method) is a graph showing the positions in time and durations of the work packages of a project as well as their logical relationship to one other. If a network plan is developed for a project, it can also be displayed as a "networked bar chart". In a networked bar chart, the logical relationships between the work packages depicted as bars are represented by connecting lines.

A distinction is made between the methods of arrow diagramming and precedence diagramming. Since the precedence diagramming method has mostly prevailed in practice, it is the only one discussed in what follows. In precedence diagramming, the work packages are represented as nodes, and the logical relationships between them are shown as arrows. In principle, the logical relationship between two work packages can be represented as a "end-start" relationship, a "start-start" relationship, an "end-end" relationship, or a "start-end" relationship (see Table 9.12).

Table 9.12: Possible Logical Relationships Between Two Work Packages

Type of relations	Relation between...
End-start	the end of an activity and the start of another activity
Start-start	the start of an activity and the start of another activity
End-end	the end of an activity and the end of another activity
Start-end	the start of an activity and the end of another activity

In contrast to developing a bar chart, in network planning a distinction can be made between the planning of the project logic and scheduling. By defining the logical relationships between work packages, a project logic is developed independently of scheduling assumptions.

As an example, a network plan for the project phase "1.7 Producing hardcover and e-book" of the project "Developing Values4Business Value" can be found in Figure 9.6 (on next pages).

Case Study: Developing Values4Business Value—Network Plan for the Project Phase "1.7 Producing hardcover and e-book"

The project phase "1.7 Producing hardcover and e-book" was performed mainly by Manz Verlag, independently of the parallel project phase "Further development of RGC Services, etc." The basis for "delivering" the individual chapters by RGC to Manz was the preparation of these chapters. These relationships are symbolized in the network plan (see Fig. 9.6) by the small circles with corresponding work package numbers.

Developing a Project Network (CPM Schedule): Objectives and Process

The objective of developing a project network is to plan the timing of the work packages and to show the logical relationships between them. Because it is a relatively complicated visualization, the network plan should only be seen as a communication tool for project management experts. However, it provides the basis for creating plans such as a networked project bar chart or a project milestones plan, which are good communication tools.

A network plan can be used as a rough or a detailed scheduling method. For example, a rough network plan can be developed for the project as a whole, and detailed network plans can be created for individual project phases.

For developing detailed network plans, work packages are to be broken down until they consist of activities for which the following hold:

- The activity is carried out without interruption.
- The usage of resources is the same for each time unit of the activity.
- There is a proportional relationship between the progress of work and the duration of the activity.

The planning of the logical relationships between the work packages is to be carried out on the basis of technological and not resource-related dependencies. However, the consideration of scarce resources can be an optimization step after the creation of the network.

The process logic of the network plan is to be developed in the project team. The use of visualization techniques (e.g., representation with Post-its) promotes communication. Only afterwards should the results achieved be documented using project

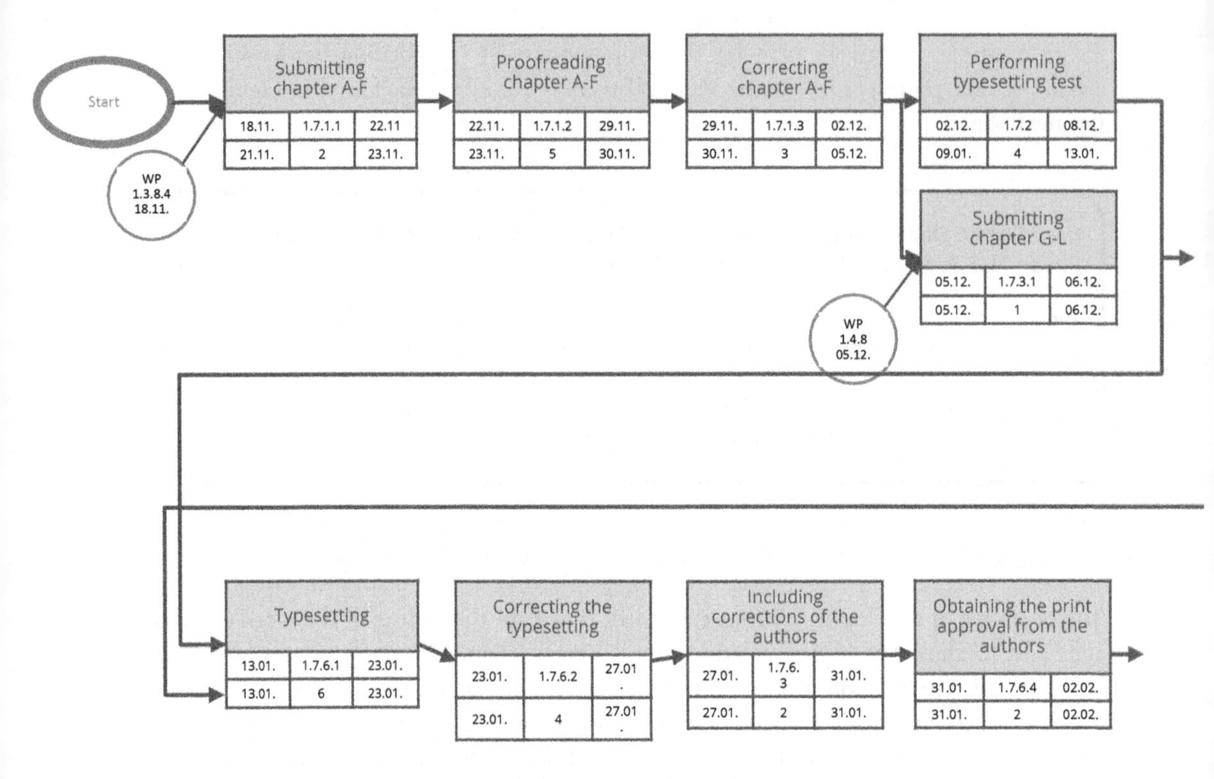

Fig. 9.6: Network plan for the project phase "1.7 Producing hardcover and e-book".
(continues on next page)

management software. This software also supports the calculation of project dates and of the project duration.

In the calculation of dates, the earliest possible or latest allowed date is calculated for each activity. If the latest allowed date and the earliest possible date are different, the activity has a (total) buffer—that is, a time reserve. If the total buffer of an activity is equal to zero, this activity is called "critical". An extension of the duration of a critical activity extends the total project duration. The chain of critical activities is called the "critical path" of a project.

If the project duration determined by scheduling proves to be too long, the duration of the project can be shortened by overlapping the activities of the critical path. If further shortening is necessary, it must be checked whether accelerations can be achieved by outsourcing or by increasing the allocation of resources to activities.

9.4.2 Planning a Project Budget

Depending on the project type, a project budget must either contain project-related costs only or also include project-related income. No income occurs in internal projects,

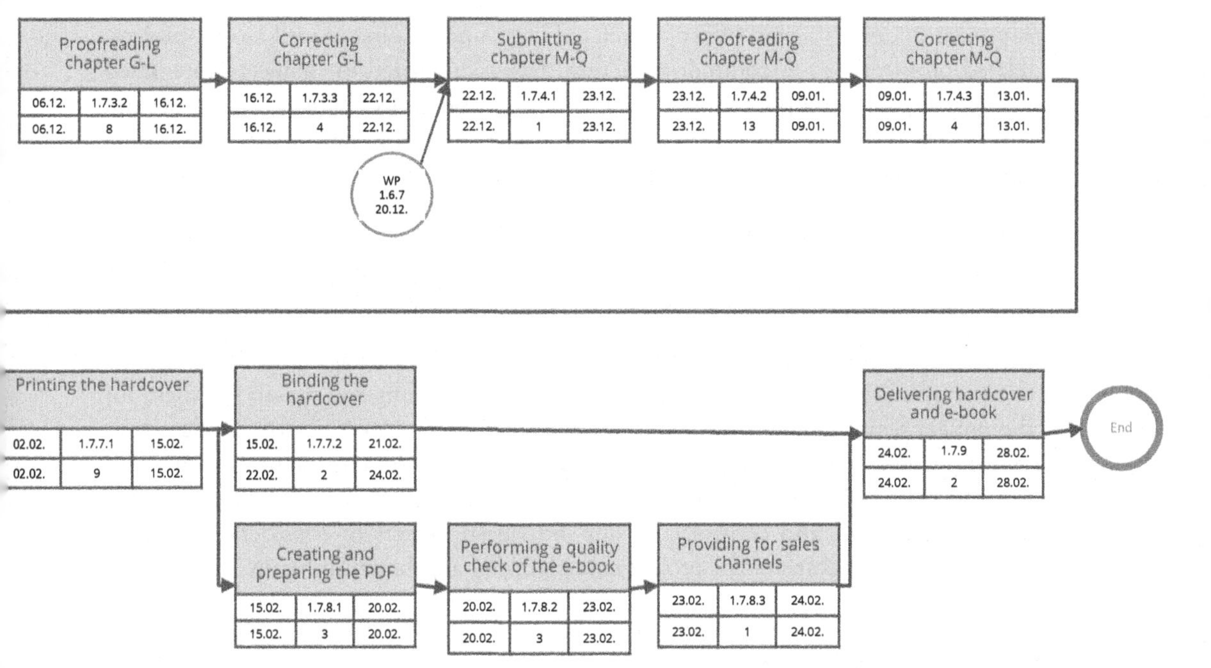

Fig. 9.6: Network plan for the project phase "1.7 Producing hardcover and e-book". *(cont.)*

so budgeting is limited to the planning of project costs. In external projects, services are provided for third parties in return for payment. The resulting income is compared with the project costs in order to calculate the profit or loss attributed to a project. According to the terminology of cost accounting, external projects are to be considered as profit centers and internal projects as cost centers. An associated challenge for project-oriented organizations is the establishment of temporary profit centers and cost centers.

In both internal and external projects, the most holistic possible view of project costs is to be striven for. This means that not only costs related to actual disbursements, but also opportunity costs are considered and that project-related costs of possible partners, suppliers, and also of the customer (in external projects) are taken into account. Thus the project costs are related to the overall project scope and make them an indicator of project size and complexity.

When planning the project budget, it is important to ensure that only costs and income incurred during the project are allocated to the project. Costs of the pre-project and of the post-project phases are not to be considered. But the costs and the income of the post-project phase are the basis for the business case analysis or the cost–benefit analysis of the investment realized by the project.

Planning a Project Budget: Definitions and Examples

A project budget is a plan of project-related costs and income as well as of a project-related profit or loss. A common structural basis is required to enable an integrated view of the project scope, the project schedule, and the project budget. This structural basis is the WBS. The units for planning a project budget are therefore the work packages and/or groups of work packages of the WBS.

The individual work packages are the units for which planned costs are estimated and actual costs are controlled. A cost structure based on the WBS creates the possibility of grouping work package costs and of influencing costs at different levels of aggregation.

Internal assignments can arise from the grouping of work packages according to areas of responsibility. Planning the costs of internal assignments and specifying them for the project team members enables "responsibility accounting". For the implementation of responsibility accounting, the breakdown of project cost planning must correspond with the structures of cost controlling. Adding up costs for objects makes it possible to plan and control costs for individual objects of consideration. These structural relationships of the project cost planning are shown in Figure 9.7.

As an example, Table 9.13 shows an excerpt of the project budget for the project "Developing Values4Business Value".

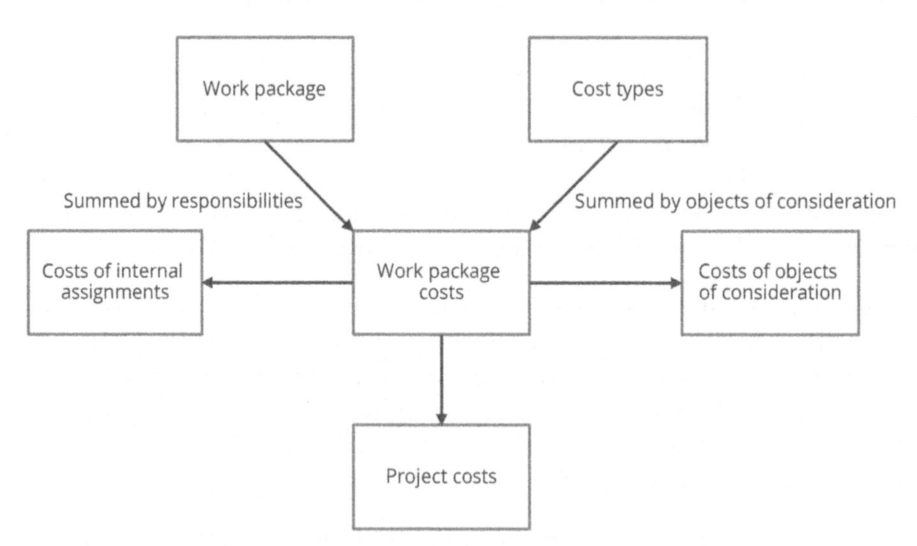

Fig. 9.7: Structural relationships of project cost planning.

Case Study: Developing Values4Business Value—Project Budget

Table 9.13: Excerpt from the Project Budget for the Project "Developing Values4Business Value"

Project Budget
Developing Values4Business Value

WBS Code	Phase / Work Package	Cost Type	Planned Quantity	Internal Price	Planned Costs
Costs					
1.1	Project managing				€ 45,000.00
1.1.1	*Project assigned*		-		
1.1.2	Project starting	Personnel: Project manager	10 PD	€ 600.00	€ 6,000.00
1.1.3	Project coordinating	Personnel: Project manager	60 PD	€ 600.00	€ 36,000.00
	...				
1.2	Prototyping and planning				€ 56,000.00
1.2.1	Controlling the initial backlog "Prototyping"	Personnel: Authors	2 PD	€ 800.00	€ 1,600.00
1.2.2	Prototyping by the conference " Agility & Processes"	Personnel: Authors	8 PD	€ 800.00	€ 6,400.00
1.2.3	Prototyping by the forum "Sustainable development"	Personnel: Authors	10 PD	€ 800.00	€ 8,000.00
1.2.4	Prototyping by the seminar "Agility & Projects"	Personnel: Authors	12 PD	€ 800.00	€ 9,600.00
	...				
1.3	Developing Chapters A-F				€ 75,000.00
1.3.1	Controlling initial backlog "book"	Personnel: Authors	1 PD	€ 800.00	€ 800.00
1.3.2	Iteration 1: Developing Chapter A				€ 8,800.00
1.3.2.1	Planning Iteration 1	Personnel: Authors	1 PD	€ 800.00	€ 800.00
1.3.2.2	Researching the content of Chapter	Personnel: Authors	2 PD	€ 800.00	€ 1,600.00
1.3.2.3	Developing a draft of Chapter A	Personnel: Authors Personnel: Assistant	2 PD 2 PD	€ 800,00 € 400,00	€ 2,400.00
1.3.2.4	Accepting a draft of Chapter A	Personnel: Authors	1 PD	€ 800.00	€ 800.00
1.3.2.5	Finishing Chapter A	Personnel: Authors Personnel: Assistant	1 PD 2 PD	€ 800,00 € 400,00	€ 1,600.00
1.3.2.6	Reflecting on and handing over Chapter A	Personnel: Authors	1 PD	€ 800.00	€ 800.00
1.3.2.7	Controlling the backlog "book"	Personnel: Authors	1 PD	€ 800.00	€ 800.00
1.3.3	Iteration 2: Developing Chapter B				€ 9,600.00
	...				
1.4	...		...		
				Sum costs	€ 381,000.00
Revenues					
1.2	Phase: Prototyping and planning				€ 30,000.00
				Sum revenues	€ 30,000.00
Difference costs-revenues					-€ 351,000.00

V. 1.001 v. S. Füreder per 5.10.2015

Since the project "Developing Values4Business Value" was an internal innovation project, mainly costs occurred, and only a low revenue (from quick wins) was produced. Revenue was produced by the seminar "Agility & Projects", which was carried out during the prototyping phase. For the project phases presented in Table 9.13, the costs incurred were mainly personnel costs. Other cost types in the project were, for example, catering costs and costs for suppliers. These, however, were incurred during later phases not shown here.

Planning a Project Budget: Objectives and Process

The objectives of planning a project budget are to estimate project costs and possibly project revenue and to calculate a project profit or loss. Project cost plans provide a basis for the decision to carry out a project or not, for determining the price for performing a contract as an external project, and for performing cost–effectiveness controlling.

The breakdown of the project cost plan must correspond to the structure of the WBS. It has to be decided at which WBS level the cost planning is to be carried out. In most cases, a relatively high WBS level is appropriate for project cost planning and project cost controlling. The work packages at the lowest WBS level, and activities which may be used for scheduling, are usually too detailed.

The cost types to be taken into account for each work package must be specified. In most cases it will be enough to restrict attention to personnel costs, costs of materials, costs of external services, and "miscellaneous" costs such as traveling costs. Quantities for each cost type of the relevant work packages must be determined. The costs for each work package can be calculated by multiplying the planned quantities by the planned prices for each cost type. To determine quantities for personnel costs requires planning of personnel resources used for the performance of a work package. This approach assures an integrative view in planning project resources and project costs.

The estimation of work package costs is to be carried out by the project team members involved in performing the individual work packages. The function of detailed project cost planning cannot be transferred to central estimating departments, because they cannot estimate the required quantities or the prices to be expected without a process-specific know-how. The involvement of the project team members in project cost planning also creates a motivational basis for agreeing on target costs.

Graphical representations of project costs over time can be shown in the form of project cost histograms and project cost curves. Project cost histograms depict project costs for time periods. In project cost curves, project costs are cumulated over time. This assures an integrated view in planning project costs and the project schedule.

The typical S-curve of project costs (see Figure 9.8) arises from the fact that the costs incurred in early and late project phases are generally lower than in the high productivity phases. Different project cost curves arise from different assumptions about the timing of the work packages (earliest possible dates, latest allowed dates, or planned dates).

The objective of planning the project revenue is to estimate the project-related profit or loss. The project team's responsibility for ensuring the project-related income is made clear. For all the different types of revenue that can be attributed to a project (e.g., payments for services, subsidies, etc.), the corresponding amount and timing must be estimated and presented in a project revenue plan.

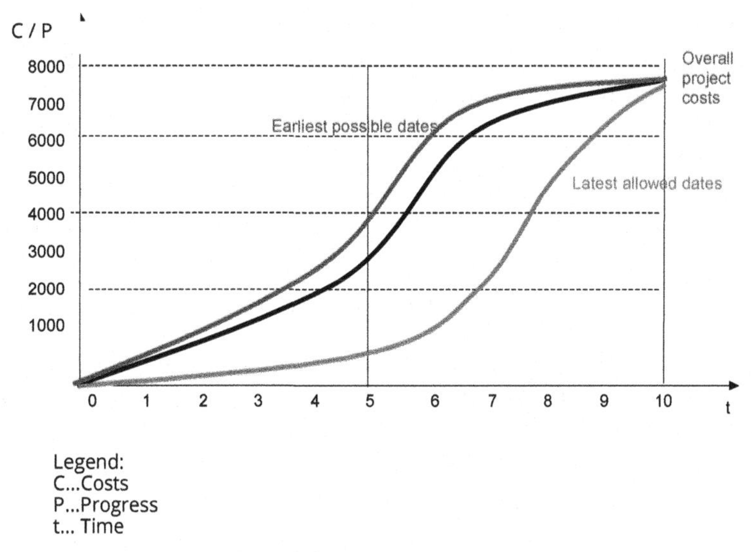

Fig. 9.8: Cumulative project cost curves.

The "project success" can be defined as the difference between project revenue and project costs, whereby the term "project costs" is defined differently in full costing and in contribution accounting. When full costing is used, the project success is calculated as the difference between project revenue and the project's "self costs". In addition to the variable project costs, "self costs" include an assumed percentage for assigning the fixed overheads of the project-oriented organization to the project. A positive project success is referred to as a project profit, and a negative project success is referred to as a project loss. When "contribution accounting" is used, the project success is calculated as the difference between project revenue and variable project costs. The resulting project success is referred to as the "contribution". It serves to cover fixed costs and provide an overall profit for the organization.

9.4.3 Planning Project Resources

Planning Project Resources: Definitions

A project resource plan is a tabular and/or graphical representation of the demand of a resource for a project over time. In project resource planning, not all resources used in a project are considered, but only so-called "bottleneck resources".

Bottleneck resources are resources that are scarce for the project and thus affect the achievement of project objectives. Bottleneck resources can be, for example, personnel, equipment, financial resources, storage areas, etc. Personnel bottleneck resources must be differentiated according to non-interchangeable qualifications. So, for example, in an engineering department, one can differentiate between design engineers and draftsmen. Individuals as well as qualification groups can be considered bottleneck resources.

A basis for project resource planning is created when estimating quantities in project cost planning.

Planning Project Resources: Objectives and Process

Supply and demand in a project for a bottleneck resource can be represented by project resource histograms and project resource curves. By comparing resource demands with available resources, project-related over- or under-coverage of project resources can be made visible. This may lead to increases or reductions in the supply of resources or can be compensated for by time shifts in work packages.

Network-supported project resource plans can be developed for different positions in time of the work packages. Two extreme positions in time of the work packages are taken as a basis for optimization decisions—namely, the earliest possible and the latest allowed positions.

Since the planning of project resources is limited to the consideration of bottlenecks, it is not an instrument for general resource planning for the project-oriented organization. Thus it is, for example, no substitution for department-specific resource planning.

Project resource plans are to be created for bottleneck resources. Therefore, each resource for which a project resource plan is to be created must be defined. Work packages are the bearers of the resource demands. The demand for a bottleneck resource must be estimated for each work package for each planning period. Quantity frameworks for resources are either directly available from project cost planning or can be derived from it. By adding up the demand for a resource for of all the work packages of a project for each period, a resource histogram can be created. The accumulation of the resource need for each period makes it possible to develop a resource sum curve.

The quantities of the bottleneck resources available in the organization must be determined and compared with the resource demands. The available resources can remain the same throughout the duration of the project, or they can vary for different periods. The resource demands for each period can be less than or greater than this available resource quantity. Over- or under-coverage can be identified (see Fig. 9.9). Using project management software, it is possible to differentiate between the timing of resource requirements at the start, at the end, or remaining the same throughout the duration of a work package.

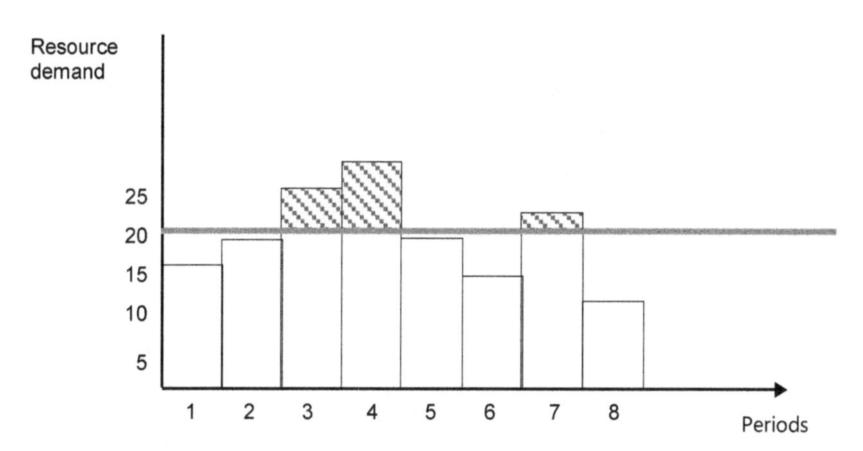

Fig. 9.9: Over-coverage and under-coverage of a bottleneck resource (histogram).

An adaption of the project plans is necessary if there are resource bottlenecks. The allocation of bottleneck resources can be optimized by rescheduling work packages, by changing the process logic, or by dividing work packages. Furthermore, the possibility of increasing or reducing the supply of resources must be taken into account. Possible measures for increasing, for example, the supply of human resources are creating incentives for higher performance, making use of overtime, or employing temporary staff.

Excursus: Planning Project Cash Flow

Financial resources must be provided in internal projects for project-related cash outflows. In external projects financial resources are required for outflow surpluses, defined as cumulative cash outflows minus cumulative cash inflows. The financial resources to be provided can be a bottleneck resource for a project. In determining the necessary financial resources, a distinction must be made between project-related cash outflows and project costs. The differentiation between costs and cash outflows is shown in Figure 9.10.

Costs without cash outflows	Costs equal cash outflows	
	Cash outflows equal costs	Cash outflows not considered as costs

Fig. 9.10: Differentiation between costs and cash outflows.

Examples of project costs requiring cash outflows are costs for services and products to be procured, project-related personnel and materials costs, project-related travel costs, etc. Cash outflows that are not cost–effective in a period are, for example, cash outflows for an investment in equipment procured. Costs not requiring cash outflows are, for example, calculated interest for own capital, depreciations for capital investments, and calculated risk premiums.

The position in time of cash outflows is usually dependent on the position in time of the costs. Costs occur continuously over time and in relation to performance. Cash outflows, on the other hand, occur at distinct points in time, independent of performance (see Fig. 9.11). Project-related cash inflows in external projects occur on the agreed payment dates.

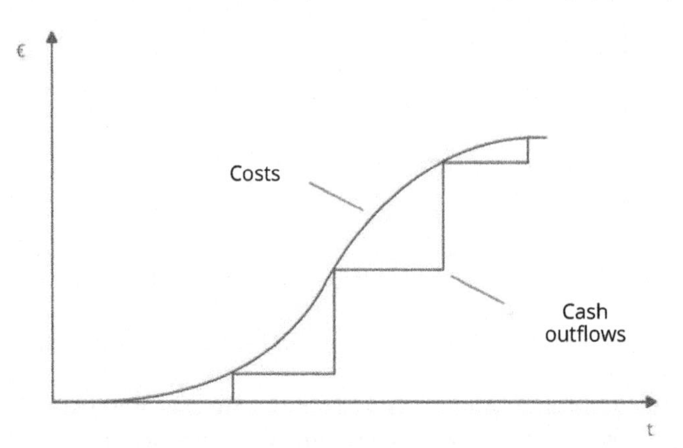

Fig. 9.11: Accumulated costs and cash outflows over time.

Project cash flow plans are used for project-related liquidity planning. A project-related need for financial resources exists if there are surpluses of cash outflows in individual periods. These are calculated from the difference between project-related cash outflows and project-related cash inflows. Interest costs incurred by the need of a project for financial resources are costs of the project and need to be considered in the project cost plan. If the project-related cash inflows exceed the cash outflows, there is an inflow surplus. Income from interest that is obtained as a result of inflow surpluses is to be considered as income of a project.

9.5 Methods: Analyzing Project Context Relations

Project contexts are understood as interrelations within which a project is performed. Contexts of a project are:

> the strategic objectives of the organization performing a project,
> other projects related to a project,
> the pre-project and post-project phases of a project,
> the project stakeholders, and
> the investment implemented by a project.

These project contexts can be analyzed, and strategies and measures for managing the project context relations can be planned. A business case analysis and a cost–benefit analysis are methods for analyzing the connection of a project to the investment implemented by a project. These methods have already been covered in Chapter 3. Methods for analyzing the other project contexts are described below.

9.5.1 Analyzing Relations to Strategic Objectives of the Organization

Analyzing the Context "Strategic Objectives of the Organization": Definitions and Example

When analyzing the connection between a project and the strategic objectives of the organization, it is necessary to clarify whether the objectives of the organization have led to the implementation of the project and in which way the project contributes to the realization of the strategic objectives of the organization (see Chapter 3). It is also possible for a project to influence the objectives of an organization (see Fig. 9.12).

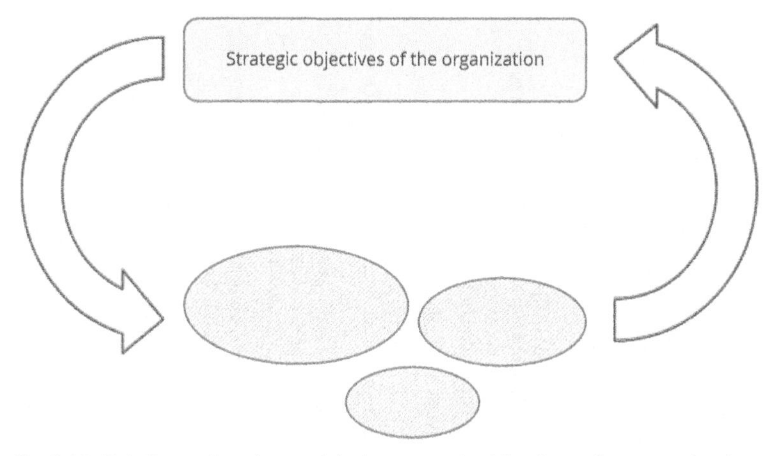

Fig. 9.12: Relations of projects with the strategic objectives of an organization.

External projects should always contribute to securing market share and to achieving revenues for the organization on a relatively short-term basis. Internal projects primarily contribute to the long-term survival of an organization by developing products, the organizational structures, the personnel, and the infrastructure of the organization.

The strategic objectives of an organization can either be explicitly planned and documented, or they can exist only implicitly. In the latter case, appropriate assumptions must be made about the objectives of the organization. As an example, the contributions of the project "Developing Values4Business Value" to the realization of the strategic objectives of RGC are presented in Table 9.14.

Case Study: Developing Values4Business Value—Contributions of the Project to the Realization of the Strategic Objectives of RGC

Analyzing the Context "Strategic Objectives of the Organization": Objectives and Process

The objective of analyzing the contributions of a project to the realization of the strategic objectives of an organization is to clarify its "sense", its significance for the organization.

Understanding the contribution of a project to the realization of the strategic objectives of an organization promotes its acceptance by the members of the project organization and stakeholder representatives.

If there are no relations between a project and the strategic objectives of the organization, then either the objectives must be changed or the project has to be seriously questioned.

Table 9.14: Relations of the Project "Developing Values4Business Value" with the Strategic Objectives of RGC

Relations between the Project and Strategic Objectives

Developing Values4Business Value

Strategic objectives of RGC	Relations to the project
To a sustainable project, program and change management in the community contributed	RGC management approaches further developed by innovations and documented in the book PROJECT.PROGRAM.CHANGE
Sustainable business value for RGC clients ensured	RGC services and products further developed
Evolutionary growth of RGC in the DACH-region realized	Optimized services and book marketed in the DACH-region
Thematic leadership regarding process, project, program and change management embraced in the scientific and consulting community	RGC management approaches further developed by innovations and documented in the book PROJECT.PROGRAM.CHANGE
Management of RGC further developed	Project management, program management and change management of the RGC further developed
Satisfaction of the employees ensured by value-oriented leadership	Employees developed by participation

V. 1.001 v. S. Füreder per 5.10.2015

9.5.2 Analyzing and Planning Relations to Other Projects

Analyzing the Context "Other Projects": Definitions and Example

Projects carried out at the same time as a project under consideration can lead to opportunities or to problems—for example, regarding the realization of the project objectives or regarding the methods and resources used. The realization of (intermediate) objectives of other projects can be prerequisites for further work in a project under consideration. For example, the implementation of new software can be a prerequisite for the continuation of work in an organizational development project. Opportunities for a project can arise, for example, from the use of common methods or of common suppliers. Resource problems for a project can arise as a result of other projects using the experts required as project team members.

In contrast to networking of projects (see Chapter 16), here the effects of "other" projects on the success of a project under consideration are analyzed and measures for optimizing the success of the project are planned. An example of the analysis and planning of measures for managing the relations of a project with other projects is shown in Figure 9.13 for the project "Developing Values4Business Value".

Case Study: "Developing Values4Business Value"—Relations to Other Projects

The projects ongoing at the time of analysis in October 2015 had a variety of effects on the project success of the project "Developing Values4Business Value"; for example, the project "Sales 16" had a large effect, as this project promoted the realization of the market-related objectives of "Developing Values4Business Value".

Also, relations to projects with medium and small influence were considered and managed.

Analyzing the Context "Other Projects": Objectives and Process

The objective of analyzing the relations between a project and other projects performed at the same time is to identify possible opportunities or problems for the project under consideration. Measures to exploit the opportunities or to avoid the problems can be planned on this basis.

Projects which are related to a project are to be listed and their influences on the project under consideration are to be analyzed. Relevant projects are identified on the basis of the information in the project portfolio database and information from members of the project organization. Possible opportunities or problems for the project under consideration have to be analyzed, and measures to deal with them have to be planned and communicated to the members of the project organization.

Relations to other Projects
Developing Values4Business Value

Project	Relevance for project success	Intensity of interaction	Quality of relation	Interpretation of evaluations	Measures for the design of the relation	Responsibility
Client project A	medium	medium	supporting	Content-related cooperation with a client representative in the peer review group	Inviting a client representatives into the peer review group.	Project Owner L. Gareis
Client project B	low	low	neutral	Innovations can be applied in the cooperation with the client, experiences can be gained.	Applying management innovations, reflecting the experiences gained.	Project team member: Service development L. Gareis
Sales 16	high	high	supporting	Support of the market-related project objectives.	Clarifying which sales measures support the market-related objectives and coordinating these measures	Project Owner R. Gareis
Event HP 16	medium	medium	supporting	Lectures and workshops for prototyping	Planning lectures and workshops for prototyping. Reflecting the experiences from lectures and workshops.	Project team member: Prototyping S. Füreder

Fig. 9.13: Relations of the project "Developing Values4Business Value" to other projects.

9.5.3 Analyzing the Pre-Project Phase and Planning the Post-Project Phase

Analyzing the Context "Pre-Project Phase and Post-Project Phase": Definitions and Example

The temporal context of a project can be determined by analyzing relevant actions and decisions from the pre-project phase and by analyzing the expectations of the post-project phase (see Fig. 9.14).

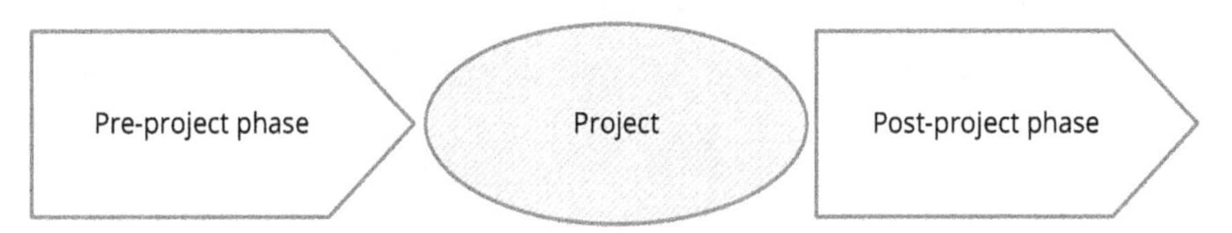

Fig. 9.14: A project in its temporal context.

Projects often have a long history. Information about the cause leading to a project, about actions before the project start, and about stakeholders who made decisions in the pre-project phase are essential to understanding a project. Although actions and decisions from the pre-project phase can no longer be influenced during the performance of a project, they can promote or inhibit the project success.

Expectations about the post-project phase influence the structuring of projects and should therefore be made transparent. Such expectations can be, for example, the use of project results for reference purposes, the establishment of a long-term customer relationship, career progression of project team members, etc. The fulfillment of these expectations can be influenced by the project.

Decisions from the pre-project phase and expectations of the post-project phase can influence, for example, the selection of project team members, the planning of project values, the planning of communication formats, the planning of the project management methods to be used, etc. As an example, the analysis of the pre-project and the post-project phase of the project "Developing Values4Business Value" is shown in Table 9.15.

Analyzing the Context "Pre-Project Phase and Post-Project Phase": Objectives and Process

The objective of analyzing the relations of a project to the pre-project phase and the post-project phase is to make the actions and decisions from the pre-project phase and the expectations about the post-project phase transparent. This information makes it possible to structure the project adequately.

Case Study: Developing Values4Business Value—Analysis of the Pre-Project and Post-Project Phase

Table 9.15: Analysis of the Pre-Project and Post-Project Phase of the Project "Developing Values4Business Value"

Analysis of the Pre-Project & Post-Project Phase Developing Values4Business Value
Analysis of the pre-project phase
Important stakeholder relationships in the pre-project phase
RGC: Need for further development of RGC management approaches identified
Clients: Good relationship as the basis for an invitation to the peer review group
Manz (publisher): Existing cooperation as a result of several publications
Decisions made in the pre-project phase
Cooperation with Manz as well as with publishers in Germany and Switzerland
No further development of the RGC process management approach; will be done in the follow-up "stabilization project"
Consideration of new values and clear epistemic positioning of the management approaches
Relevant documents developed in the pre-project phase
Concept as a result of the working group "Values4Business Value"
Initial backlog, initial project plans, initial change plans
Existing consulting and training materials
Analysis of the post-project phase
Expected development of stakeholder relationships
Publishers: Cooperation extended/developed
Peer review group members: Perception of their role also as change agent
Measures in the post-project phase
Performing the project "Stabilizing Values4Business Value"
Further marketing of the new RGC management approaches
Developing an english version of the book
Utilization of experiences from the project
Applying optimized services and products in consulting and trainings
Using the new management approaches in the RGC management

V. 1.001 v. S. Füreder per 5.10.2015

Information about the relevant actions and decisions of the pre-project phase and expectations about the post-project phase are to be exchanged by the members of the project organization during project starting. This ensures a common level of information. Existing documents (minutes, plans, specifications, etc.) that were created in the pre-project phase are to be analyzed for relevant information. Interviews with promoters of the project idea and decision makers may also be carried out. Possible questions for the analysis are shown in Table 9.16. Consequences of the analysis for the structuring of the project can be derived and documented.

Table 9.16: Questions for the Analysis of the Pre-Project and Post-Project Phases of a Project

Analysis of the pre-project and post-project phases of a project	
>	What concrete cause has given rise to the project?
>	Who supported the project initiation? Who impeded it?
>	What relevant documents and documentation were produced?
>	What decisions were made which have to be taken into account?
>	What similar projects have already been implemented in the organization?
>	Who has what expectations about results in the post-project phase?
>	What actions and decisions are necessary after the project end?
>	What follow-up projects should be undertaken?

9.5.4 Analyzing and Planning Relations to Project Stakeholders

Analyzing the Context "Relations to Project Stakeholders": Definitions and Example

The social context of a project is constructed by defining its stakeholders. In a project stakeholder analysis, the relations of a project with its stakeholders can be analyzed and planned. It can be assumed that the stakeholders of a project cannot be changed but that relations with project stakeholders can be designed. The design of these relations is a project management task.

Stakeholders who can positively or negatively influence the project success are "relevant" for a project. Project stakeholders can be differentiated into project-internal, organization-internal, and organization-external stakeholders. Organization-external stakeholders are, for example, customers, partners, suppliers, residents, etc. Organization-internal stakeholders can be profit centers, departments, and employees of the organization carrying out the project. The project owner, the project manager, and the project team are project-internal stakeholders whose relations to a project also influence its success. All of these relations must be designed.

The documentation of the project stakeholder analysis can be limited to a graphical representation of the relations of a project with its stakeholders, or it can include analyses of expectations between a project and individual stakeholders. In the latter case, analyses of the opportunities and conflicts in the relations are needed, and strategies and measures for designing the relations have to be decided upon. As an example of a project stakeholder analysis, Figure 9.15 shows a stakeholder analysis graph for the project "Developing Values4Business Value", developed during project starting.

Case Study: Developing Values4Business Value—Project Stakeholder Analysis

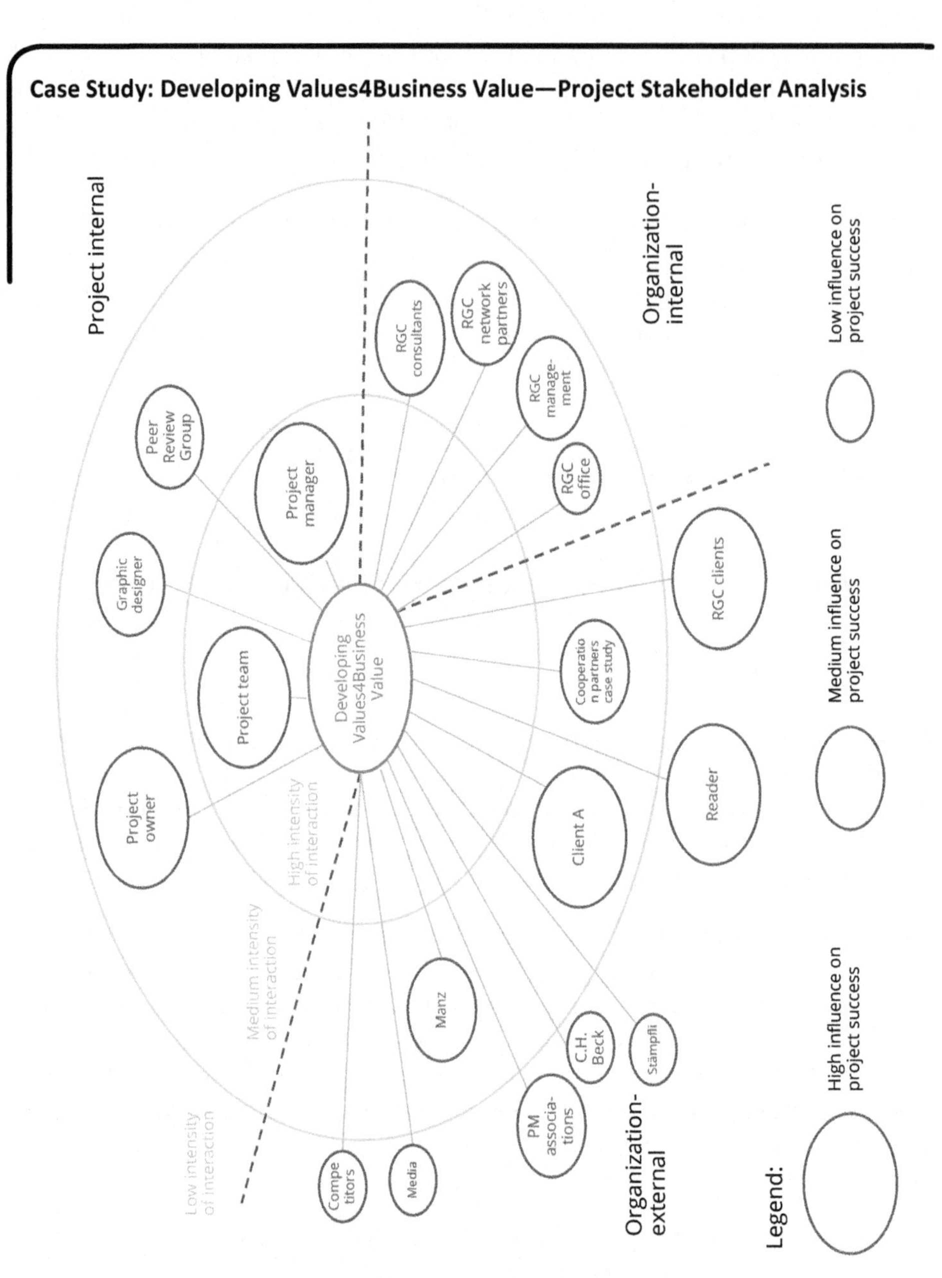

Fig. 9.15: Project Stakeholder Analysis for the Project "Developing Values4Business Value" (Status: October 2015).

The social complexity of the project due to stakeholder relations was perceived as average. A lot of support was expected from within the organization. In the early phases of the project, the intensity of the interactions with the organization-external stakeholders, such as publishers, project management associations, and customers, was relatively low.

Analyzing the Context "Project Stakeholders": Objectives and Process

The objective of a project stakeholder analysis is to provide a basis for appropriate management of the project stakeholder relations. This analysis ensures an external orientation of the project. The project stakeholder analysis is also a basis for project marketing.

To create a project stakeholder analysis, "relevant" project stakeholders must be identified. The identification of project stakeholders is a social construct. The project identifies the stakeholders who should receive management attention. Differentiation of project stakeholders is necessary if different strategies and measures are needed to manage a project's relations with different stakeholders. For example, it can be useful to see the individual managers of an organization which is to be changed as individual stakeholders, instead of considering the group of managers as a stakeholder.

When identifying project stakeholders, they must be distinguished from change stakeholders. Project stakeholders have interests in the project and have opportunities to influence the project success. Change stakeholders have an interest in the overall change and have opportunities to influence the success of change. Organizations, groups, and individuals can be both project and change stakeholders at the same time, but in any case, they have different expectations of the project and the associated change.

The importance of individual project stakeholders for a project and the intensity of interaction between the project and the corresponding stakeholder can be analyzed and evaluated. The results of this analysis can be documented in the form of a list. An example for the project "Developing Values4Business Value" is presented in Table 9.17. This assessment of the importance and intensity of the interactions refers to an analysis at a specific time point. Project stakeholder relations change during the course of the project. Therefore "social" project controlling is necessary to manage them accordingly (see Chapter 10).

Case Study: "Developing Values4Businesss Value"—Importance and Intensity of Interactions with Project Stakeholders

Table 9.17: Analysis of the Importance and Intensity of the Interaction of the Project with Individual Project Stakeholders

Stakeholder	Importance 1...great, 2 ... average 3...little	Interaction 1...frequent, 2 ... average 3...rare
Project owner team	1	2
Project team	1	1
RGC Management	2	2
Client A	2	2
Graphic designer	2	2
Media	3	3

In the project stakeholder graph, the importance and the intensity of interactions between the project and each stakeholder are expressed by symbols. The size of each stakeholder represents their importance for the project at the date of analysis. The distance of each stakeholder to the project represents the intensity of the interactions between the project and the stakeholder.

Selected project stakeholder relations can be subjected to more detailed analysis and planning. The nature of the relation between a project and a stakeholder can be based on a description of mutual expectations. A distinction can be made between expectations related to the process of cooperation and expectations related to the project results. Fears are also expectations—namely, negative ones.

The advantage of formulating mutual expectations is that project team members put themselves in the place of stakeholders and thereby understand their views. Potential conflicts between the project and the stakeholder can then, for example, be recognized as being structural and not personal.

The analysis of the opportunities and/or conflicts in a project stakeholder relation enables the definition of strategies and measures for managing the corresponding project stakeholder relation (see the example in Table 9.18). A strategy and measures for designing relations with Project Management Associations could be derived from an analysis of mutual expectations. It became clear that measures were only useful in the later project phase of marketing.

A detailed project stakeholder analysis can be carried out in a project start workshop by the project team on the basis of the initial analysis results from the project initiation. Representatives of stakeholders can be actively involved. The communication of the results obtained, in particular the evaluation of stakeholder relations, must be carried out carefully.

Analyzing the Context "Project Stakeholders: Management for Stakeholders

Freeman et al. distinguish between "management *of* stakeholders" and "management *for* stakeholders".[1] The "breadth" of the definition of stakeholders and the type of interaction between an organization and stakeholders are relevant to this distinction. The approach to project stakeholder management described above corresponds fundamentally to "management of stakeholders", as the project view and the project interests are the main focus.

Consideration of the principles of sustainable development, however, leads to a further development of the "traditional" stakeholder management approach. In the following, possibilities for the further development of project stakeholder management are presented, based on an explicit "management for stakeholders".

[1] Freeman et al., 2007.

Case Study: Developing Values4Business Value—Analysis of the Relation Between a Project and a Project Stakeholder

Table 9.18: Analysis of the Relation Between a Project and the Project Stakeholder "Project Management Association"

Analysis of a Project Stakeholder Relation

Developing Values4Business Value

Project to Project Management Association	Project Management Association to project
Performance-related expectations	
Positive view of the optimization of the management approaches	No competition regarding project management approaches
Recommendation of the book as standard reference work	
Perception of the management innovations as add-on to project management standards and not conflicting with the certification standards	
Cooperation in events for presenting the management innovations	
Process-related expectations	
Assistance in marketing the book	No participation in developing the publication
	Usual support in marketing for the book

Strategy
Informing about the book when it is finished
Measures
Providing the book to board members of the association
Inviting to the book presentation
Joint planning of the support in marketing

V. 1.001 v. S. Füreder per 5.10.2015

Differentiation between stakeholders and non-stakeholders determines the "breadth" of the definition of stakeholders. Freeman defines stakeholders as "any group or individual who can affect or is affected by the achievement of the organization's objectives".[2] Being perceived as project stakeholders and gaining management attention is therefore dependent not only on the interests of a project, but also on the interests of organizations, groups, or individuals in a project. Project stakeholders have their own interests and demands and are not just means for achieving project purposes.

It is therefore necessary to identify as project stakeholders not only organizations, groups, and individuals who can support a project or harm a project, but also those who can benefit from a project or who can be damaged by a project. Organizations, groups, and individuals who indirectly support or damage a project by influencing other stakeholders must also be considered. "Management for stakeholders" therefore requires a broad definition of stakeholders.

Stakeholders have a right to receive management attention. The extent of management attention given to a project stakeholder depends on the form of the interaction between the project and the stakeholder. A distinction can be made between "informing" and "conducting a dialog". In informing, facts about a project and its consequences are transmitted in the form of one-way communication. In communicating in dialog, information is exchanged: workshops and events are held by representatives of the project and project stakeholders to ensure shared views. Stakeholders are involved in the management structures of projects in order to get to know and understand their needs. Active stakeholder commitment is pursued. "Engaging stakeholders" means taking measures that enable stakeholders to participate in an active fashion.

As an example of involving and communicating with project stakeholders, the "Stakeholder Engagement Plan" of the project "Implementing Dorobantu Windparc" by OMV Petrom in Romania is shown in Table 9.19. The "Grievance Procedure" for resolving complaints, which is listed in the Table of Contents under 7.2, is shown in Figure 9.16 (on page 232). This makes the formal structures for managing stakeholder relations clear.

There are limited resources for stakeholder management. "Management of stakeholders" must therefore be combined with "management for stakeholders". An appropriate balance is to be found. One must not lose focus. Prioritizations must be made. The attention given to the management of a stakeholder relation is, according to Mitchell et al.,[3] dependent on the power of the stakeholder and the legitimacy and urgency of the demand. Mitchell et al. distinguish stakeholders according to "the stakeholder's power to influence the firm, the legitimacy of the stakeholder's relationship with the firm, and the urgency of the stakeholder's claim on the firm". A comparison of the "management of stakeholders" and "management for stakeholders" approaches can be found in Table 9.20.

[2] Freeman, 1984.

[3] Mitchell et al., 1997

Table 9.19: "Stakeholder Engagement Plan" for the Project "Implementing Dorobantu Windparc" (Table of Contents)

TABLE OF CONTENTS

Table 9.20: Comparison of the "Management of Stakeholders" and "Management for Stakeholders" Approaches

	Management of Stakeholders	Management for stakeholders
Perceptions of stakeholders	> Stakeholders are instruments for achieving objectives > Stakeholder interests are regarded as disruptive for the achievement of the project objectives	> Stakeholders are sources of ideas > Stakeholders are involved as co-creators to achieve objectives from which many stakeholders benefit
Consideration of stakeholders	> Only important stakeholders are considered > The most important stakeholder is the investor	> Many stakeholders are considered > The different interests of the stakeholders are considered
Understanding of conflict	> Conflicts are bad and must be avoided	> Conflicts are inherent. > A culture that can deal with contradiction is necessary
Values	> Strong economic orientation > Rather short-term orientation > Little consideration of ethical principles	> Consideration of the principles of sustainable development, such as economic, ecological and social orientation, short-, medium- and long-term orientation > Consideration of ethical principles, such as fairness, transparency and co-operation
Challenges	> Achieve sustainable solutions	> High management complexity, slow decision-making process

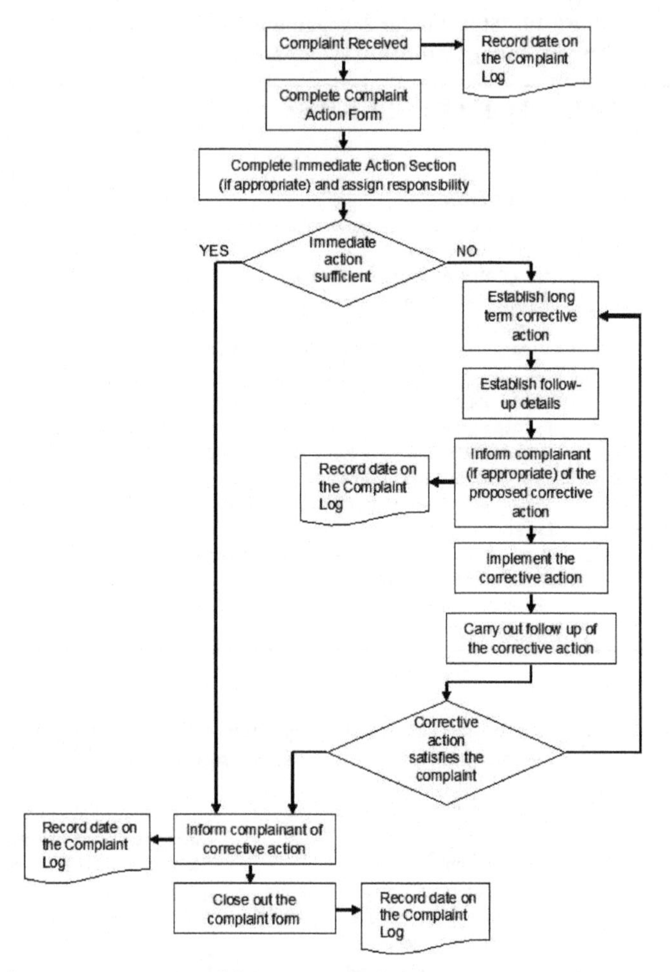

Fig. 9.16: "Grievance Procedure" for the project "Implementing Dorobantu Windparc".

9.6 Methods: Designing a Project Organization, Developing a Project Culture, Managing the Project Personnel

9.6.1 Designing a Project Organization

Designing a Project Organization: Overview

Traditional forms of project organization and new elements for designing project organizations, developing project organization charts, and describing project roles are discussed in Chapter 7. Concepts relating to teamwork and project leadership are discussed in Chapter 8. The following methods are based on these concepts:

> developing a project assignment,
> defining project roles,
> developing a project organization chart,
> developing a project responsibility chart,
> designing project communication formats, and
> defining project rules.

Designing a project organization is a creative process. An appropriate organization for carrying out a project creates a competitive advantage. The project organization is to be designed when starting a project. The initial plans from the project initiation can be used.

The project organization changes during the course of the project. Additional project team members may be required, the frequency of project meetings may change, project rules may need to be supplemented, etc. Changes like this are made to increase the efficiency of a project. They are the result of reflection processes of a project. The project organization is therefore the object of "social" project controlling. Fundamental changes to a project organization are necessary when transforming a project.

Developing a Project Assignment: Definitions and Example

A project assignment summarizes the agreements between the project owner and the project manager and the project team. The project assignment should contain the following information: Project start date and project end date; project objectives and non-objectives; project phases; project costs and possibly project profits; project owner, project manager, and project team members; relations with other projects and key project stakeholders. As an example, the project assignment for the project "Developing Values4Business Value" is shown in Table 9.21.

Developing a Project Assignment: Objectives and Process

The objective of the project assignment is to document the agreements made between the project owner, the project manager, and the project team in the project starting process.

The project assignment is finalized by detailing and extending the initial project plans from project initiating. Revision of the project plans by the actual project team necessitates an adaptation of the project assignment at the end of project starting. The iterative creation of the project assignment during the project initiating and project starting ensures its quality.

The project assignment should be drawn up in writing. In order to symbolize the empowerment of the project team, the project owner should assign the project team and not just the project manager with the project.

Case Study: Developing Values4Business Value—Project Assignment

Table 9.21: Project Assignment for the Project "Developing Values4Business Value"

Project Assignment

Developing Values4Business Value

Project start date	10/5/2015	Project end date	3/31/2017
Project start event	1.1.1 Project assigned	Project end event (formal)	1.1.7 Project approved
Main objectives			
Service- and market-related project objectives			
RGC management approaches further developed by innovations (values and management approaches, sustainable development, agile approaches, benefits realization management, requirements management, business modelling)			
RGC management approaches documented in the book PROJECT.PROGRAM.CHANGE (hardcover and e-book)			
Practical relevance of the developed RGC management approaches and readability of the book ensured by peer review groups and layout optimizations			
Services (consulting, training, events, lectures) and products (sProject, books) further developed by optimizing the processes and tools			
Optimized services and book PROJECT.PROGRAM.CHANGE marketed in Austria, Germany and Switzerland; additional marketing planned			
Organization-related and personnelrelated project objectives			
RGC management further developed			
RGC personnel further developed			
Infrastructure-related and budget-related project objectives			
Establish a basis for an increase in sales of RGC services and RGC products			
Stakeholder-related project objectives			
Cooperation with Manz continued			
Publishing partners C.H. Beck (Germany) and Stämpfli (Switzerland) involved			
Clients bound by the optimized management approaches			
Peer review group involved			
Non-objectives			
English version of the book developed			
Advanced seminars further developed			

Project phases:	Project budget:
1.1 Project managing	Project costs
1.2 Prototyping and planning	381.000.-
1.3 Developing Chapters A-F	
1.4 Developing Chapters G-L	
1.5 Planning the marketing activities, the e-book, developing other contents	Project profits
1.6 Developing Chapters M-Q	30.000.-
1.7 Producing the hardcover and e-book	
1.8 Further developing the services, products and management	
1.9 Initial marketing of the book and services	
Project owner:	R. Gareis, L. Gareis
Project manager:	S. Füreder
Product owner team:	R. Gareis, L. Gareis
Project team member:	
PTM Prototyping	S. Füreder
PTM Service development	L. Gareis
PTM Chapter development	R. Gareis
Relations to other projects	**Key project stakeholders**
Client project A	Project owner
Client project B	Project manager
Sales 16	RGC consultants
Event HP 16	Publishers
	Graphic designer

Project owner (team)	Project manager

V. 1.001 v. S. Füreder per 5.10.2015

Defining Project Roles: Definition and Example

A project role is an element of the organizational structure of a project. One can distinguish between individual roles (for example, project manager or project team member) and team roles (for example, project owner team and project team). Project roles must be defined and staffed. The expectations regarding the fulfillment of the various project roles are independent from their staffing. As an example for the definition of project roles and their staffing, the list of project roles for the project "Developing Values4Business Value" is shown in Table 9.22.

Case Study: Developing Values4Business Value—List of Project Roles (for the Early Project Phases)

Table 9.22: List of Project Roles for the Project "Developing Values4Business Value"

List of Project Roles
Developing Values4Business Value

Project roles	Name
Project owner	R. Gareis; L. Gareis
Project manager	S. Füreder
Product owner team	R. Gareis, L. Gareis
Project team member: Prototyping	S. Füreder
Project team member: Service development	L. Gareis
Project team member: Chapter development	R. Gareis
Subteam: Prototyping	
Project contributor: Consulting and products	R. Gareis
Project contributor: Lectures	L. Gareis
Project contributor: Seminars	M. Stummer, W. Seidler
Project contributor: Documentation	K. Ludat
Project contributor: Organisation	V. Riedling
Subteam: Chapter development	
Project contributor: Visualization	L. Weinwurm
Project contributor: Research	P. Ganster
Project contributor: Chapter development	L. Gareis
Project contributor: Case study cooperation partner 1	F. Mahringer
Project contributor: Case study cooperation partner 2	M. Paulus
Peer Review Group	
Project contributor: Book production, book sales	C. Dietz
Project contributor: Graphic design	M. Riedl

The project organization of the project "Developing Values4Business Value" changed over time. The list shows the roles for the early project phases "prototyping and planning" and "creating Chapter A–F". The roles and their relations to one another are shown in the project organization chart in Figure 9.17 (on page 238).

Defining Project Roles: Objectives and Process

The objective of defining project roles is to identify and to describe the project roles needed to carry out a project. Individual roles and team roles must both be considered. The completeness of the definition of project roles must be ensured. Project roles must be defined and named project specifically. The "labels" of the roles fulfilled by people in the permanent organization must not be used for project roles.

It must be possible to assign responsibilities for performing all work packages on the basis of the defined project roles. This means that even company-external partners are able to perform project roles. In the early stages of the project "Developing Values4 Business Value", for example, the roles "Project Contributor Graphic Design" and "Project Contributor Book Production, Sales" were performed by a supplier and a project partner.

A project role can be described by presenting the objectives of the role, its organizational integration, and the tasks to be performed by the role. By describing project roles, project role bearers gain clarity about their tasks and cooperation with other project roles.

Descriptions of project roles can be standardized. Chapter 7 contains standard role descriptions for all types of project roles. Although these descriptions are generally valid, they can be adapted as required by a project. These adaptations are to be carried out by the project team in order to collectively clarify the expectations of specific roles. The defined project roles are the basis for the development of a project organization chart.

Designing a Project Organization Chart: Definition and Example

A project organization chart is a graph showing the organizational structure of a project. In a project organization chart the roles in the project organization and their relations with each other are represented. As an example, project organization charts for the project "Developing Values4Business Value" are shown in Figures 9.17 and 9.18.

Case Study: Developing Values4Business Value—Project Organization Charts

The evolution in time of the project organization can be seen in the two organization charts. For the late project phases, additional project roles were defined, and roles from the early phases, such as the subteam "prototyping and planning", were dissolved.

In both project organization charts, integrated project organizations were designed based on the involvement of external project partners.

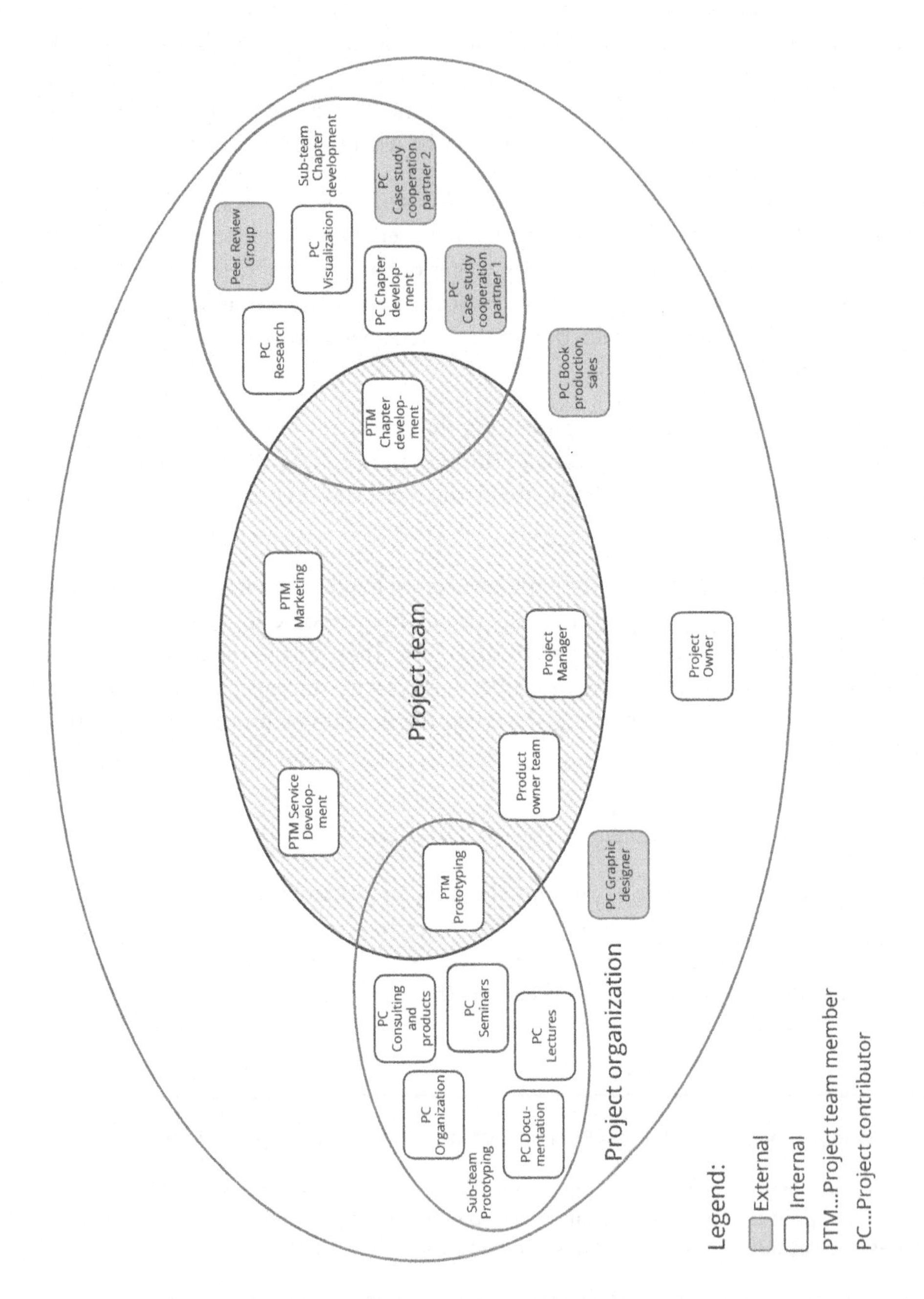

Fig. 9.17: Project organization chart of the project “Developing Values4Business Value”—early project phases.

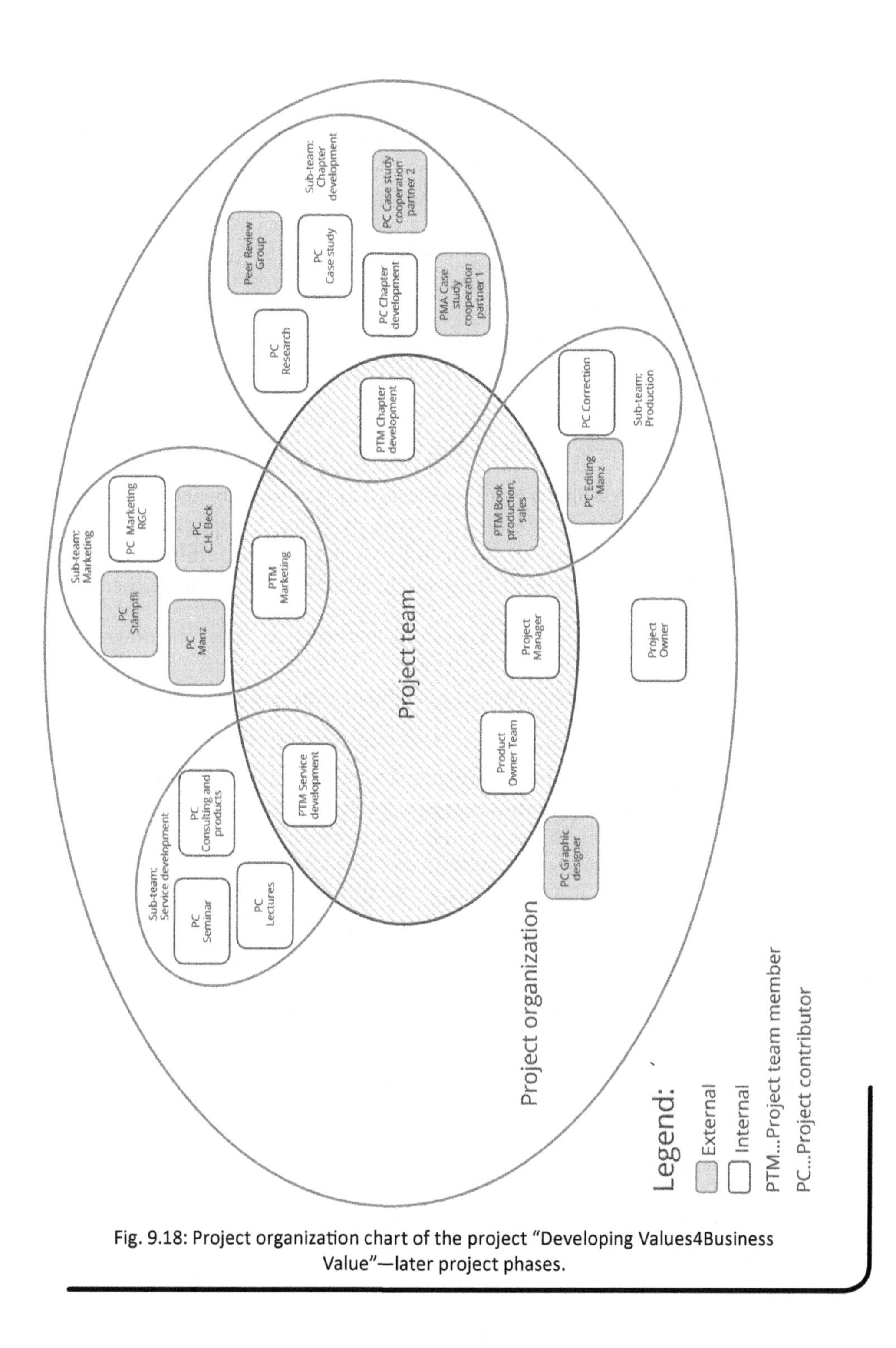

Fig. 9.18: Project organization chart of the project "Developing Values4Business Value"—later project phases.

Designing a Project Organization Chart: Objectives and Process

The objective of designing a project organization chart is a visualization of the organizational structure of a project. The project roles and essential relations between the roles are depicted. The names of project role bearers can also be made visible in the project organization chart. A project organization chart should give orientation for project work to members of the project organization.

Standard project organization charts (see, for example, Fig. 7.10 on page 145) can be used as a basis for designing a project-specific organization chart. Relations between project roles can be exemplified in the organizational chart. There is no claim that this is exhaustive. Communication relations can be seen in a project communication plan, which must also be created.

Developing a Project Responsibility Chart: Definitions and Example

Developing a project responsibility chart is a method for planning cooperation between project roles for the fulfillment of individual work packages. A responsibility chart is a method for conflict management. Any conflicts regarding the fulfillment of work packages can be identified at an early stage and handled before the conflict occurs.

The responsibility chart is an integrative planning method. The work breakdown structure, the project role definitions, and the project stakeholder analysis are the basis for developing it. The completeness and level of detail of these project plans can thereby be checked.

Project responsibility charts are matrix representations. The rows of the matrix represent work packages, and the columns represent project roles or project stakeholders. Functions to be performed by project roles or stakeholders are presented in the intersection fields of the matrix. As an example of a project responsibility chart, the responsibility chart for the project "Developing Values4Business Value" is shown in Table 9.23.

Developing a Project Responsibility Chart: Objectives and Processes

The objective of developing a project responsibility chart is to agree on the cooperation between the members of the project organization and representatives of stakeholders in the fulfillment of individual work packages.

The work packages to be considered when creating the responsibility chart must be selected. Work packages for which the organizational responsibilities are basically clear or which do not require any cooperation between project roles need not be considered.

Case Study: Developing Values4Business Value—Project Responsibility Chart

Table 9.23: Excerpt from the Project Responsibility Chart for the Project "Developing Values4Business Value"

Responsibility Chart
Developing Values4Business Value

	Legend P...Performing C...Contributing I...Being informed Co...Coordinating	Roles								
WBS-Code	Phase / Work package	Project manager	Project owner	Product owner team	PTM Prototyping	PTM Chapter Development	Subteam Chapter Development	PTM Marketing	PTM Service Development	PTM Book production, Sales
...	....									
	Planning Iteration 1	I		C		Co	P			
1.3.2.2	Researching the content of Chaper A	I				Co	P			
	Developing a draft of Chapter A	I				Co	P			
	Accepting the draft of Chapter A	I	I	P		Co	C			
	Finishing Chapter A	I				Co	P			
1.3.2.6	Reflecting on and accepting Chapter A	I	I	P		Co	C			
1.3.2.7	Controlling the backlog "book"	I		P		Co	C			
...	...									

V. 1.001 v. S. Füreder per 5.10.2015

As a result, the planning effort can be kept low. The focus should be on work packages which are unclear as regards organizational responsibilities.

A responsibility chart is created by

> listing the work packages to be considered as rows of the responsibility chart,
> listing project roles and project stakeholders as columns of the responsibility chart, and
> entering the functions for the fulfillment of work packages as codes in the intersection fields of the matrix.

Different codes can be used for different functions. For ease of understanding, it is recommended to use expressive letter codes and not numeric codes. No more than four to five different codes should be used.

By analyzing the rows of the responsibility chart, it can be seen what cooperations are needed to fulfill work packages. By analyzing the columns, the project-related workload of the different project roles can be determined.

In developing the responsibility chart, the project team members agree on the contents of the work packages and the organizational responsibilities. Potential conflicts which become visible can be dealt with. Thus the essential role bearers must be involved in the development process. For repetitive projects, standard responsibility charts can be developed and used. These standard charts are to be adapted according to project-specific needs.

Designing Project Communication Formats: Definitions and Example

A distinction can be made between communications within the project organization and communications with project stakeholders. In principle, these project communications can be carried out either digitally or in analog formats. Digital media that can be used are the intranet, e-mails, electronic newsletters, postings in social media, etc.

Formats for personal communication, which can be either face-to-face or virtual, are individual meetings, group meetings, workshops, and presentations (see Chapter 6). These formats can be combined as required. The type, objectives, participants, and frequency of the communication formats used in a project can be documented in a project communication plan.

In Table 9.24, the communication plan of the project "Developing Values4Business-Value" is presented, differentiated by communications within the project organization and communications with project stakeholders.

Designing Project Communication Formats: Objectives and Process

The objective of designing the project communication formats is to define the required communication formats differentiated by communication within the project organization and communication with project stakeholders. The project communication formats have an important integrative function in projects. They also make it possible to direct the "energy" in a project.

A project communication plan can be used to plan and agree on the objectives, participants, and frequency of the different communication formats. The dates for meetings and workshops should be agreed to in the medium term to provide orientation for all participants. Additional meetings may be held if necessary.

Case Study: Developing Values4Business Value—Project Communication Plan

Table 9.24: Communication Plan for the Project "Developing Values4Business Value"

Project Communication Plan
Developing Values4Business Value

Project-internal communication				
Meeting	Content	Participants	Frequency	Responsibility
Project start workshop	Starting the project; developing the project manual, developing the "big project picture"	Project owner, project manager, project team members	1x	Project manager
Project team meeting	Controlling the project, developing directive measures	Project manager, project team members	1x/month	Project manager
Project owner meeting	Discussing project status, (strategic) decision making	Project owner, project manager, project team members upon request	1x/month	Project manager
Subteam meeting	Coordinating the subteam, discussing content-related topics	Project team member(s), project contributors	upon request	Project team member(s)
Projekt standup meetings	Short-term coordinating, sharing of information	Project team member(s), subteams	upon request, at least 1x/week	Project manager
Project closing workshop	Closing the project; reflecting & giving feedback, dissolving the project team	Project owner, project manager, project team members	1x	Project manager
Peer review group workshop	Discussing and reflecting the contents of the chapters	Project manager, project owner, project team, peer review group	3x	Project manager
Project stakeholder communication				
Stakeholder	Meeting & content	Participants	Frequency	Responsibility
RGC Clients	Information about the book at prototyping events and RGC events	Project team members "Marketing" and "Service development", event participants	each seminar / event	Project team member: Marketing

V. 1.001 v. S. Füreder per 5.10.2015

In addition to the usual communication formats of projects, regular stand-up meetings were held. This proved to be particularly useful for short-term coordination.

In the early project phases, "Values4Business Value" was communicated to participants of RGC seminars and of prototyping events only, but not yet to the broader management community.

Defining Project Rules: Definitions and Example

In addition to general organizational rules for project-oriented organizations that must be followed in projects, project-specific rules can be defined. These project rules can, for example, be related to signatory rights, financial decision-making powers, documentation and filing, the use of information and communication technology, and behavior in the project.

As an example, some rules of the project "Developing Values4Business Value" are documented in Table 9.25.

Case Study: Developing Values4Business Value—Project Rules

Table 9.25: Excerpt from the Project Rules of the Project "Developing Values4Business Value"

Project Rules Developing Values4Business Value
Rules regarding applying ICT
sPROJECT is used for all project management documentation
MS Office 2013 is used for emails and other documentations
Graphics are created in InDesign
Rules regarding filing project data
All project documents are centrally stored in the Dropbox file "PROJECT.PROGRAM.CHANGE"
Electronic formats are used due to ecological reasons
Rules regarding meetings
Minutes of the meeting are taken for every project meeting and workshop
A Scrum Board is used for documentation during standup meetings
No mobile phones, no phone calls during project meetings
Everyone arrives in time for project meetings
Participants have the decision-making authortiy
Disruptions are prioritized
No deputies allowed
Rules regarding the behaviour in the project
New ideas and different points of view are welcome
There is a common responsibility for the results

V. 1.001 v. S. Füreder per 5.10.2015

Project Rules: Objectives and Process

Project rules should give orientation to the members of the project organization. Cooperation in the project is to be efficient and cooperative.

The initial draft of the project rules must be created by the project team during project starting and agreed to with the project owner. Project rules are to be documented. The rules are to be adapted as necessary as a result of regular reflections of the project team which form part of the "social" project controlling.

9.6.2 Developing a Project Culture

Developing a Project Culture: Definitions and Examples

As a temporary organization, a project has a specific culture. A project culture can be observed in the behavior of the members of the project organization as well as in the methods applied and communication formats used in the project. Observable elements of the project culture are the project name, the project logo, project-specific values, project slogans, and project artifacts such as project plans (see Chapter 7). As examples of this, elements of the project culture of the project "Developing Values-4Business Value" are described below.

Case Study: Developing Values4Business Value—Elements of the Project Culture

Project Name

"Developing Values4Business Value" was chosen as a project name based on the name of the investment and the change "Values4Business Value". This makes clear the relation between the investment, the change, and the first project in the project chain for realizing the change. The follow-up project had the name "Stabilizing Values4Business Value".

Project Values

The values guiding behavior in the project "Developing Values4Business Value" are shown in Table 9.26.

Table 9.26: Solution-Related and Process-Related Values of the Project "Developing Values4Business Value"

Project Values	
Solution related	> Innovative > Holistic > Practice oriented
Process related	> Iterative > Context oriented

In defining project-specific values, the project organization oriented itself according to the general management values of RGC. A distinction was made between solution-related and process-related project values. These values guided the structuring of the project, important project decisions, and the behavior of members of the project organization. The value "holistic", for example, promoted an integrative view of project, program, and change management. The value "practice-oriented" led to the involvement of the peer review group. The project values were interpreted in each situation by the project owner and the project manager in order to make project decisions transparent.

Project Slogan

As "Values4Business Value" was the name of the investment and the change, it could also be used as a slogan for the project. In a later project phase, a logo in the form of a word image was developed for this slogan.

The benefit of the solution developed for RGC customers in the project was communicated by the slogan "Values4Business Value". This was consistent with the context orientation and business value orientation.

Project Artifacts

The main artifacts of the project were the project plans and project results documents. Professionalism in the management of the project "Developing Values4Business Value" was made visible to the members of the project organization and to the stakeholders by the use of project plans. The layout of the project artifacts was defined by general RGC standards—for example, the use of project management software sPROJECT.

Project Infrastructure

Project workshops and project meetings took place in the rooms of the RGC office. Stand-up meetings were supported by visualizing the project status on a Scrum board.

Project Events

As part of the project "Developing Values4Business Value", two peer review group workshops were held, and a book presentation was planned. The half-day peer review group workshops (see photo in Fig. 9.19) took place in seminar hotels. The presentation of the book *PROJECT.PROGRAM.CHANGE* was planned in cooperation with the telecommunications provider A1.

Developing a Project Culture: Objectives and Process

The objectives of the explicit development of a project culture are described in Chapter 8. A project-specific identity should be created in order to secure competitive advantages. Competitive advantages are provided by clearly differentiating a project from other projects, by ensuring the recognizability of the project, and by promoting the identification of members of the project organization with the project. The results of developing a project culture also form the basis for project marketing.

Fig. 9.19: Good atmosphere at the second peer review group workshop.

Developing a project culture takes time and energy. Since projects are usually used to carry out short-term and medium-term business processes, project work is to start as soon as possible after the project has been assigned. This deficit in available time is to be compensated for by a high use of resources and appropriate communication formats during project starting. Initial elements of the project culture may be available from the project initiation.

9.6.3 Managing the Project Personnel

Managing the Project Personnel: Definitions and Examples

People who perform roles in projects can be considered the project personnel of a project-oriented organization. The recruitment, assignment, evaluation, development, and release of the project personnel have to occur for each single project, but also generally for the project-oriented organization overall.

A non-project–related task in the management of project personnel is the recruiting of employees, without their assignment to a specific project yet being planned. Other non-project–related tasks are the assessment and general development of project personnel as well as the release of project personnel, if the organization no longer needs their services.

Recruiting, assigning, leading, developing, and releasing project personnel in a specific project is an integral part of project management. The tasks of project personnel management to be fulfilled within the project management subprocesses are listed in Table 9.27.

Table 9.27: Managing Project Personnel as Tasks of the Project Managing Sub-Processes

Project managing: Sub-processes	Managing project personnel in the project
Project starting	> Recruiting, allocating and leading project personnel > Creating incentive system > Planning personnel development
Project coordinating	> Leading project personnel
Project controlling	> Possibly re-allocating and leading project personnel > Controlling the personnel development
Project closing	> Leading, assessing and releasing project personnel

Reflections on the project-specific management of project personnel during project starting of the project “Developing Values4Business Value” are presented as an example in Table 9.28.

Case Study: Developing Values4Business Value—Managing Project Personnel

Table 9.28: Project-Related Management of Project Personnel During Starting the Project “Developing Values4Business Value”

Managing project personnel	Tasks
Recruiting project personnel	> The project personnel was particularly recruited from the RGC pool of employees > Two persons of the project organization namely the "Graphic designer" and the expert for "Book production & sales" were representatives of a supplier and a project partner.
Allocating project personnel	> The project personnel was briefed regarding roles in the project. by the project manager with the aid of project plans and standard role descriptions > Agreements regarding time availability were made
Planning the project personnel development	> There were big learning opportunities for all members of the project organization because of the management innovations. All members have seen these opportunities. > It was planned to organize explicit learning by reflections. > Also the discussion of the management innovations in the annual RGC workshop "Knowledge Management" was agreed.
Arranging incentives	> Learning about the management innovations was an important incentive for all members of the project organization. > It was agreed that the project was the basis for the IPMA Project management certification of the project manager.
Leading project personnel	> The project manager and the project team were informed by the project owners about project contexts, investment objectives and the structures of the change "Values4Business Value" at the project start workshop > The project manager developed the project plans together with the project team and coordinated those with the project owners.

During the project starting of the project “Developing Values4Business Value”, the tasks shown in Table 9.28 for managing project personnel were performed.

Managing the Project Personnel: Objectives and Process

The objectives of managing the project personnel during project starting are to secure the necessary personnel for a specific project, to ensure their appropriate assignment to project roles, and possibly to plan their personal development, as well as to agree on project-related incentives (see Chapter 16).

Project personnel can be recruited either internally or externally. Prerequisites for this are a clear design of the project organization, defined project roles, and qualification profiles for each project role.

Project-specific incentives may be material or non-material. The incentives should be agreed upon between the project owner, the project manager, and project team members during project starting. It is also possible to agree on project-related assessments of project personnel and project-related personal development. Personal development can be carried out "on the job" and by means of "training on the project". The reallocation of project personnel after completion of a project has also to be planned.

Recruitment plans, assessment forms, and personnel development plans can be used to help fulfill personnel management tasks in a project. Project training and coaching can be used for project-related personal development of project personnel. "Training on the project" is implemented specifically for the needs of a project. Its objective is the qualification of members of the project organization. This project-based training enables individual and collective learning. The members of the project organization are given common method knowledge and learn a common language. Agreements about rules for cooperation in the project can be made within the framework of the training.

Management coaching for individuals can be useful for the project manager and for individual project team members, but also for the project owner. The objective of project-related management coaching is personal support in the fulfillment of a project role. The competences of a person should be further developed, and the implementation of guidelines for project management may be ensured (see Chapter 16). Management coaching can be organized not only for individuals, but also for teams—namely, project owner teams, project teams, and project sub-teams. Coaching a team has the objective of developing team competence. Competencies should be developed for constructing common project realities, for assuring team commitment, for resolving conflicts, and using synergies in the team.

9.7 Methods: Managing Project Risks

Project Risk Definition and Risk Types in Projects

Projects involve risks due to their relative uniqueness, complexity, and dynamics. A project risk can be defined as the possibility of a negative or positive deviation from a project objective.[4] In the case of project risk management, deviations are considered with regard to the project scope, the project schedule, project costs, and project income.

A possible differentiation of project risks according to risk types is shown in Table 9.29. This list can be used as a rough checklist for risk analysis.

Table 9.29: Differentiation of Project Risk Types

Criterion	Type of risk	Criterion	Type of risk
Project level	• Project risk • Project phase risk • Work package risk	Cause	• Market risk • Organizational risk • Etc.
Object level	• Object risk • Component 1 - risk • Component 2 - risk • Etc.	Range	• Isolated risk • Related risks
Function	• Engineering risk • Purchasing risk • Installation risk • Etc.	Measurability	• Measurable risk • Non-measurable risk
Area	• Technical risk • Economical risk • Legal risk • Etc.	Coverage	• Coverable risk • Non-coverable risk

If several risks are interdependent, these risks must be considered together in order to avoid repeated consideration.

Project risks are to be distinguished from investment risks. Investment risks relate primarily to the usage phase of an investment object. Risk management for the investment is to be performed when developing a cost–benefit analysis or a business case analysis.

Managing Project Risks: Definition and Example

Project risk management is a project management task to be performed during project starting and project controlling. It includes risk identification and analysis, the planning and implementation of risk management measures, and risk controlling (see Fig. 9.20). In project risk management, deviations from the planned project scope, the project schedule, project costs, and project income are considered at the levels of work packages, groups of work-packages, project phases, and the project as a whole.

[4] In project management practice, it can be observed that positive deviations from objectives are not considered as a risk and are therefore not taken into account in the risk analysis. Significant opportunities in projects are lost as a result.

Fig. 9.20: Tasks of project risk management.

When planning risk management measures, a distinction must be made between measures to avoid risks and measures to provide for risks. Possible measures for avoiding the risk of a car accident are, for example, the choice of a known route or careful driving. Possible measures for providing for the case of a car accident occurring are, for example, the taking out of accident insurance and the usage of safety belts.

Case Study: Developing Values4Business Value—Project Risk Analysis

The results of the project risk analysis and the planning of risk management measures carried out for the project "Developing Values4Business Value" are presented as an example in Table 9.30.

Risk management was performed roughly and intuitively in the project "Developing Values4Business Value". In the project team, potential risks and their deviations from objectives were identified by brainstorming. The most important project risks were documented at the project phase level. There was no monetary analysis, and there were no estimates of the probability of occurrence. Risk management measures were planned for the management of the identified risks, and their implementation was agreed upon.

Table 9.30: Project Risk Analysis and Planning of Risk Management Measures for the Project "Developing Values4Business Value"

Project Risk Analysis

Developing Values4Business Value

WBS-Code	Phase / Work package	Project risk	Positive deviation	Negative deviation	Measure
1.1	Project managing	Scope: Project organization		Role conflicts as a result of multiple role assignments to individuals	Differentiating clearly the project roles
1.2	Prototyping and planning	Schedule: Duration of the phases		Exceeding duration as a result of inappropriate prototyping results	Integrating competent cooperation partners
1.3, 1.4, 1.6	Developing Chapters 1–6	Scope: Practical relevance		Too theoretical contents of the chapters as a result of the innovations	Using a peer review group for feedback
1.5	Planning the marketing activities and the e-book, developing other contents	Scope: Marketing by publishers	New client contacts in DACH-region		Identifying the pubslisher's contact persons for marketing issues
1.9	Initial marketing of the book and services	Scope: Book presentation	Cooperation for book presentation with new partners		Defining benefits of a common book presentation for partners

V. 1.001 v. S. Füreder per 5.10.2015

Managing Project Risks: Objectives and Process

Project management implicitly helps to avoid negative deviations from project objectives and to promote positive deviations. However, in addition to this implicit risk management, it is advisable to manage project risks explicitly.

The objectives of project risk management are the early identification of project risks, the minimization of negative deviations from objectives, and the maximization of positive deviations.

A distinction can be made between detailed, analytical and rough, intuitive management of project risks. Which one is used will depend on the expected extent of the deviations and the expected probability of their occurrence. Detailed, analytical project risk management will only be used when large deviations from objectives are expected and the probability of occurrence is high (see Fig. 9.21).

For other situations, rough, intuitive project risk management will usually suffice. An example of rough, intuitive analysis for the "Developing Values4Business Value" project was presented above in Table 9.30.

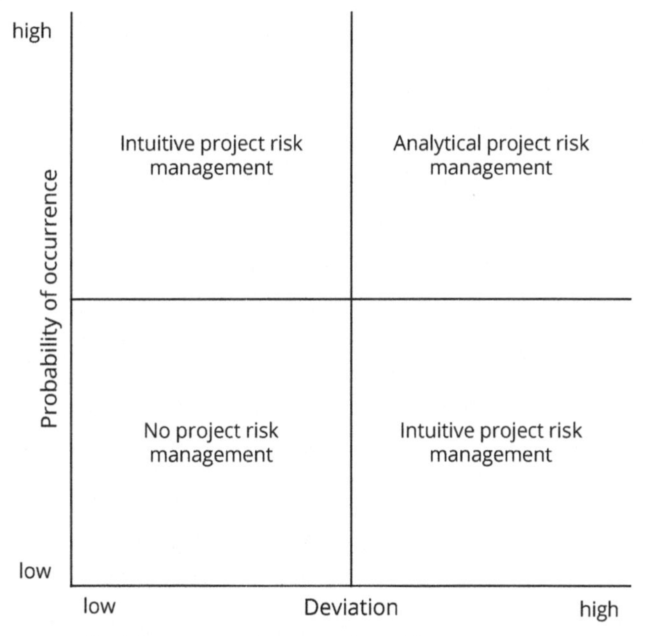

Fig. 9.21: Approaches to project risk management.

In practice, detailed, analytical risk management is most frequently used for repetitive contracting projects. In this case, empirical evidence for risk analysis is usually

Table 9.31: Excerpt from a Risk Analysis for an Organizational Development Project

WBS-Code	Phase / Work package	Risk	Positive deviation	Negative deviation	Extent of deviation	Probability of occurrence
1.1	Project managing	Relationship to consultant	High confidence in performance		4	h
1.2	Prototyping and planning	Process of developing		No common views of partners, little experience	5	m
		Compliance with corporate standards		Not enough compliance	2	l
1.3.4	Data collection	Data quality		Inadequate quality	3	m
		Process of developing	Much information available, fast data collection		4	m
1.3.6	Workshop: Agreement upon decision makers	Decision making		Slow decision making process	2	m
1.3.8	Developing a concept for IT-implementation	Compatibility to existing IT-infrastructure		Solution not compatible with existing IT-infrastructure	1	m
1.4.1	Defining roles, career paths, certification requirements	Acceptance of defined roles, etc.		Resistance of the works council and HR	3	m
1.4.3	Performing personnel assessment	Acceptance		Resistance of employees and works council	3	l
1.5	Developing a concept of the organization structure	Process of developing		Diversity of interests among R&D, organization, engineering	4	h
...	...					

Key:
Scale for the extent of deviation: From 1 (very low) to 5 (very high)
Scale for the probability of occurence: „low“, „medium“, „high“

available, and standardized risk premiums for risk provisioning, so called "contingencies", can be used. Initial risk management plans from project initiating should be available to build on during project starting.

Identifying Project Risks

Project risks should be identified by work package, because planning and controlling of scope, schedules, costs, and income are carried out at the work package level. Therefore, any deviations from these objectives can be analyzed at the work package level. For the risk analysis, only those work packages for which risks are expected are to be considered. It is not necessary to consider all work packages.

To identify project risks, other project plans, such as the objects of consideration plan and the project stakeholder analysis, can be used as project-specific checklists. In general risk checklists, which are also used as standards, there are frequently combinations of risks related to work packages, to objects of consideration, and to stakeholders. Such checklists are not compatible with the WBS, so their benefit is limited. It is advisable to recognize and use the project plans as project-specific checklists.

Analyzing Project Risks

Qualitative and possibly also quantitative analyses of possible deviations from objectives are carried out in risk analysis. Often a qualitative description of the deviations is sufficient. A quantification of project risks, in the form of estimates of deviations from scope, schedule, costs, and income, is not always necessary.

A quantitative assessment can be in monetary units for costs and income, and in time units for durations and dates. Assessments may also be based on other scales, such as "(very) low, medium, and (very) high". An example of this is shown in Table 9.31.

Quantitative analyses of risks also require estimates of the probability of occurrence of a risk. An example of a risk analysis from a plant construction project is shown in Table 9.32. The expected value of a risk is calculated by multiplying the possible risk costs ("financial damage") by the probability of occurrence of the damage. But this "expected value" does not correspond to the costs caused by the occurrence of a risk. It is only a statistical value.

Excursus: Analyzing Project Risks Using Probability Theory

Using probability theory, project costs, project income, or project duration can be represented as probability distributions. To this end, work package costs and income or work package durations are represented as stochastic variables. Costs, income, and duration are considered as random variables, which can be described by probability distributions. These probability distributions are generally constant distributions, since work package costs or durations can assume any value within certain limits.

Table 9.32: Example of Monetary Risk Analysis from a Plant Construction Project

WBS-Code	Phase / Work package	Risk	Deviation	Costs in € 1.000.-	Probability of occurrence	Expected value in € 1.000.-
1160	Construction of threads of scew	Engineering effort	Additional engineering effort required, as necessary pulling force not feasible	50,-	50%	25,-
2110	Delivery of components	Delivery period	Delay of delivery, as no suitable supplier available	200,-	30%	60,-
3250	Production slewing gear	Completion date	production cannot be kept because of high reject rates; compliance with tolerances cannot be ensured	30,-	70%	21,-
Expected Value: Overall financial risk						106.-

Because usually only limited statistical data is available for project planning, probabilities are usually determined by subjective estimates. These can be represented by standard probability distributions (e.g., normal distribution, beta distribution, or uniform distribution). The advantage of using standard distributions is that only a few estimated values are required to specify the entire distribution.

Standard distributions can easily be determined on the basis of three estimates: optimistic, most likely, and pessimistic estimates. In addition, the probabilities that the optimistic value is not exceeded and the pessimistic value is not fallen short of must be defined. These figures are sufficient to calculate the expected value and the variance or standard deviation of the random variables. The relevant formulas are shown in Figure 9.22.

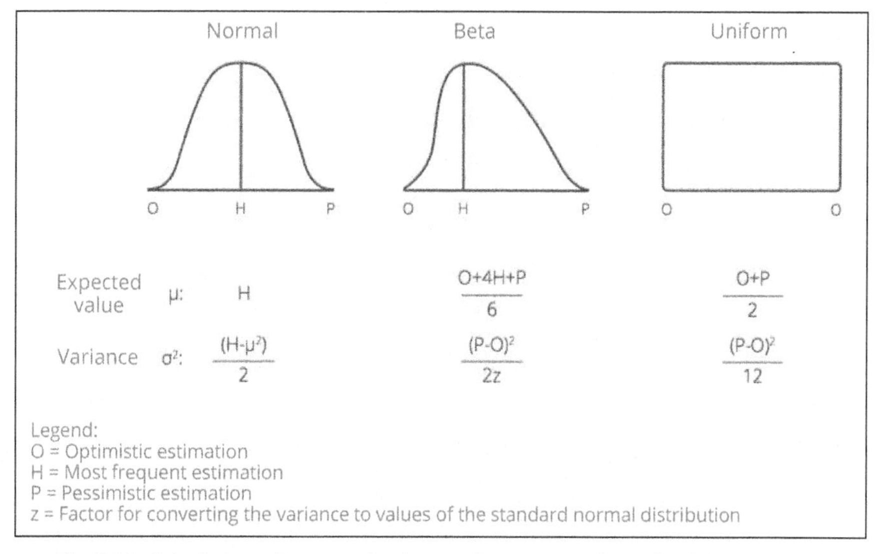

Fig. 9.22: Calculation of expected values and variances of standard distributions.

Expected values and variances of project costs, project income, or project duration can be calculated using a Monte Carlo simulation. The variances calculated represent key risk figures.

The distribution of project costs, project income, or project duration is approximately normally distributed, since the central limit theorem says that the sum of many independent, arbitrarily distributed random variables is approximately normally distributed.[5] The probability that a given value of the project costs, project income, or project duration will be exceeded or fallen short of can be calculated from standardized normal distributions.

Risk Management Measures in Projects

Risk management measures in projects should avoid the occurrence of negative deviations from objectives and mitigate their consequences through provision measures. On the other hand, positive deviations from objectives and their consequences should be promoted. Risk management measures must be planned and implemented.

The planning of risk management measures requires selecting risks for which measures are to be taken. Relevant risks can be selected by using a project risk matrix. A project risk matrix with standard management policies is shown in Figure 9.23. The

[5] Cf. Jann, B., 2002, p. 126.

standard policies refer to different combinations of deviations from objectives and probabilities of occurrence.

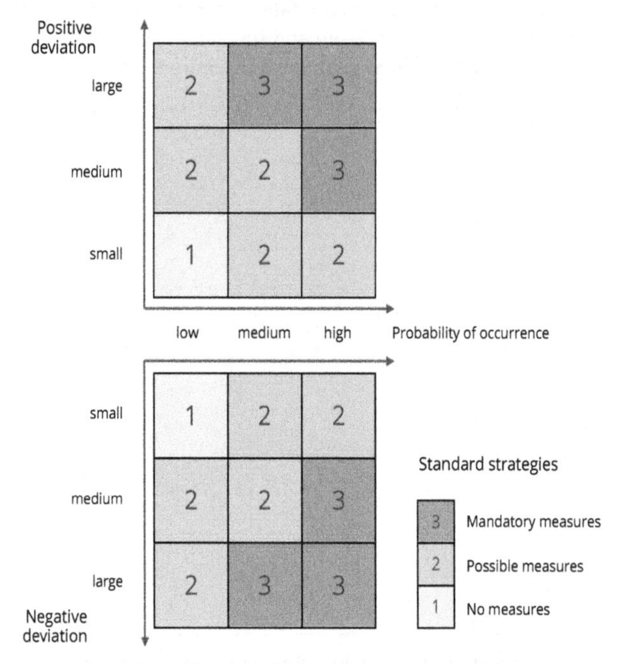

Fig. 9.23: Standard risk management policies.

Risk management measures can be differentiated into risk avoiding, risk promoting, and risk measures for providing for risks. Avoidance measures should avert the occurrence of negative deviations from objectives. Provision measures should reduce negative consequences in the event of a risk occurrence. Examples of risk avoidance or promotion and risk provision measures are given in Table 9.33.

Avoidance, promotion, and provision measures incur costs. Costs and benefits of risk measures must therefore be estimated before the measures are implemented. Benefits can be the elimination of a project risk or merely a reduction of risk.

Planned risk management measures must be included in the project plans. Additional work packages in the WBS, additional costs in the project costs plan, etc. have to be considered.

Managing Project Risk: Organization

Project risk management is a project management task and as such is to be fulfilled by the members of the project organization—namely, the project owner, project manager, and project team. The involvement of representatives of stakeholders in risk management extends the points of view and creates design possibilities in the agreement of risk management measures.

Table 9.33: Examples of Avoidance, Promotion of, and Provision for Project Risks

Avoidance and/or promotion measures	Provision measures
> Ensuring creditworthiness in supplier selection > Using experienced project personnel > Using proven practices and products > Involving of stakeholders affected by project results > Promoting creativity in problem solving > Etc.	> Transferring risks to customers, consortia or suppliers in contract design > Considering risk surcharges in costing > Creating reserves > Establishing insurance contracts > Creating redundancies in project structures > Defining an escalation model for problem situations > Etc.

Risk management requires appropriate communication formats to secure the necessary experience and creativity for risk identification, risk analysis, and planning of risk management measures. Teamwork and the carrying out of risk management workshops are recommended. Possible participants in a risk management workshop are shown in Figure 9.24.

Fig. 9.24: Possible participants in a project risk management workshop.

Risk management software solutions are available for the analysis of project risks and for project risk documentation. These software packages can, for example,

> represent costs, income, resources, and durations as stochastic variables,
> carry out Monte Carlo simulations,
> calculate probability distributions for start events of work packages, for the project duration, for the project costs and the project income, and
> develop a "Criticality Index" for noncritical activities of the project network plan.

These specialized software solutions can be used as a supplement to project management software. Some project management software solutions, such as MS Project, also have basic functions for project risk management.

Literature

Freeman, R. E.: *Strategic Management: A Stakeholder Approach,* Pitman, Boston, MA, 1984.

Freeman, R. E., Harrison, J. S., Wicks, A. C.: *Managing for Stakeholders: Survival, Reputation, and Success,* Yale University Press, London, New Haven, CT, 2007.

Gareis, R., Huemann, M., Martinuzzi, A.: *Project Management & Sustainable Development Principles,* Project Management Institute (PMI), Newton Square, PA, 2013.

Greenwood, M.: Stakeholder Engagement: Beyond the Myth of Corporate Responsibility, *Journal of Business Ethics,* 74(4), 315–327, 2007.

Jann, B., *Einführung in die Statistik,* Oldenbourg, München, Wien, 2002.

Julian, S. D., Ofori-Dankwa, J. C., Justis, R. T.: Understanding Strategic Responses to Interest Group Pressures, *Strategic Management Journal,* 29(9), 963–984, 2008.

Mitchell, R. K., Agle, B. R., Wood, D. J.: Toward a Theory of Stakeholder Identification and Salience: Defining the Principle of Who and What Really Counts, *Academy of Management Review,* 22(4), 853–886, 1997.

Rowley, T. J.: Moving Beyond Dyadic Ties: A Network Theory of Stakeholder Influences, *Academy of Management Review,* 22(4), 887–910, 1997.

10 Sub-Processes: Project Coordinating and Project Controlling

In contrast to periodic project controlling, project coordinating takes place continually during the project. The objectives of project coordinating, besides providing ongoing information for members of the project organization and representatives of project stakeholders, are the coordinating of project resources, the ensuring of project progress, the ensuring of the quality of work package results, and project marketing.

The need for periodic project controlling is a result of the dynamics of a project and the dynamics of relations of the project with its stakeholders. Any deviations from plans must be identified. Controlling measures are required for exploiting new potentials or for correcting undesired deviations. From a project change management point of view, project controlling can be perceived as "organizational learning of a project" or "further developing a project".

Project controlling applies to all dimensions of a project. This means for example, that when performing "social" controlling of the project organization, the project culture and relations with project stakeholders and other projects are also considered.

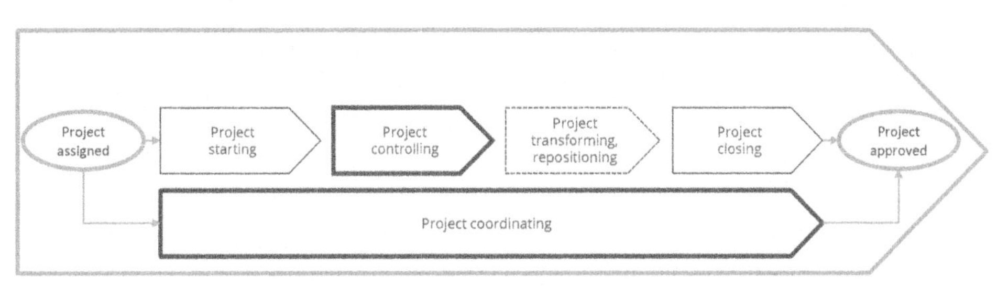

Overview: "Project coordinating" and "Project controlling" as sub-processes of the business process "Project managing".

10 Sub-Processes: Project Coordinating and Project Controlling

10.1 Sub-Process: Project Coordinating

The sub-process "project coordinating" starts when the project is assigned. The fulfillment of the first work packages must be coordinated with the help of the initial project plans. This takes place in parallel with project starting. Detailed project plans for further project coordinating are available as a result of project starting. Project coordinating ends with the formal project approval. So it takes place throughout the entire project duration.

The process "project initiating" takes place before project coordinating. Processes occurring partly in parallel are "project starting", "project controlling", "developing solutions", "project administrating", "change managing", "benefit realization controlling", and "project closing".

In contrast to periodic project controlling, project coordinating takes place continuously. It consists of the ongoing activities of project managing. The application of project coordinating methods is not demanding. Nevertheless, owing to intensive communications, this sub-process is important for project success.

10.1.1 Project Coordinating: Objectives and Process

Project Coordinating: Objectives

The objectives of project coordinating, besides providing information for members of the project organization and representatives of project stakeholders, are the coordinating of project resources, the ensuring of project progress, the ensuring of the quality of work package results, and project marketing. Project progress is ensured by coordinating relations between work packages and by the acceptance of work package results by the project manager. If project risks become acute, measures must be taken to deal with these situations. The measures are to be planned on an ad hoc basis or have already been planned as risk provision measures.

The objectives of the sub-process "project coordinating", differentiated into economic, ecological, and social objectives, are described in Table 10.1.

Project Coordinating: Process

Tasks, roles, and responsibilities for project coordinating are shown in the responsibility chart of Table 10.2.

The tasks of project coordinating are to be performed continuously by the project manager. Communication with the project owner, project team members, and project contributors, as well as with representatives of project stakeholders, can take place in

Table 10.1: Objectives of the Sub-Process "Project Coordinating"

Objectives of the sub-process: Project coordinating
Economic objectives > Members of the project organization and representatives of project stakeholders are continuously informed > Project resources coordinated > Relations between work packages coordinated > Project progress and quality of work package results ensured > Measures taken to manage a situation due to an active project risk > Work package results accepted > Project coordinating performed efficiently
Ecological objectives > Ecological consequences of project coordinating optimized
Social objectives > Project personnel led appropriately during project coordinating > Project stakeholder involved in project coordinating

meetings, but also by e-mail, telephone, or video conferencing. Project marketing can also be done both face to face and digitally. The acceptance of work package results by the project manager can be formalized by means of an acceptance certificate.

Table 10.2: Sub-Process "Project Coordinating"—Responsibility Chart

Tool/document:
1 ... To-do list
2 ... Work package acceptance certificate
3 ... Minutes of the meeting

Legend
P .... Performing
C...Contributing
I... Being informed
Co...Coordinating

	Roles						
Process tasks	Project owner	Project manager	Project team member	Sub-team	Expert pool managers	Representatives of stakeholders	Tool/Document
Communicating with the project owner	C	P	C				1
Communicating with the project team members and project contributors		P			I	C	1
Coordinating project resources		P	C		C	C	
Accepting work package results	I	P	C	C			2
Communicating with representatives of project stakeholders	C	P	C			C	1
Performing project marketing continuously	C	P	C			C	
Attending sub-team meetings		C		P			3

Since project coordinating is carried out on an ongoing basis, there is no need for differentiation between preparing, implementing, and following up. The communication situations of the project manager, however, must be structured as a method of self-management at a micro level.

10.1.2 Project Coordinating: Organization and Quality

Project coordinating is a central task of the project manager. In the case of larger projects, the project manager can be supported by a project management assistant or a project controller.

The quality of project coordinating is dependent on the quality of the project manager's communication with members of the project organization and representatives of project stakeholders. Appropriate social competence is necessary for this, and appropriate communication methods must be used.

These are above all the project plans, but also specific project coordinating methods—namely, To-Do lists, minutes of meetings, and acceptance certificates. The quality of project coordinating is determined by the binding nature of the communication results.

10.1.3 Methods: Communicating with Project Plans, Providing Minutes of Meetings, Using To-Do Lists, Accepting Work Package Results

In project coordinating, the project plans developed during project starting and adapted in project controlling have to be communicated. In addition, the methods "providing minutes of meetings", "using to-do lists", and "accepting work packages" can be used.

Communicating with Project Plans

The ongoing communication of the project manager with members of the project organization and representatives of project stakeholders during the project is supported by project plans such as the project objectives plan, the work breakdown structure (WBS), the project bar chart, the project costs plan, and the project stakeholder analysis. This reduces the complexity of the project, because one refers to what has already been agreed upon, to known terms, and places discussion topics into the corresponding project context.

Providing Minutes of Meetings

Minutes of meetings are to be created to ensure commitment and transparency within the project. This formalism is justified, as people, some of whom are unknown to one another, work on relatively unique tasks in different locations. Documenting agreements is therefore an essential coordinating method.

Minutes of meetings should be formulated as briefly as possible. As an appendix to minutes, agreed-upon measures, dates, and responsibilities should be documented in a To-Do List. The objectives of the meeting, the items on the agenda, the date and duration of the meeting, the meeting participants, and the meeting place should all be recorded. To enable transparency, the documentation of discussion contributions and decisions should be structured by project phases and work packages.

Using To-Do Lists

Using a To-Do List is an important method for project coordinating. The project manager should only use one list for a project. It is to be used as a supplement to the WBS, since it contains detailed activities (to-dos) for fulfilling the work packages.[1]

The To-Do List should be structured by project phases and work packages. In addition to listing activities and their allocation to work packages, it should also include the responsibilities for the activities, the completion dates, and the status of the activities. Completed activities should be given the status "completed"; additional agreed-upon activities should be included in the list and provided with a target date.

A To-Do list is to be used progressively. Completed activities should be kept in the To-Do List in order to document the process. This gives the list the character of a "project diary".

If the To-Do List is adapted during a meeting, it must be attached to the minutes as an appendix. An example of a To-Do List of the project "Developing Values4Business Value" is shown in Table 10.3.

Accepting the Work Package Results

The progress and the quality of the results of single work packages must be monitored at all times. This is carried out by project coordinating.

After the completion of a work package by a project team member or a subteam, the project manager has to accept the work package results. This can take the form of a formal approval carried out by the project manager and the project team member.

A work package acceptance certificate can be issued. In this way, the project team member is formally relieved. In an acceptance certificate, the completion of the work according to the defined requirements is confirmed.

[1] In project practice, a To-Do List is sometimes used without the Big Project Picture provided by the WBS.

Case Study: Developing Values4Business Value—To-Do List

Table 10.3: Excerpt of the To-Do List of the Project "Developing Values4Business Value"

To-Do List

Developing Values4Business Value

WBS-Code	Process	Date of arrangement	Responsible	Completion date	Status
...	...	...	...	...	...
1.6.5	Coordinating figures for Chapter Q with graphic designer	2/22/2017	P. Ganster	3/1/2017	done
1.6.7	Developing bibliography	2/27/2017	T. Koch	3/1/2017	In progress
	Developing index	2/27/2017	T. Koch	3/15/2017	In progress
	Writing the introduction	2/27/2017	R. Gareis	3/1/2017	In progress
1.6.8	Updating the work breakdown structure for the case study	2/27/2017	L. Weinwurm	2/28/2017	pending
	Updating the project bar chart for the case study	2/27/2017	L. Weinwurm	2/28/2017	pending
	Proofreading Chapters M-Q	2/27/2017	R. Gareis	3/1/2017	In progress
...	...	...	...	...	...
...	...	...	...	...	...

V. 1.006 v. P. Ganster per 27.02.2017

From the To-Do list, it can be seen that several activities are required to fulfill a work package. To ensure flexibility, these can be agreed upon at short notice.

10.2 Sub-Process: Project Controlling

The need for periodic project controlling is a result of the dynamics of projects. Project meetings and workshops enable reflections of projects. Weaknesses and strengths can be identified, and risks which have become acute can be recognized. Any deviations from plans must be identified, and directing measures must be defined for exploiting new opportunities or for correcting undesired deviations.

Project controlling relates to all project dimensions—that is, not only to "hard facts", such as project objectives, project scope, project schedules, and project costs, but also to "soft facts" such as the project organization, the project culture, and relations with project stakeholders. These soft facts are considered in "social project controlling".

The directing measures of project controlling lead to project changes. Project controlling can be perceived as "organizational learning of a project" in the case of minor changes, and "further developing a project" in the case of more extensive changes.[2]

[2] The methods of change management can also be applied to temporary organizations (see Chapter 11).

10.2.1 Project Controlling: Objectives and Process

Project Controlling: Objectives

The objective of project controlling is to agree periodically on the project status and on appropriate measures for achieving the project objectives. This requires collective reflections, constructions of project realities, and decisions by members of the project organization.

Project controlling must be carried out periodically. Its frequency depends on the duration of the project. In the case of a six-month product development project, formal project controlling every three weeks is recommended. In a construction project with a duration of some 18 months, project controlling meetings every six weeks might be sufficient for project controlling. A project controlling cycle takes about one week depending on the size of the project.

The objectives of the sub-process "Project controlling", differentiated into economic, ecological, and social objectives, are described in Table 10.4.

Table 10.4: Objectives of the Sub-Process "Project Controlling"

Objectives of the sub-process: Project controlling
Economic objectives > Project status determined, common project reality constructed by the project organization > Directive measures agreed > Project plans adapted > Project controlling reports prepared (status report, project score card) > Project controlling performed efficiently
Ecological objectives > Ecological consequences of project controlling optimized
Social objectives > Project personnel led appropriately during project controlling > Project stakeholder involved in project controlling > Organizational learning of the project performed

Project Controlling: Process

Project controlling is a periodic process, to be carried out several times in a project. A project controlling cycle is complete when the project controlling report is distributed to the members of the project organization. The tasks, roles, and responsibilities for fulfilling the task of a project controlling cycle are shown in the responsibility chart of Table 10.5.

The structures for project controlling—the frequency, the content, and the form of the report; the communication formats; etc.—are to be planned during project starting. The project owner must decide what minimum requirements the project controlling has to fulfill. Different controlling standards can be functional for different types of projects.

The following tasks must be performed by the project owner, the project manager, and the project team members in a project controlling cycle:

> Project monitoring: Recording actual data, performing planned vs. actual comparisons, analyzing deviations
> Project directing: Planning and agreeing on directive activities
> Re-planning the project: Updating project plans
> Preparing project controlling reports: Preparing a project status report, project scorecard, deviation trend analysis

Changes are to be encouraged during project controlling. Deviations should be seen as learning opportunities. It requires a "rough" instead of a "fine" approach, and thinking in alternatives.

Table 10.5: Sub-Process "Project Controlling"—Responsibility Chart

Legend P .... Performing C...Contributing I... Being informed Co...Coordinating Process tasks		Roles						
		Project owner	Project manager	Project team	Project team members	Expert pool managers	Representatives of project stakeholders	Tool/Document
1	Preparing project controlling							
1.1	Gathering actual data and comparing actual versus planned		P		C			
1.2	Analyzing deviations, planning directive measures		P		C	C	C	
1.3	Adapting project plans		P		C		C	
1.4	Preparing project controlling reports		P		C			
1.5	Preparing project controlling meetings		P					1
2	Performing project controlling meetings							
2.1	Distributing info-material to participants	I	P	I		I	I	
2.2	Performing project team meetings	C	Co	P			C	
2.3	Performing project owner meeting	P	Co		C			
3	Following up project controlling							
3.1	Completing the adaption of project plans		P		C			2
3.2	Completing the project controlling report		P		C			3
3.3	Arranging updates in the project portfolio database		P					
3.4	Performing project marketing	C	Co		P		C	
3.5	Distributing the project controlling report	I	P		I	I	I	
4	Performing content work (parallel)				P		P	

Tool/document:
1 ... Invitation of the participants for the project team meeting
2 ... Adapted project plans
3 ... Project controlling reports

Project controlling is carried out for every project dimension—that is, for the project objectives, the project scope, the project schedule, the project costs, etc. The earned value analysis provides a holistic view of progress, costs, and schedule. Project directing refers to the project as a whole as well as to individual work packages. The directive activities decided upon must be considered in the project plans. The preparation of project controlling reports is carried out for the project as a whole.

The communication formats of project controlling usually include project team meetings, project owner meetings, and, if required, sub-team meetings. In these meetings a common project reality is to be constructed by the members of the project organization. This provides the basis for agreeing on directive activities. Already prepared project controlling reports are further developed in the meetings.

10.2.2 Project Controlling: Organization and Quality

Project Controlling: Organization

Project controlling is a sub-process of project management. It is performed by the project owner, project manager, and project team members. In large projects, the project manager can delegate some functions to a project controller or a project management assistant.

The controlling of a project-oriented organization—for example, of a Profit Center—is a context for projects (see Fig. 10.1). The project managers or the project controllers inform the profit center controlling about the status of individual projects. This information can be summarized by the profit center controlling into overall organizational reports.

Specific methods are used for project controlling (see Table 6.2 on page 106). In addition to updating project plans, developing project status reports, earned-value analyses, and project trend analyses is relevant. The project scorecard is a specific type of project status report. These methods are described below.

Project Controlling: Quality

Appropriate project management quality is ensured by applying consistent communication formats and project management methods during project starting, project coordinating, project controlling, and project closing. The project plans which were developed during project starting are to be used to support communications and are to be adapted if necessary. The use of project plans in the sub-processes of project management justifies the communication effort for the development of the project plans during project starting.

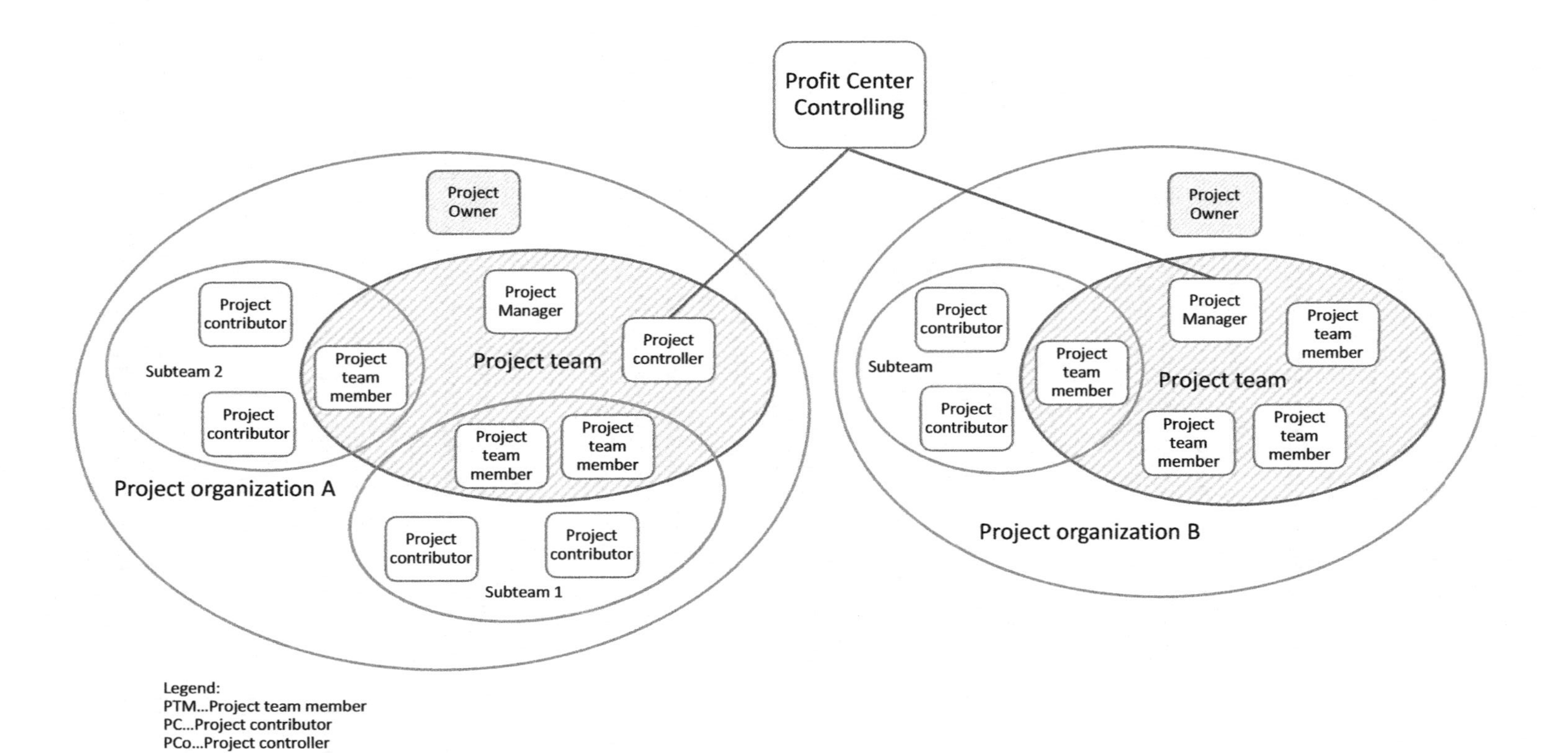

Fig. 10.1: Relation between project controlling and profit center controlling.

In practice, often too little importance is ascribed to project controlling. The effort of data collection, analysis, and reporting is often not seen to be justified. The clarity that was achieved during project starting seems to be enough. This causes many learning and optimization opportunities to be lost.

10.2.3 Methods: Monitoring, Directing, and Re-Planning

Project monitoring, directing, and possible re-planning are to be performed for all project dimensions—for the solution requirements, the project objectives, the project schedule, etc. The relevant methods are described in the following, with the exception of the controlling of requirements, which has already been dealt with in Chapter 4.

Monitoring, Directing, and Re-Planning of Project Objectives

Analyzing the status of the project objectives on the controlling date is difficult, as the project objectives are not formulated with respect to controlling dates but in terms of the results to be achieved at the end of the project. It is therefore necessary to analyze whether fulfillment of the objectives defined for the end of the project is still realistic in the light of developments up to the controlling date.

When directing and re-planning the project objectives, objectives which are no longer relevant are to be deleted, still-relevant objectives are to be preserved, and any new objectives are to be added. Causes for re-planning the project objectives are quantitative or qualitative changes in the solution requirements. The change of the project objectives can be performed on the basis of so-called "change requests" (see Chapter 4). If more extensive changes in the solution requirements and thus in the project objectives are made, this situation can be perceived as a change "Further developing a project" (see Chapter 11).

A change in the project objectives necessitates changes in most of the project plans. The scope of work has to be adapted in the work breakdown structure. But the project schedule, project costs, and probably the project organization and the project stakeholder relations also have to be adapted.

As an example of an adapted project objectives plan, the project objectives plan for the project "Developing Values4Business Value" is shown in Table 10.6.

Monitoring, Directing, and Re-Planning of Objects of Consideration

In monitoring the objects of consideration of a project, it is necessary to check whether the objects of consideration plan is complete and whether planned objects of consideration may be omitted or are to be renamed. The consistency of the objects of

Case Study: Developing Values4Business Value—Adapted Project Objectives Plan

Table 10.6: Adapted Project Objectives Plan for the Project "Developing Values4Business Value" as of the Controlling Date 20.12.2016

Adapted Project Objectives Plan
Developing Values4Business Value

Service-related and market-related project objectives	Objectives adapted per 20.12.2016
RGC management approaches further developed by innovations (values and management approaches, sustainable development, agile approaches, benefits realization management, requirements management, business modelling)	Business modeling not considered
RGC management approaches documented in the book PROJECT.PROGRAM.CHANGE (hardcover and e-book)	
Practical relevance of the developed RGC management approaches and readability of the book ensured by peer review groups and layout optimizations	2 peer review workshops instead of 3 carried out
Services (consulting, training, events, lectures) and products (sProject, books) further developed by optimizing the processes and tools	
Optimized services and book PROJECT.PROGRAM.CHANGE marketed in Austria, Germany and Switzerland; additional marketing planned	
	Concept for English translation of the book developed
	Marketing materials (e.g. logo "Values4Business Value", marketing leaflet, testimonials) developed
Organization-related and personnel-related project objectives	
RGC management further developed	
RGC personnel further developed	
Infrastructure-related and budget-related project objectives	
Establish a basis for an increase in sales of RGC services and RGC products	
Stakeholder-related project objectives	
Cooperation with publisher Manz continued	
Publishing partners C.H. Beck (Germany) and Stämpfli (Switzerland) involved	
Clients bound by the optimized management approaches	
Personnel developed contentually by participating	
Peer review group involved	
	PR measures for "Values4Business Values" taken
Additional objectives	
Basis for certification of the project manager established	
Case study for RGC trainings developed	
Non-objectives	
English version of the book developed	
Advanced seminars further developed	

V. 1.003 v. P.Ganster per 20.12.2016

In the project "Developing Values4Business Value", project objectives were both supplemented (e.g., "Concept for English translation of the book developed") and adapted (e.g., "2 peer review workshops instead of 3 carried out").

Because of changes in the project objectives, the WBS, the project cost plan, the stakeholder analysis, etc. had to be adapted. The changes were agreed on in the project team and with the project owner and communicated to selected project stakeholders (e.g., publishers, graphic designers).

consideration plan with the other project plans must be assured. If deviations exist, re-planning is necessary.

If additional solution requirements or project objectives have been planned, the associated objects of consideration must be taken into account when re-planning. A further differentiation of already planned objects of consideration can also be useful. As an example, the additional objects of consideration of the project "Developing Values-4Business Value" as of the controlling date 20.12.2016 are described in the case study.

Case Study: Developing Values4Business Value—Adapted Objects of Consideration Plan

As a result of the additional project objectives of the project "Developing Values4Business Value", the following additional objects of consideration were considered in the plan:

> Concept for the English version of the book *PROJEKT.PROGRAMM.CHANGE*
> Various marketing tools (e.g., "Values4Business Value" logo)
> Public relation for "Values4Business Value"

Monitoring, Directing, and Re-Planning of the Project Scope: Measuring Progress

The objective of controlling the project scope is to determine the progress of individual work packages, project phases, and the project as a whole. The actual progress is to be compared with the planned progress up to the controlling date. This comparison makes it possible to identify deviations from the plan.[3]

The following techniques can be used to measure the progress of a work package:

> 0% or 100% assumption,
> intuitive estimation,
> output measurement, and
> measurement of defined progress milestones.

In the "0% or 100% assumption" technique, the work packages in progress are assumed to be either not yet started (progress = 0%) or to be finished (progress = 100%). This technique requires a high level of detailing of the work packages. It is the most imprecise of the techniques mentioned here. In the "intuitive estimation" technique, the project team member responsible for performing the work package provides an estimate of the progress. Precision can be increased by three-point estimates (optimistic, probable, and pessimistic values). In the "output measurement" technique, progress is determined by measuring the output achieved. This method is suitable for a relatively continuous performance of a work package. Table 10.7 shows examples of such work packages from the construction industry.

[3] Qualitative controlling of work package results in an "empowered" project organization is performed in the project team by project team members. Coordination of the content-related results can take place by means of presentations and discussions in project team meetings, as well as in project coordination.

Table 10.7: Examples of Indicators of Work Package Progress

Work	Indicators
Excavation	m^3 earth excavated
Pouring concrete	m^3 concrete poured
Pipe laying	m pipes laid
Equipment assembly	t equipment assembled

For many work packages, such as design, training, testing, etc., progress is not continuous. In such cases "progress milestones" can be defined to measure progress. Progress percentages can be assigned to these by the project manager and the relevant project team member. An example of progress milestones and the associated cumulative progress percentages is shown in Table 10.8.

Table 10.8: Progress Milestones of the Work Package "Developing Construction Plans"

Progress milestone of the work package: Developing construction plans		
Number of the progress milestone	Name of the progress milestone	Progress (cumulative)
1	Construction plans for internal review developed	40 %
2	Construction plans for discussion with the client developed	50 %
3	Construction plans for tendering completed	80 %
4	Construction plans provided	100 %

The progress of project phases and the project as a whole can be calculated from work package progress using relevance tree methods. A relevance tree can be developed by weighting the work packages and the work package groups in the work breakdown structure. The weightings determine the relevance of work packages for their work package group ("relative relevance") and for the project ("absolute relevance"). The relevance of a work package for the project can be determined by multiplying the relative relevance of the work package by the absolute relevance of the work package group.

The sum of the weights of the work packages belonging to a work package group is 100%. A uniform weighting criterion (e.g., number of person days) must be used for all work packages belonging to a work package group. If different resources are used in a project, the costs are the standard weighting criterion at higher levels of the relevance tree. An example of a relevance tree is shown in Figure 10.2.

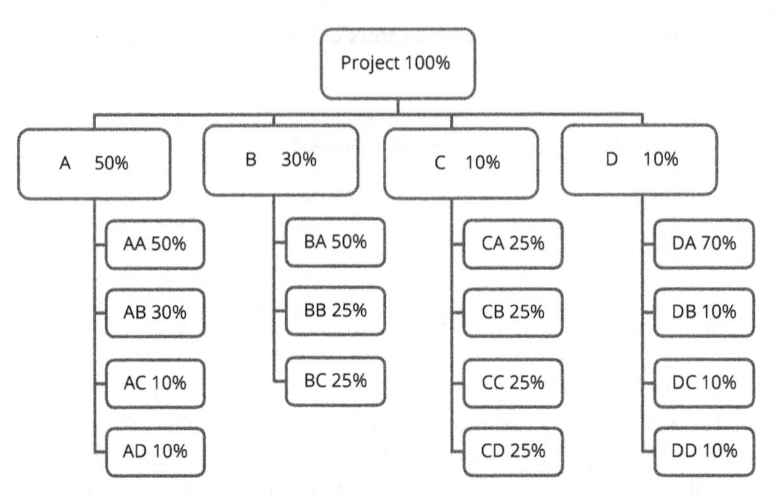

Fig. 10.2: Weightings of work packages of a project in a relevance tree.

Monitoring, Directing, and Re-Planning of Project Scope: Adapting the Work Breakdown Structure

Changes in the project objectives and in the objects of consideration necessitate changes in the scope. A change in the scope of a project can be documented in the work breakdown structure. Any additional work packages to be performed are to be added. Work packages which are no longer necessary are to be deleted.

The progress of the project can also be visualized in the work breakdown structure. Figure 10.3 shows the progress of the project "Values4Business Value" as of the controlling date 20.12.2016.

Monitoring, Directing, and Re-Planning the Project Schedule

The objective of monitoring the project schedule is to record the status of the dates of the work packages and the project. In monitoring the dates, the work packages completed or in progress during the period between two controlling reports are considered. Progress controlling is the basis for date monitoring, since an analysis about deviations from planned dates can only be made in relation to the progress achieved. Date monitoring can be carried out in milestone plans, lists of dates, bar charts, or network plans.

The actual dates of milestones or work packages are recorded in milestone plans or in lists of dates and compared with the planned dates. Any deviations between planned and actual dates must be interpreted to provide corresponding information for directive activities.

(text continues on page 279)

Case Study: Developing Values4Business Value—Adapted Work Breakdown Structure

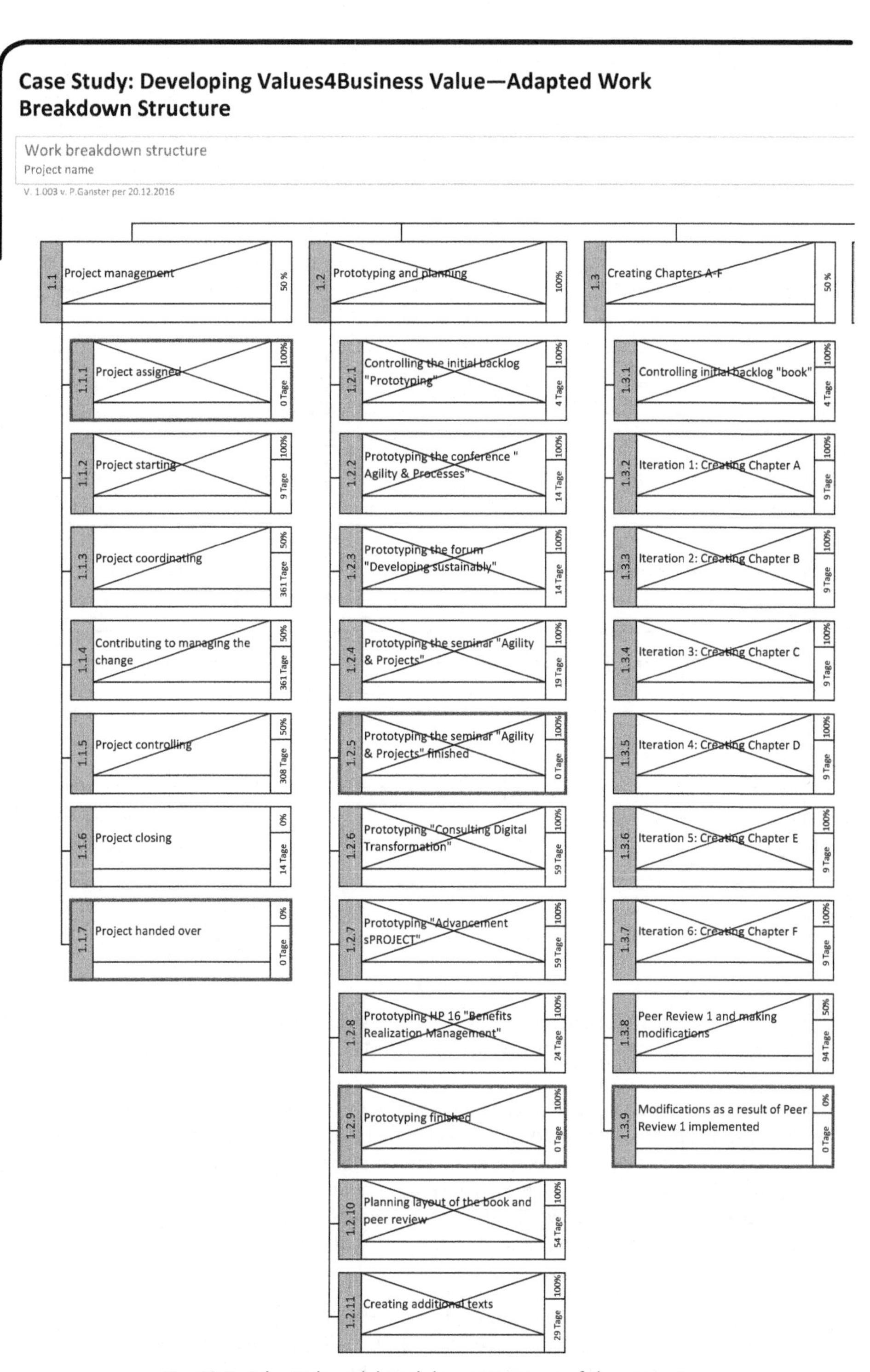

Fig. 10.3: Adapted work breakdown structure of the project "Developing Values4Business Value" as of 20.12.2016 *(continues on next pages)*.

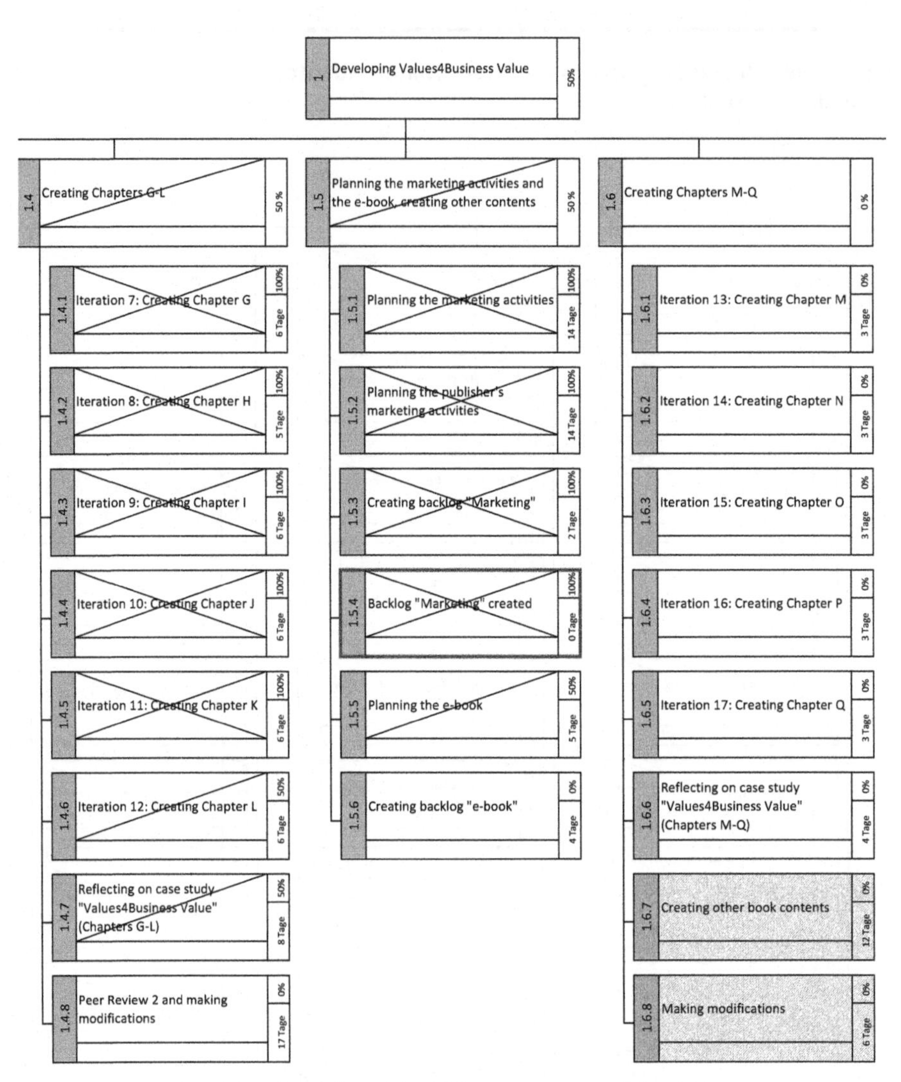

Fig. 10.3: Adapted work breakdown structure of the project "Developing Values4Business Value" as of 20.12.2016. *(continues on next page)*

Because of the additional project objectives and the results of the creation of the marketing backlog, the last phase of the work breakdown structure "1.9 Initial marketing book and services" was supplemented by work packages and re-planned. The work package "1.6.7 Producing other book contents", which is shown grayed out in Phase 1.6, was moved from Phase 1.5 to Phase 1.6. The scope of work package 1.6.8 was reduced, because no third workshop of the peer review group was planned. All additional work packages are shown grayed out in the WBS for better transparency.

The completed work packages are identified by two diagonal slashes, work packages in progress by one diagonal slash. If, as it is here, the work breakdown structure is structured in a process-oriented manner, the work packages should be processed from left to right.

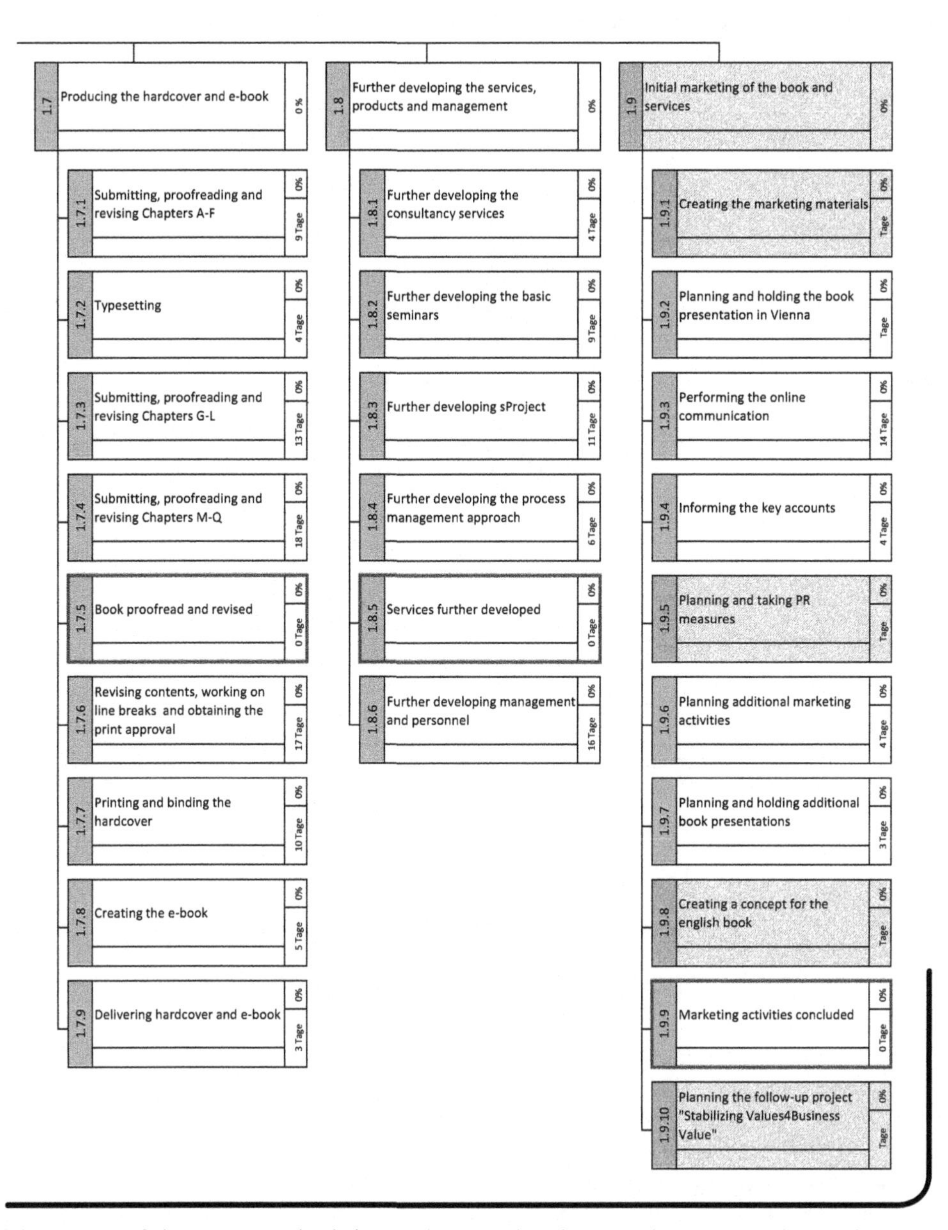

The status of the project schedule can be visualized using the project plans. The timing of a completed work package can be made visible in the bar chart by the position of the actual bar. The start dates and the duration up to the control date can be displayed for work packages that are in progress. In network plans the process logic and the project schedule can be controlled.

Case Study: Developing Values4Business Value—Adapted Milestones Plan

Table 10.9: Adapted Project Milestones Plan as of 20.12.2016

Adapted Project Milestone Plan

Developing Values4Business Value

WBS-Code	Milestone	Basic plan	Actual date	Adapted plan
		per 05.10.2016	per 20.12.2016	per 20.12.2016
1.1.1	Project assigned	05.10.2015	05.10.2016	
1.2.5	Prototyping by the seminar "Agility & Projects" finished	18.03.2016	18.03.2016	
1.2.9	Prototyping finished	27.05.2016	27.05.2016	
1.5.4	Backlog "Marketing" developed	18.11.2016	18.11.2016	
1.3.9	Modifications as a result of Peer Review 1 implemented	23.11.2016		16.01.2017
1.7.5	Book proofread and corrected	13.01.2017		06.03.2016
1.8.5	Services further developed	24.02.2017		26.04.2016
1.9.8	Marketing activities performed	10.03.2017		12.05.2016
1.1.7	Project approved	31.03.2017		31.05.2016

V. 1.003 v. P. Ganster per 20.12.2016

The first peer review group workshop was very creative and constructive. Because the peer review group noted that the central messages were not easy enough for a critical reader to follow, Chapters A–F had to be completely revised. As a result, the end of Phase 1.3 was shifted from 23.11.2016 to 16.1.2017. This delay resulted in an extension of the project duration by two months.

Directing and re-planning of the project schedule is necessary if the logic has changed, if changes in the durations have occurred, or if dates have to be planned for additional work packages.

In a milestones plan, the planned dates of the "baseline", adapted planned dates, and actual dates can be displayed (see Table 10.9). In a bar chart, bars can be displayed with the baseline dates, adapted planned dates, and actual dates for comparison.

Monitoring, Directing, and Re-Planning Project Costs and Project Income

The objectives of monitoring the project costs and the project income are to record actual costs and income and to identify deviations in project costs, project income, and expected financial project success on the controlling date. Monitoring of the project resources can be performed by monitoring the resource quantities implicitly as part of project cost controlling or explicitly.

A planned versus actual comparison of project costs and work package costs reveals whether there are deviations from the plan. Possible errors in planning project costs, changes in quantities, and changes of prices can influence project costs and the financial project success. Opportunities for cost savings and increases in yields should also be identified in project cost controlling.

Adequate project structures are a prerequisite for efficient project cost controlling. The structures of project cost planning must correspond to the structures of progress measurement and the recording of actual costs. Recording of actual costs is performed according to the process of performing work packages. Thus the project cost plan should also be structured in a process-oriented manner.

The actual costs for work packages are recorded as follows: for personnel costs, on the basis of proof of the working hours of members of the project organization; for costs of external services, on the basis of supplier invoices; material costs on the basis of invoices for materials; and for other costs, such as travel expenses, on the basis of relevant evidence. Potential weaknesses in controlling project costs are listed in Table 10.10.

If the scope of a project is changed in the course of project implementation, the planned costs must be updated. By comparing the planned costs of the baseline with

Table 10.10: Potential Weaknesses in Controlling Project Costs

Potential weaknesses in controlling project costs
> No common structures of planned costs and actual costs > No clear distinction between costs and cash-outflows > Invoices and records for cost controlling not available to the project manager > Opportunity costs (e.g. internal personnel costs) are not considered as part of the project costs > No recording of the costs of supplier services according to the progress of work

the updated planned costs, the cost effects of changes in the scope can be determined (see Table 10.11).

Table 10.11: Calculation of the Cost Deviation Due to a Change in the Scope

Updated planned costs (based on a changed scope of work)
– Planned costs (according to the basic plan)
Cost deviation (due to the changed scope of work)

Re-planning of project costs is performed by determining the remaining costs of work packages in progress and by adapting the costs of work packages yet to be fulfilled. The determination of remaining costs is not an extrapolation of already incurred work package costs, but rather a systematic re-planning of the costs, taking into account the improved level of information and the consequences of directive activities, such as changes in the use of resources, new pricing information, etc.

A relatively imprecise re-planning of work package costs can be performed by estimating the remaining costs still to be expected, without analyzing the progress already achieved. For appropriate planning of the remaining costs of a work package, information on the progress already achieved, as provided by the earned value analysis, is necessary (see Excursus: Earned-Value Analysis).

Re-planning of project costs must be performed jointly by project team members and the project manager. Each project team member determines the remaining or total costs for the work packages. Possible cost deviations for future work packages must be presented as early as possible.

Excursus: Earned-Value Analysis

Objectives of Earned-Value Analysis

The objective of earned-value analysis is a monetary evaluation of the progress of a work package, project phase, or project. This is achieved by an integrative consideration of progress, costs, and dates. The calculation of the "earned value" provides clear information about the project status and serves as a basis for estimating remaining costs and remaining duration.

Earned-Value Analysis for a Project

A comparison of planned and actual project costs is only useful if the costs are compared on a common basis. A comparison of "planned costs" (for the planned progress) with "actual costs" (for the actual progress) has little meaning. The actual progress is to be used as the common basis for comparing planned and actual costs.

The planned project costs are based on the assumption of the planned progress up to the controlling date. However, it is not the planned progress but the actual progress which led to the actual costs. Defining the actual progress makes it possible to calculate the originally planned costs for this actual progress. The planned costs for the actual progress are called "earned value". The earned value is to be compared with the actual costs to determine whether there is a cost deviation.

The recording of the actual progress is performed at the lowest level of the work breakdown structure. Earned-value analysis requires the possibility of aggregating data using the relevance tree method. The following assumptions are made about the relationship between progress, costs, and dates:

> There is a direct proportionality between progress and costs at all levels of consideration (project, project phase, work package).
> Proportionality between progress and duration is not assumed.

The data from the planning and controlling of progress, costs, and dates makes it possible to develop cumulative curves for planned costs, planned progress (this corresponds to the curve of the planned costs), actual costs, actual progress, and earned value. These curves can be displayed in a coordinate system with project duration on the *x* axis and progress as well as project costs on the *y* axis. On the basis of the assumption of proportionality between progress and costs, both progress and costs can be shown on the *y* axis (see Fig. 10.4).

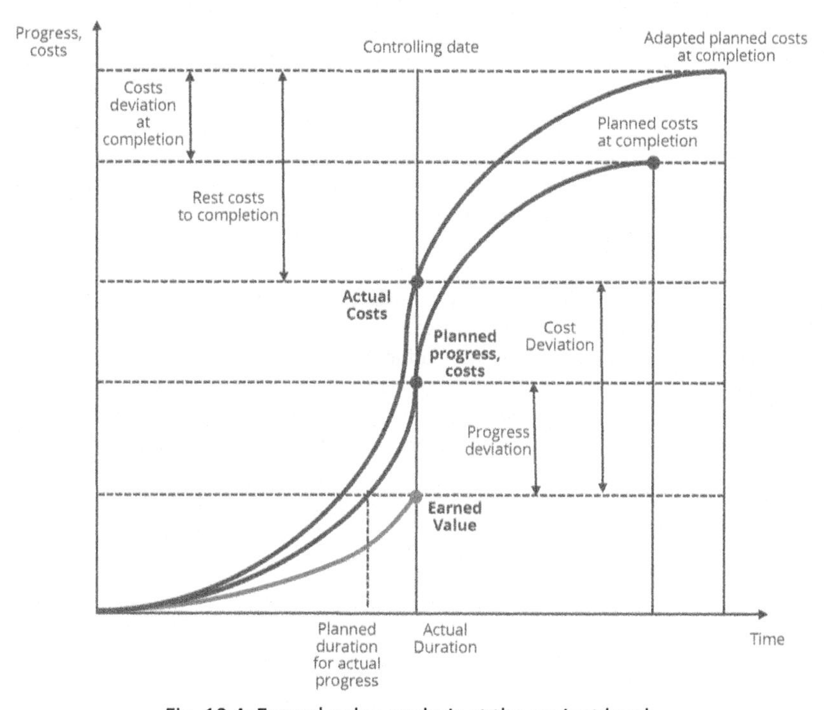

Fig. 10.4: Earned-value analysis at the project level.

On the controlling date, the actual progress of the project can be compared with the planned progress. This shows the deviation in progress as of the controlling date, and the actual costs of the project can be compared with the earned value. This shows the cost deviation as of the controlling date.

The planned duration (for the actual progress of the project) can be determined graphically by projecting the actual progress onto the planned progress curve. A (theoretical) deviation in the project duration can be identified by comparing this planned duration with the actual duration.

Earned-Value Analysis for Work Packages

The earned value of a work package can be determined by multiplying the planned costs of the work package by the actual progress (as a percentage). As an example, an application of earned-value analysis at the work package level is shown in Table 10.12 for two work packages.

Table 10.12: Earned-Value Analysis for Work Packages

WBS-Code	Work package	Planned costs (baseline)	Progress at controlling date	Earned Value	Actual costs at controlling date	Costs deviation at controlling date	Rest costs to completion	Adapted planned costs at completion	Planned total costs deviation at completion
		(1)	(2)	(3) = (1) x (2)	(4)	(5) = (4) - (3)	(6)	(7) = (4) + (6)	(8) = (7) - (1)
2.2.1	Planning	€ 30,000.00	50%	€ 15,000.00	€ 15,500.00	€ 500.00	€ 15,000.00	€ 30,500.00	€ 500.00
2.2.2	Developing a specification	€ 20,000.00	20%	€ 4,000.00	€ 7,000.00	€ 3,000.00	€ 16,000.00	€ 23,000.00	€ 3,000.00
...									

Earned-value analysis enables the determination of cost deviations as of the controlling dates. It also provides an adequate basis for determining the remaining costs of work packages by determining its progress at the reporting date. The work yet to be performed is to be clarified as a basis for this.

By comparing actual costs with actual prices, and actual costs with planned prices and earned value, the cost deviation of a work package can be divided into a price deviation and a quantity deviation (see Fig. 10.5). The price deviation is that part of the cost deviation resulting from changes in invoiced prices. The quantity deviation is that part of the cost deviation resulting from changes in the quantities. An analysis of the causes of cost deviations and a cost–responsibility calculation is only made possible by differentiation of price and quantity deviations.

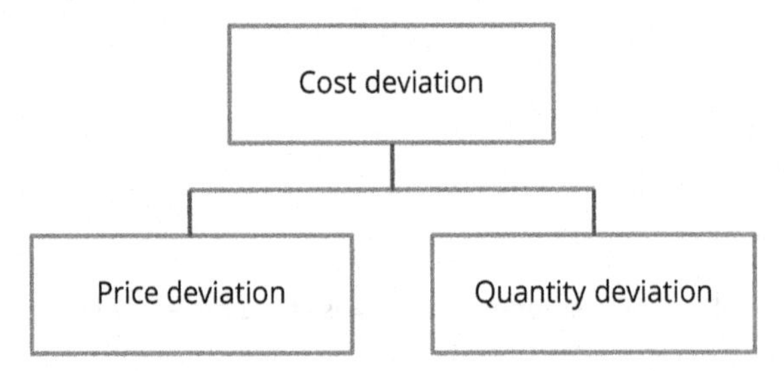

Fig. 10.5: Differentiation of the cost deviation into a price deviation and a quantity deviation.

Monitoring, Directing, and Re-Planning of Project Risks

Controlling of project risks includes controlling of risk management measures, identification of new risks, analyzing of newly identified risks, adaptation of the analyses of still active risks, and planning and implementing of new risk management measures.

The methods used during project starting must be used to fulfill these tasks (see Chapter 9).

"Social" Project Controlling: Monitoring, Directing, and Re-Planning of the Project Organization

"Social" project controlling involves controlling the project organization, the project culture, and the relations of a project with project stakeholders and other projects.

The functionality of the organizational structures of a project must be checked, the developed project culture must be reflected upon, and the appropriateness of the measures for managing relations with project stakeholders and other projects must be controlled. Changes to the project objectives and objects of consideration of the project or improved information can necessitate adaptations.

Controlling of a project organization applies to the design of the project organization chart, the description of project roles, the composition of the project team, the staffing of the project roles, the communication formats, and the project rules. The relations in the project team can also be subjected to "social" controlling by means of reflections and mutual feedback between project team members (see Chapter 8).

As an example of change of the project organization over the course of a project, the project organization charts of the project "Developing Values4Business Value" are shown in Figures 9.17 (page 238) and 9.18 (page 239).

From the organization charts shown in Figure 9.17 and 9.18, it can be seen that after the end of the phase "1.2 Prototyping and planning", the sub-team was dissolved and the new sub-teams "marketing", "further development of services", and "book production and distribution" were established. This also changed the composition of the project team. The roles and the sub-teams for the later project phases and their staffing are shown in Table 10.13 (in Case Study on next page).

"Social" Project Controlling: Monitoring, Directing, and Re-Planning of the Project Culture

Project controlling also includes controlling the project culture. Adaptations of the project culture are the result of reflection processes in the project.

Case Study: Developing Values4Business Value—Adapted List of Project Roles

Table 10.13: List of Project Roles for the Later Phases of the Project "Developing Values4Business Value"

Adapted List of Project Roles
Developing Values4Business Value

Project roles	Name
Project owner	R. Gareis; L. Gareis
Project manager	P. Ganster
Product owner team	R. Gareis, L. Gareis
Project team member: Service development	L. Gareis
Project team member: Marketing	V. Riedling
Project team member: Chapter development	R. Gareis
Project team member: Book production, sales	R. Gareis
Subteam: Service development	
Project contributor: Consulting and products	R. Gareis
Project contributor: Lectures	L. Gareis
Project contributor: Seminars	M. Stummer
Subteam: Marketing	
Project contributor: Manz	C. Dietz
Project contributor: Stämpfli	H. Gruber
Project contributor: C. H. Beck	M. Maier
Project contributor: Marketing RGC	P. Ganster
Subteam: Chapter development	
Project contributor: Case studies	L. Weinwurm
Project contributor: Research	P. Ganster
Project contributor: Chapter development	L. Gareis
Project contributor: Case study cooperation partner 1	F. Mahringer
Project contributor: Case study cooperation partner 2	M. Paulus
Peer review group	
Subteam: Book production, Sales	
Project contributor: Editing Manz	C. Dietz
Project contributor: Revision	P. Ganster
Project contributor: Graphic design	M. Riedl

V. 1.003 v. P. Ganster per 20.12.2016

The identity-creating elements of the project culture, however—namely, a project name, a project logo, and project values—should not be fundamentally changed. Minor adaptations of the values or the development of phase-specific slogans in order to meet the specific requirements of different project phases are possible. For an IT project, for example, different values may be relevant for the phases of software development from those for the next phase of the roll-out.

A fundamental change in the project culture is necessary when transforming or repositioning a project.

"Social" Project Controlling: Monitoring, Directing, and Re-Planning of Project Stakeholder Relations

Controlling relations with project stakeholders involves the following tasks:

- analyzing the project's relations with individual project stakeholders,
- planning strategies and measures to redesign existing relations,
- identifying project stakeholders who are no longer to be considered,
- identifying additional stakeholders who are to be considered, and
- planning strategies and measures for managing relations with the additional stakeholders.

When analyzing the project's relations with individual project stakeholders, the quality of the relation, possible changes in the importance of a stakeholder, and the intensity of communications with the stakeholder can be considered, and a possible need for an adaptation of the relation can be identified.

As an example of re-planning of a project's relations with project stakeholders, the project stakeholder analysis of the project "Developing Values4Business Value" is presented in Figure 10.6 (in Case Study on next page) on the basis of the project controlling on 20.12.2016.

"Social" Project Controlling: Monitoring, Directing, and Re-Planning Relations with Other Projects

The relations of a project with other projects must also be monitored during project controlling and re-planned as necessary. The basis for this is those projects from the project portfolio of the project-oriented organization which are to be considered. However, projects by customers and partners may also be relevant to the success of a given project.

Monitoring, Directing, and Re-Planning of the Investment

In project controlling, a possible investment which is realized by a project must also be controlled as necessary. Relevant assumptions about the costs and benefits of an investment must be questioned and changed if necessary, and the cost–benefit analysis or business case analysis must be adapted.

The controlling of an investment, however, can also be understood as the task of controlling the benefits realization (see Chapter 3).

Case Study: Developing Values4Business Value—Adapted Project Stakeholder Analysis

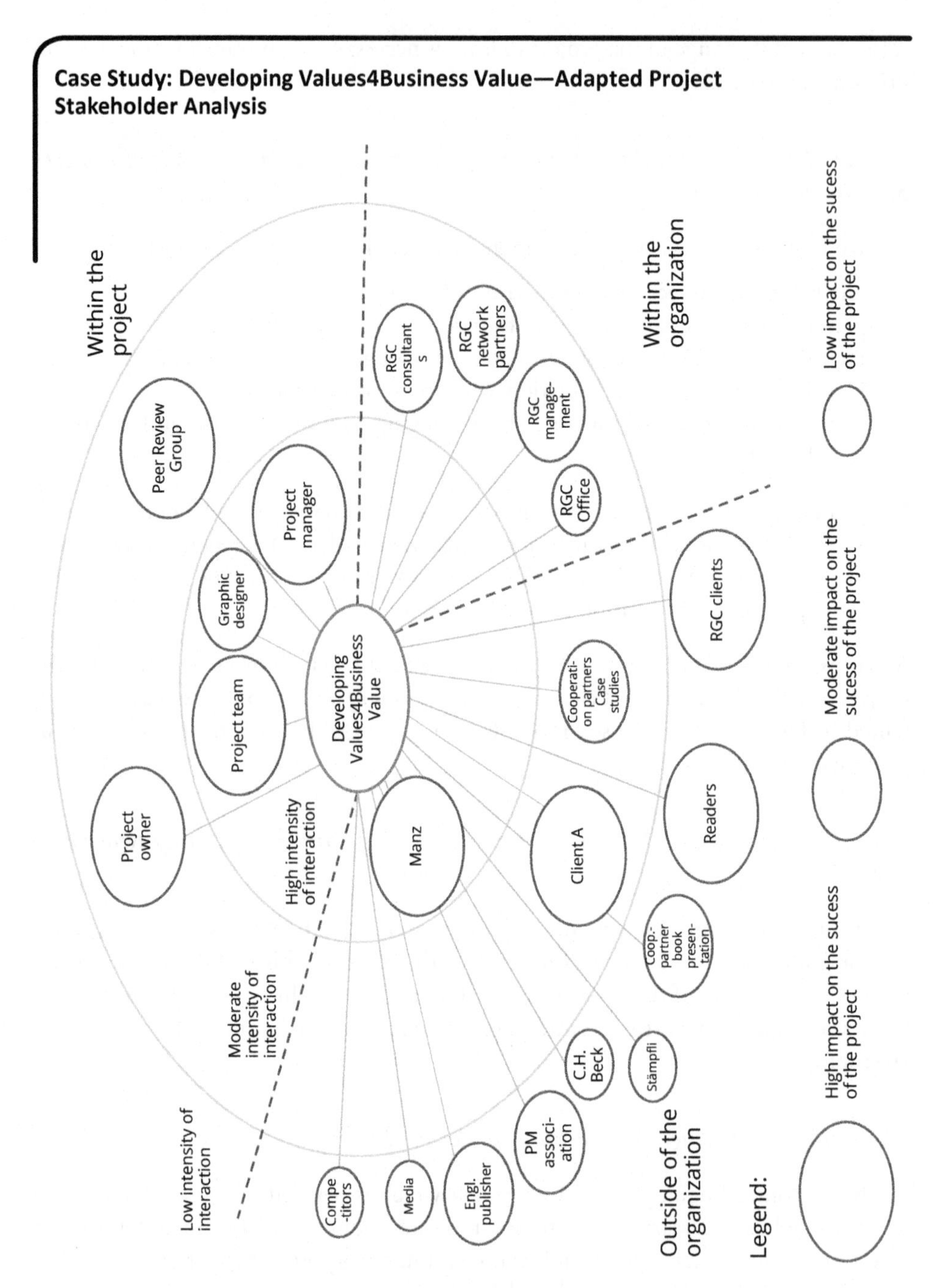

Fig. 10.6: Project stakeholder analysis of the project "Developing Values4Business Value" as of 20.12.2016.

Compared to the project stakeholder analysis presented in Chapter 9, which was drawn up during project starting, the following changes may be observed:

> English publishers and cooperation partners were considered as additional project stakeholders.
> Manz Verlag has increased in importance, and the interaction with this stakeholder has also become more intense.
> The peer review group also increased in importance.
> The interactions with the graphic designer became more intense.
> The importance of readers and RGC customers for project success remained high. It was planned, however, to distribute information only in the "initial marketing" phase.

10.2.4 Methods: Preparing Project Controlling Reports

Project Status Report

The central report which is the result of a project controlling cycle is the project status report. This describes the status of a project on a controlling date. The structure of a project status report is oriented towards the project dimensions managed. After evaluating and interpreting the overall project status, the status of the project objectives, the project progress, the project schedule, the project resources, the project costs, the project income, the project organization, the project culture, and the project stakeholder relations must be evaluated and interpreted.[4]

The scope of the project status report should be kept short; the formulations can take the form of keywords. The adapted project plans must be referred to in the appendix as a supplement to the verbal descriptions. A project status report is a formal, relatively elaborate report, which should therefore be prepared at larger intervals, about once per month.

Target groups for the project status report are the project owner and project team members. The project status report is an important project-internal communication tool. In the case of contracting projects, the project status report also serves to provide information to the customer.

Project Scorecard

A project scorecard is a specific form of a project controlling report. A project scorecard contains essentially the same information as a project status report, but in a more visualized form.

In analogy to the balanced scorecard model of Kaplan and Norton, the project scorecard uses a number of quantitative and qualitative criteria for assessing the project status.[5] The criteria to be considered depend on the project management approach

[4] The cost–benefit analysis must also be monitored periodically during the implementation of the project. This can, however, be done within the framework of benefits realization (see Chapter 3).

[5] See Kaplan, R., Norton, P., 1997.

used. The project scorecard of the project "Developing Values4Business Value", shown in Figure 10.7, is based on the RGC project management approach.

Depending on the project type, individual criteria can be considered or omitted. For example, the business case analysis can be omitted for contracting projects. The project stakeholders to be considered must always be specified in a project-specific manner.

Case Study: Developing Values4Business Value—Project Scorecard

The status of the project "Developing Values4Business Value" as of 20.12.2016 was presented in a project scorecard (see Fig. 10.7).

Interpretation of the scores in the project scorecard from 20.12.2016:

Planning, Controlling:

> Project progress as of the control date was not according to plan, as the first chapter group, A–F, had to be revised after the workshop with the peer review group.
> This additional work also led to an extension of the project duration by two months and a corresponding increase in use of resources and project costs.

Project Stakeholder Relations:

The relations with Manz Verlag and the peer review group were constructive and contributed significantly to quality assurance.

The RGC network partners were still too little involved in this project phase.

The relations with the companies that provided case studies were sometimes challenging, because it was necessary to clarify the extent to which company data could be published. This clarification had to be obtained.

Project Objectives and Contexts:

> The achievement of the project objectives seemed to be guaranteed on the controlling date. The relevant objects of consideration had been taken into account, and the investment benefits and the corresponding contributions to company strategies seemed to be ensured.
> There were no content-related conflicts with other projects, but there was competition for scarce human resources.

Project Organization:

> The project organization was perceived as basically adequate. Project communication and project teamwork functioned well. Project stakeholders such as Manz Verlag, the graphic designer, and Scrum experts were appropriately involved.
> Since RGC is a small company, several project roles were carried out by the same people. This increased the project complexity.
> In December 2016, a change in the role "project manager" took place because Susanne Füreder left RGC and Patricia Ganster took her place.

The project as a whole was rated "good" as the completion of the solution was not time-critical. It was assumed that the quality of the solution would have a lasting effect on the company's sustained business value.

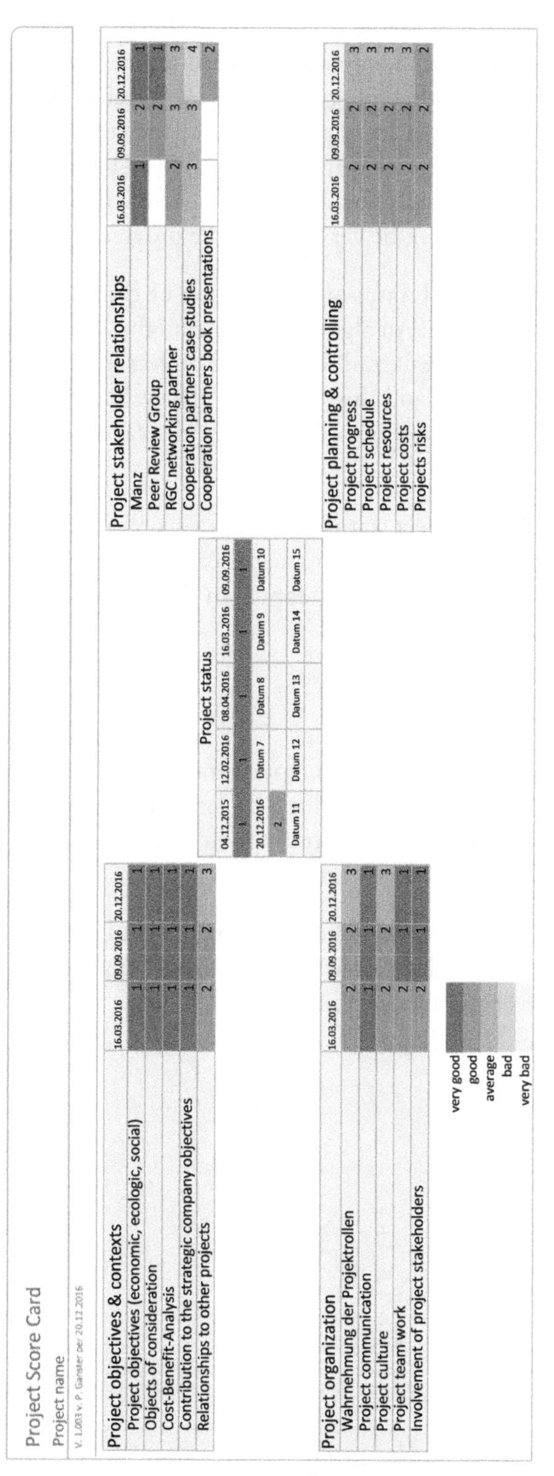

Project Score Card

Project name

V. 1.003 v. P. Ganster per 20.12.2016

Project objectives & contexts	16.03.2016	09.09.2016	20.12.2016
Project objectives (economic, ecologic, social)	1	1	1
Objects of consideration	1	1	1
Cost-Benefit-Analysis	1	1	1
Contribution to the strategic company objectives	1	1	1
Relationships to other projects	2	2	3

Project status				
04.12.2015	12.02.2016	08.04.2016	16.03.2016	09.09.2016
1	1	1	1	1
20.12.2016	Datum 7	Datum 8	Datum 9	Datum 10
2				
Datum 11	Datum 12	Datum 13	Datum 14	Datum 15

Project stakeholder relationships	16.03.2016	09.09.2016	20.12.2016
Manz	1	2	1
Peer Review Group		2	1
RGC networking partner	2	3	3
Cooperation partners case studies	3	3	4
Cooperation partners book presentations			2

Project organization	16.03.2016	09.09.2016	20.12.2016
Wahrnehmung der Projektrollen	2	2	3
Project communication	1	1	1
Project culture	2	2	3
Project team work	2	1	1
Involvement of project stakeholders	2	1	1

Project planning & controlling	16.03.2016	09.09.2016	20.12.2016
Project progress	2	2	3
Project schedule	2	2	3
Project resources	2	2	3
Project costs	2	2	3
Projects risks	2	2	2

Fig. 10.7: Project scorecard of the project "Developing Values4Business Value" from 20.12.2016.

The scoring of the criteria of the project scorecard can be with the traffic light colors red, yellow, and green or with a five-color scale. The latter allows more differentiation in the "scores". The project scorecard provides a holistic view of the project status. The status of each project dimension managed is assessed in an integrated form. Connections between the scores of individual criteria can be considered. For example, the relationship with the project owner cannot be very good if the progress of the project is very bad.

The assessment of the overall status of a project can be carried out either intuitively or with the aid of an algorithm. An algorithmic rule might, for example, be: "If two criteria are scored red, then the overall score of the project is red". The individual scores are to be interpreted briefly, and the reasons for the evaluations have to be explained. The scores from several controlling dates may be displayed, so that the development of the project can be observed.

The visualization of the project status in a project scorecard is an important communication tool of a project. The project manager can produce a first draft of the project scorecard. This is to be discussed and adapted in the project team and then presented to the project owner.

Deviation Trend Analysis

Deviation trend analysis is a graphical representation of the development of a project ratio as determined by several controlling cycles. Trend analyses can be carried out based on updates during project controlling—for instance, for the project costs, the project income, or the project end date.

Figure 10.8 shows a deviation trend analysis for the end date of a project. It can be seen that a later project end date was planned in each controlling cycle and that the project duration was continuously prolonged.

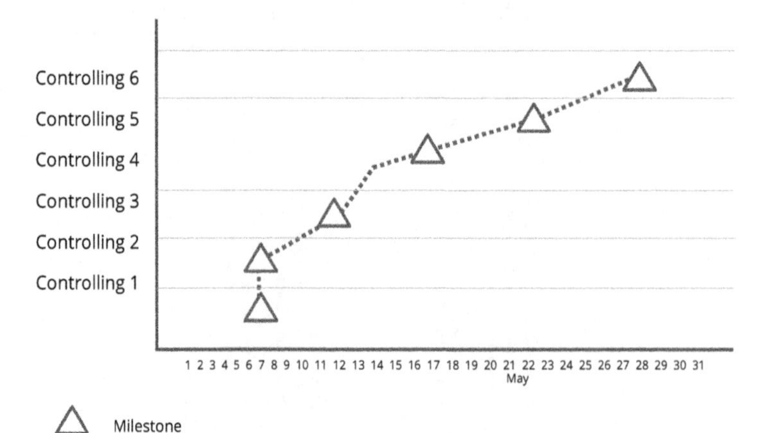

Fig. 10.8: Deviation trend analysis for the project end date.

Literature

Kaplan, R., Norton, P.: *Balanced Scorecard, Strategien erfolgreich umsetzen,* Schäffer-Poeschel, Stuttgart, 1997.

11 Sub-Processes: Project Transforming and Project Repositioning

Projects develop, both continuously and discontinuously, during their performance. These developments of projects can be perceived as project changes. Therefore, not only permanent organizations, but also temporary organizations, are change objects. Change management methods can also be applied for project changes.

It is possible to distinguish between the changes "Learning of a project", "Further developing a project", "Project transforming", and "Project repositioning". The changes "Learning of a project" and "Further developing a project" were already discussed in Chapter 10 in the context of project controlling.

The management of discontinuous developments of projects involves the avoidance or promotion of project discontinuities, provision for their possible occurrence, and coping with a project discontinuity. To cope with a project discontinuity, the changes "Project transforming" or "Project repositioning" can be performed. As a case study for "project transforming", the transforming of a project into a program for establishing a hospital will be presented.

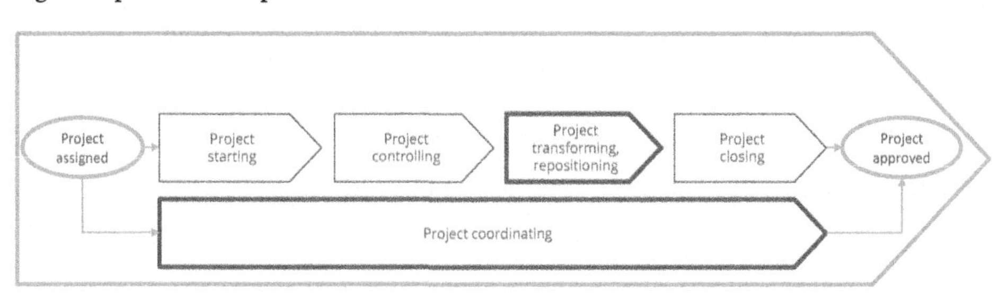

Overview: "Project transforming" and "Project repositioning" as sub-processes of the business process "Project managing".

11 Sub-Processes: Project Transforming and Project Repositioning

11.1 Project Changes

11.1.1 Continuous and Discontinuous Developments of Projects

Projects can develop both continuously and discontinuously. The causes for continuous or discontinuous developments of projects can be their self-organization processes or interventions by project stakeholders. The frequent occurrence of discontinuities in projects is due to their complexity and dynamism.

Continuous Development of a Project: Definition

A continuous development of a project is a result of small changes in one or more project dimensions. A continuous development can be reflected upon during project controlling. Continuous development of a project does not lead to a change in the project identity.

Discontinuous Development of a Project: Definition

A discontinuous development of a project is a phase of instability in a project that leads to a change in the project identity. One can distinguish between the following types of project discontinuities (see Table 11.1):

- a project crisis and a project chance,
- a structurally determined change of the project identity, and
- an "ad hoc" change of the project identity.

Table 11.1: Definition of Project Discontinuities

Definition of project discontinuities	
Project crisis	> An existential threat of a project > Consequence: Change "Project repositioning"
Project chance	> Major, new potentials of a project > Consequence: Change "Project repositioning"
Structurally determined change of the project identity	> Expected, fundamental change of a project > Consequence: Change "Project transforming"
Ad hoc change of the project identity	> Unexpected, fundamental change of a project > Consequence: Change "Project transforming"
No discontinuities are ...	> Conflicts or catastrophes

Project crises and project chances come unexpectedly despite possible measures to avoid or promote them. This also applies to "ad hoc changes of the project identity". "Structurally determined changes of the project identity" are foreseeable and can therefore be planned for.

Examples of project crises, project chances, and structurally determined changes of the identity of projects are shown in Table 11.2. The occurrence of one or more of the factors listed in Table 11.3 can lead to the definition of a project crisis or a project chance.

Table 11.2: Examples of Discontinuities of Projects

Examples of project crises
> The reorganization of the service department of a magistrate had to be interrupted because of outstanding basic organizational decisions, which had to be made by politicians and a municipal council external to the project.
> A reactor accident led to a threat to the personnel of a large-scale engineering company on a nearby construction site. Crisis management measures had to be undertaken on the construction site and for family members of the construction site personnel.
Examples of project chances
> A developed software application could be used not only by the parent company but also by subsidiaries. As a result, the scope of the project was significantly extended during implementation. It was redefined as a cooperative project with the subsidiaries.
> In the course of implementing a logistics project for a ministry, a new technology came onto the market. The specialist departments decided to switch to the new technology. The structures of the project had to be completely changed.
Examples of structurally determined changes of project identities
> After the development of an organizational solution for the branches of a trading company, a roll-out was performed in the branches. A reorientation in the project was necessary to create appropriate external orientation after the first phases of creativity and internal orientation. The project organization had to be redesigned for the roll-out.
> A redesign of an engineering project at the beginning of the construction phase after the phases of technical planning, procurement and production was required.
Example of an "ad hoc necessary" change of a project identity
> The project "Developing Values4Business Value" was initially defined and understood as a book publication project. During the early phases of the project, a potential for further developing RGC management approaches was identified. As a result, greater benefits were created for RGC customers and employees. This changed perception of the project led to an "ad hoc necessary" change in the identity of the project. The benefits of the investment, the project objectives, the scope, the costs, the duration, etc. were changed.

Table 11.3: Possible Criteria for Defining a Project Crisis or Project Chance

Factors that can lead to the definition of a project crisis:
> Significant deterioration of the results of the business case analysis or the cost-benefit analysis of the investment implemented by a project
> Substantial increase of the project scope within the same project budget
> Substantial increase of the project costs within the same project scope of work
> Substantial increase of the project duration within the same project scope of work
> Loss of a major project partner
> Insolvency of the customer (for contracting projects)
Factors that can lead to the definition of a project change:
> Significant savings and/or reference potential as a result of using a new technology in the project
> Significant enlargement of the scope of work (and related revenues) of a contracting project

11.1.2 Projects as Change Objects, Types of Project Changes

Developments of projects which lead to changes and not to cancellations or suspensions can be perceived as project changes. It is possible to distinguish between the changes "Learning of a project", "Further developing a project", "Project transforming", and "Project repositioning".

Projects as Change Objects

Projects develop over time, both continuously and discontinuously. These developments can be perceived as project changes. A project change is a continuous or discontinuous development of a project in which one or more dimensions of the project are changed.[12] It is assumed that in a project change the social structures of the project and the project stakeholder relations are changed. It is therefore a "social" and not, for example, just a technological change.

The perception of continuous and discontinuous developments of projects as project changes allows change management concepts, such as defining a change vision, creating a sense of urgency for a change, planning quick wins, and practicing appropriate change communication, to be used for managing project chances.

Social systems are the objects of changes. The social systems that are considered here are the temporary organizations.[3] Both projects and programs can be the objects of changes (see Fig. 11.1).

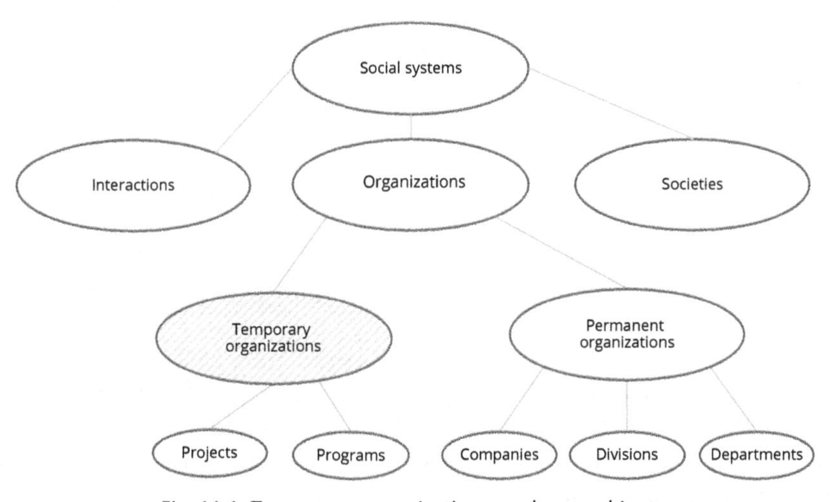

Fig. 11.1: Temporary organizations as change objects.

[1] Lewin defines change as the transformation of an existing dynamic balance of an organization into a new one.
[2] See Lewin, K., 1947.
[3] Chapter 14 discusses changes of permanent organizations.

Dimensions of a project that can change are its objects of consideration, objectives, strategies, dates, costs, organization, culture, personnel structures, infrastructures, stakeholder relations, relations with other projects, relation with the objectives of the project-oriented organization, and relation to the investment realized by the project.

A project change is a continuous or discontinuous development of a project in which one or more dimensions of a project are changed.

Types of Project Changes

Taking into account the need for change of a project and the number of change dimensions to be considered, one can distinguish between the first-order changes "Learning of a project" and "Further developing a project", as well as the second-order changes "Project transforming" and "Project repositioning" (see Figure 11.2). First-order changes lead to continuous changes in projects, second-order changes lead to changes in project identities (see Chapter 14 for the definition of first- and second-order changes).

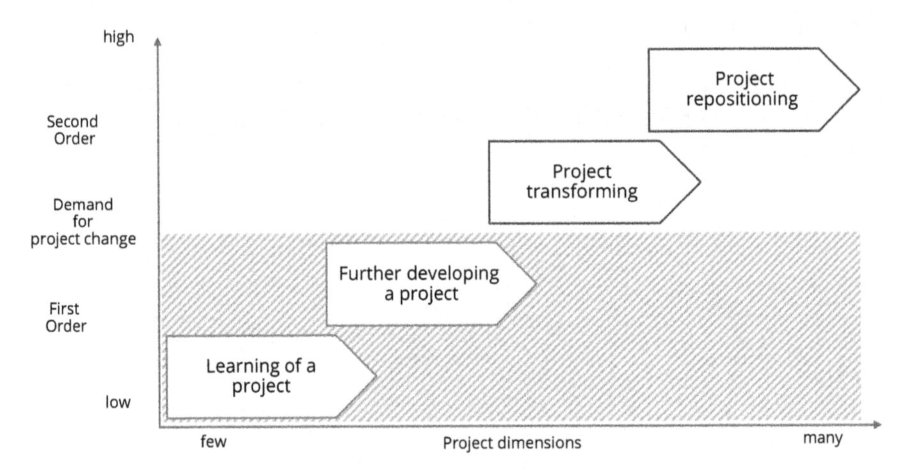

Fig. 11.2: Types of project changes.

A relatively low demand for change, which necessitates the change "Learning of a project", results from the daily project work. This need is met by periodic project controlling. The improvements agreed upon during project controlling are integrated into the existing project structures.

A medium demand for change, which necessitates the change "Developing a project", is usually due to a change in the requirements for the solution to be provided by a project. New solution requirements are often formulated as "change requests".

Signs of a future threat, but also of new potentials for a project, can lead to a medium to high demand for change. A strategic and cultural reorientation of a project may become necessary; many or all dimensions of a project are to be considered. "Project transforming" is the appropriate change type for this.

The highest demand for change for a project exists when it is in a crisis or if there is a chance. Both necessitate a fundamental change of the project. The loss of the project owner or an important project partner, a new technology that one is not qualified to use, new objectives and strategies of the project-oriented organization, or greatly increased project costs might be indicators of a project crisis. A project chance exists, for example, if a customer increases the volume of an order that is processed as a project by 100 percent, or if a fundamentally new technology, which considerably reduces the project costs and the duration of the project, can be used. In these cases, all project dimensions must be considered in the change. A change "Project repositioning" is required. The basic objective of the change "Project repositioning" is to continue the project. To achieve this, positive project ratios have to be achieved, and major project innovations have to be implemented.

11.2 Managing Continuous Developments of Projects

Continuous developments can be managed by the changes "Learning of a project" and "Further developing a project".

11.2.1 Managing the Change: Learning of a Project

The cause for the change "Learning of a project" is the objective of continuously improving the quality of the project results and the project processes. Available potentials for learning are used, and small innovations are promoted.

Only a few project dimensions are affected by the change "Learning of a project". A stakeholder relationship may be changed by optimizing a common tool, or a communication format can be improved.

This organizational learning of a project takes place implicitly in the sub-process "Project controlling" (see Chapter 10). Existing practices can be reflected upon in controlling meetings and agreements made to improve them. Learning is often associated with the need to unlearn. It is difficult to learn how to get rid of familiar practices. Symbolic acts, such as tearing up a previously used template in a project meeting, support the unlearning.

To ensure learning in projects, learning and small innovations are to be formulated as an explicit project objective. Appreciation should be given for improvement proposals, and (non-monetary) incentives are possible.

11.2.2 Managing the Change: Further Developing a Project

The change "Further developing a project" is also performed implicitly in the sub-process "Project controlling". In contrast to the learning of a project, however, several project dimensions are affected, and formal approval of a change request is usually required.

The cause for the change "Further developing a project" is often a "change request". In this case, the change is associated with a "change" in the requirements of the solution to be provided by the project. However, the optimization of project structures and major innovations can also be objectives. This shows that in practice the term "change" is often just used for a change in solution requirements. It is not understood in an organizational sense, as applied here.

As a result of changed solution requirements or other possible improvements, the project structures have to be adapted and the change has to be communicated. In the change "Further developing a project", the project structures are adapted, but there is no change in the identity of the project.

In Figures 11.3 and 11.4, the process for approving a "change request" and a change request form are shown as examples.

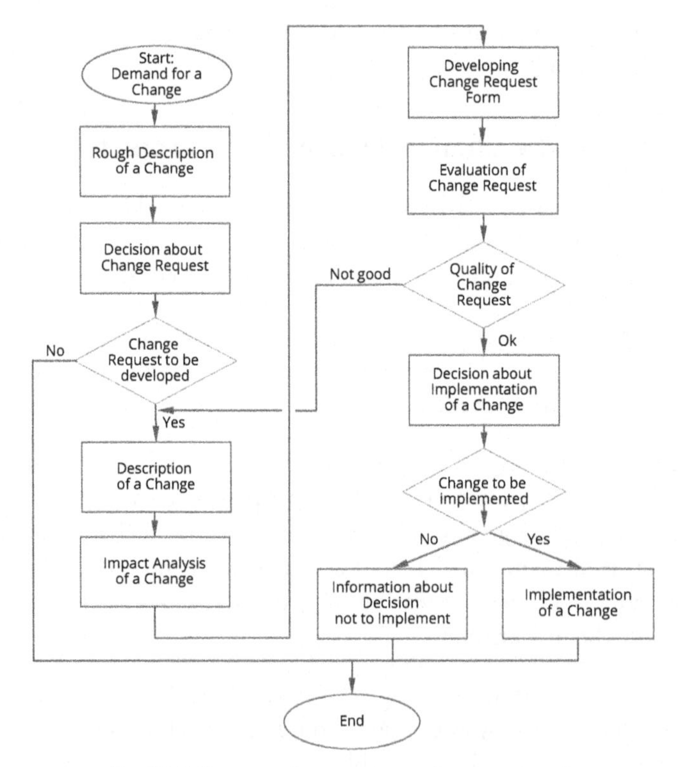

Fig. 11.3: Process of approving a change request.

Project:			
Change request ID:	Requestor:	Priority: Low / Medium / High	Initiation date:
Change title:			
Reason for change:			
Short description of the change :			

Impact Analysis	
Change impact related to:	Project objectives:
	Project scope:
	Project schedule:
	Project budget:
Date:	Person in charge of implementation:

Fig. 11.4: "Change request for projects" form.

11.3 Managing Discontinuous Developments of Projects

The tasks of managing project discontinuities are avoiding project crises or promoting project chances, provisioning for project discontinuities, and coping with a project discontinuity (see Fig. 11.5).

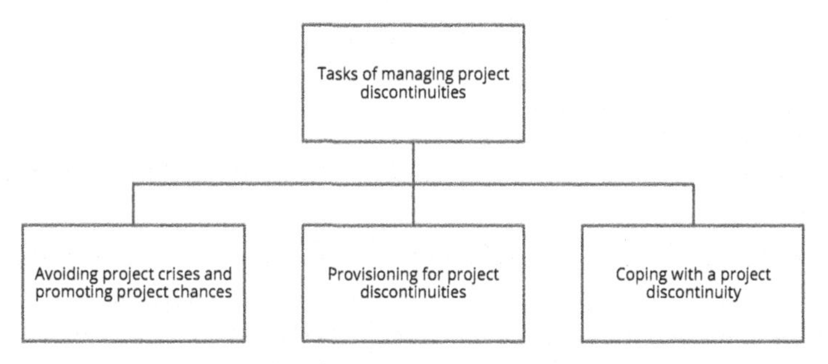

Fig. 11.5: Tasks of managing project discontinuities.

The avoidance of project crises or the promotion of project chances and provision for project discontinuities are not separate project management sub-processes, but tasks that are to be performed during "Project starting" and "Project controlling".

Coping with a structurally determined change differs fundamentally from coping with a project crisis or a project chance. The objectives and tasks of its process are similar to project starting, and it is therefore not presented here as a separate business process. The changes "Project transforming" and "Project repositioning" are considered as separate project management sub-processes.

11.3.1 Avoiding Project Crises and Promoting Project Chances

The tasks for avoiding project crises are

> early recognition of potentials for project crises,
> analysis of already implemented measures for avoiding project crises,
> development of strategies and measures for avoiding project crises,
> implementation of strategies and measures for avoiding project crises, and
> communication of the implemented strategies and measures.

In any case, relevant measures for avoiding project crises are professional project management and explicit project risk management.

Similar tasks, only in the opposite direction, are to be performed to promote project chances. Planning and implementing the strategies and measures for avoiding project crises or for promoting project chances must be carried out by the project team in agreement with the project owner.

11.3.2 Provisioning for Project Crises or Project Chances

The objective of provisioning for project crises or project chances is to develop strategies and measures for the event of a project discontinuity. This is to enable efficient management in the event of occurrence.

The tasks of provisioning for project crises or project chances are

> early recognition of potentials for project crises and project chances,
> analysis of measures which have already been implemented for providing for project crises and project chances,
> planning strategies and measures for providing for project crises and project chances,
> implementation of strategies and measures for providing for crises and chances, and
> communication of these strategies and measures for providing for crises and chances.

Measures for provisioning for project crises and project chances which are relevant in all cases are creating standardized structures for the management of a project crisis and developing alternative project plans.

The need for measures for avoiding or promoting and for provisioning for project discontinuities depends on the type of project. For external customer projects, crisis-avoidance measures are planned in the tendering process. On the one hand, detailed object plans, project plans, and contracts are developed. On the other hand, risk management measures such as insurance contracts or the imposition of risk premiums are implemented. As the risks and complexity of the tendering process are high, organizations even perform projects for tendering processes. This is also to be seen as a risk management measure.

Due to their social complexity and dynamism, internal projects are exposed to a higher degree of discontinuity than external projects. Project cancellations are more frequently observed in internal projects than in external projects.

11.3.3 Avoiding or Promoting and Provisioning for Project Discontinuities: Methods

The basis for avoiding or promoting and provisioning for project discontinuities is the identification of potentials for project crises or project chances. Appropriate methods, differentiated by methods of the first, second, and third "generation", are listed in Table 11.4. Both first-generation and second-generation methods are past-oriented methods, while third-generation methods are forward looking.

Table 11.4: Methods for Identifying Potentials for Project Crises and Project Chances

Methods for identifying project crises and project chances potentials	
First generation methods: Project ratios	> Project scope > Project costs > Project duration > Project profit
Second generation methods: Indicators	> Fluctuation of members of the project organization > Formalism in the project > Commitment of suppliers
Third generation methods: Weak signals, scenarios	> Weak signals: Bad structured, hard-to-interpret information > Project scenarios

Weak signals are often overlooked or unheard. They require specific monitoring. Examples of weak signals for projects are

> a sudden accumulation of similar events,
> the opinions, comments, and statements of project stakeholders, and
> changes in laws or regulations.

The development of project scenarios is another future-oriented method.[4] This is used to describe possible future states of a project. Alternative project scenarios (e.g., a positive and a negative one) are developed in addition to a desired target scenario. Thinking in alternatives is deliberately performed, in order to expand the spectrum

[4] See, for example, Reibnitz, U. v., 1992.

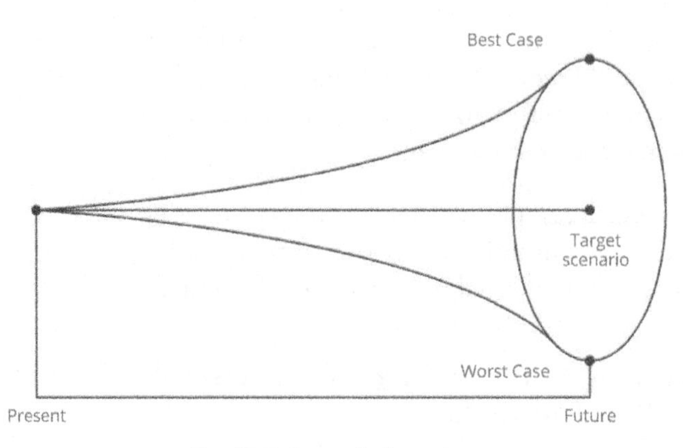

Fig. 11.6: Scenario funnel.

of action (funnel-shaped) (see Fig. 11.6). The project scenarios provide a basis for strategies and measures for providing for these scenarios. For example, measures are defined for the achievement of the target scenario, and alternative project plans can be prepared to provide for different scenarios.

11.3.4 Coping with a Project Crisis or a Project Chance: Options

The following basic strategic options exist for coping with project discontinuities:

> project cancelling,
> project interrupting, and
> project changing either by "project transforming" or by "project repositioning".

The existence of the social system "project" is terminated (at least for a period of time) by project cancellation or project interruption. In project transforming and project repositioning, the social system is changed. An attempt is made to keep the project alive, but its identity is changed.

The following tasks must be performed when cancelling a project:

> dissolving project stakeholder relations,
> safeguarding achieved project results,
> giving feedback to members of the project organization, and
> supporting members of the project organization in their reorientation.

The following tasks must be fulfilled when interrupting a project:

> informing project stakeholders about the project interruption,
> safeguarding achieved project results,
> giving feedback to members of the project organization,
> planning the restarting of the project, and
> ensuring the availability of members of the project organization for the restarting.

The cancellation of the project represents a catastrophe for the social system "project", since survival is no longer possible. From the perspective of the project-oriented organization, the decision to cancel a project can, however, make sense. When a project is interrupted, it is assumed that the project can be continued later on. The objectives and processes of project cancellation and project interruption are similar to those of project closing (see Chapter 12). They are therefore not described here. The project management sub-processes "Project transforming" and "Project repositioning" are dealt with below.

11.4 Managing the Change: Project Transforming

The change "Project transforming" is a second-order change. This means that it leads to a change in the identity of a project.

11.4.1 Project Transforming: Objectives and Process

Causes for the change "Project transforming" can be signs of an existential threat or a fundamental potential for optimizing a project. The objectives of a transformation are a strategic and cultural reorientation of the project, achieved by optimizing project structures and implementing major project innovations.

All project dimensions are considered in project transforming. This requires results-oriented top-down management. The objectives of the project change have to be formulated, and members of the project organization and project stakeholders who are not convinced of the urgency of the project change have to be convinced. Quick wins are to be ensured in order to create appropriate change momentum. The contradiction between the demand for hard cuts and the promotion of new potentials in the project is to be managed. The change "Project transforming" is to be explicitly defined and explicitly ended after performance. Information is to be provided about going "back to daily business" (with new project structures).

Project transforming is a project management sub-process with the phases "planning project transforming", "implementing project transforming", and "stabilizing the project" (see Fig. 11.7). The tasks of these project phases and the responsibilities for their fulfillment are shown in Table 11.5 (on page 309).

11.4.2 Project Transforming: Organization and Quality

Project Transforming: Organization

Project transforming can basically be carried out by the project organization. It must be ensured that the project owner is heavily involved. His or her decision-making power is needed. The members of the project organization and representatives of

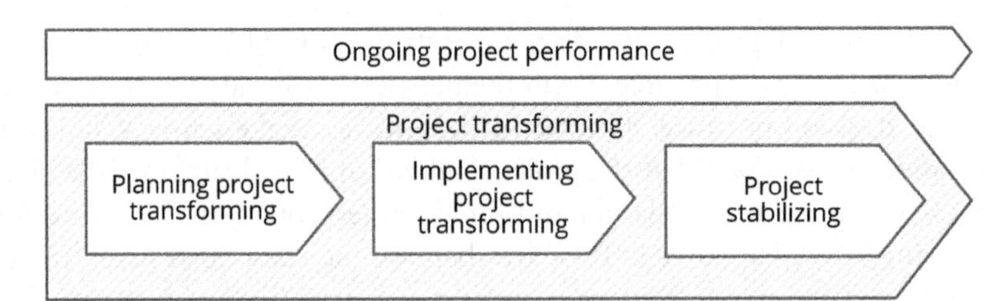

Fig. 11.7: Phases of the sub-process "Project transforming".

project stakeholders must also be involved so as to benefit from their creativity. If necessary, a project-project management consultant can support the process. Intensive project communication is necessary.

Project Transforming: Quality

One challenge in project transforming is simultaneously performing the tasks of Project transforming and the ongoing project tasks. The members of the project organization are responsible for both.

The stabilization phase is important for the success of Project transforming. It must be implemented so as to ensure the benefits of new project structures and contextual relations.

11.5 Managing the Change: Project Repositioning

The change "Project repositioning" is also a second-order change. This means that it leads to a change in the identity of a project.

11.5.1 Project Repositioning: Objectives and Process

In the case of a project crisis, an existential threat of to a project is the cause for the change "Project repositioning". This threat can arise as a result of project errors, financial losses, bad relations with project stakeholders, etc.

The objectives of repositioning a project are a fundamental improvement of the project results, the securing of positive project ratios, and the recovery of the potential for an efficient continuation of the project. This requires a strategic and cultural reorientation of the project.

All project dimensions are to be considered in project repositioning. It is necessary to focus on essential project objectives and project tasks. This also makes it possible to reduce project costs and duration and contribute to the reduction of project complexity. A short-term results orientation is essential for repositioning. The frequency and intensity of project controlling is to be increased, which means that key project ratios are controlled on a daily basis in order to assess the success of the implemented measures. The project organization usually has to be redesigned. Ensuring the continued cooperation

Table 11.5: Sub-Process "Project Transforming"—Responsibility Chart

Tool/document:
1 ... Project change vision
2 ... Project change communication plan
3 ... Project manual

Legend
P .... Performing
C...Contributing
I... Being informed
Co...Coordinating

Process tasks		Roles							
		Project owner	Project manager	Project team	Project team members	PM-Consultant	Expert pool managers	Representatives of project stakeholders	Tool/Document
1	Planning project transforming								
1.1	Interrupting the project routine	P	C	C		C	I	I	
1.2	Analyzing the project situation	C	Co	P		C			
1.3	Drafting a project change vision	C	Co	P		C			1
1.4	Building awareness for the urgency of the project change	P	C	C		C	C	C	
1.5	Establishing competences for project transforming		Co		P	C		C	
1.6	Planning project change communication	P	Co	C		C	I	I	2
1.7	Planning the implementation of project transforming	C	Co	P		C	I		
2	Implementing project transfomring								
2.1	Replanning the project objectives, project strategies	C	Co	P		C			
2.2	Replanning the project scope, schedule, costs, etc.		Co	P		C	C		
2.3	Redesigning the project organization	C	Co	P					
2.4	Redeveloping the project culture	C	Co	P					
2.5	Redesigning the project context relations	C	Co	P		C	C	C	
2.6	Documenting new project structures, context relations	I	Co	P					3
2.7	Performing project change communication	P	Co	C		C	I	I	
2.8	Planning project stabilizing	C	Co	P					
3	Project stabilizing								
3.1	Further developing project structures	I	Co	P					
3.2	Further developing project context relations	I	Co	C					
3.2	Further developing project context relations	I	Co	C					
3.3	Performing project change communication	P	Co	C		C	I	I	
3.4	Closing project transformation	P	Co	C		C	I	I	
4	Performing content work		Co		P			P	

of qualified members of the project organization is a key challenge. Qualified project team members are often lost, and there is a large integration effort for new members.

Project crisis communication is also challenging. It is to be decided when which stakeholders are to be provided with information regarding the project crisis and by which communication media. The change "Project repositioning" is to be explicitly defined and explicitly ended. Repositioning a project is a project management sub-process with the phases "defining a project crisis, ad-hoc measures", "analyzing causes, additional measures 1", "additional measures 2", and "stabilizing the project" (see Fig. 11.8). The tasks of these project phases and the responsibilities for their fulfillment are shown in Table 11.6.

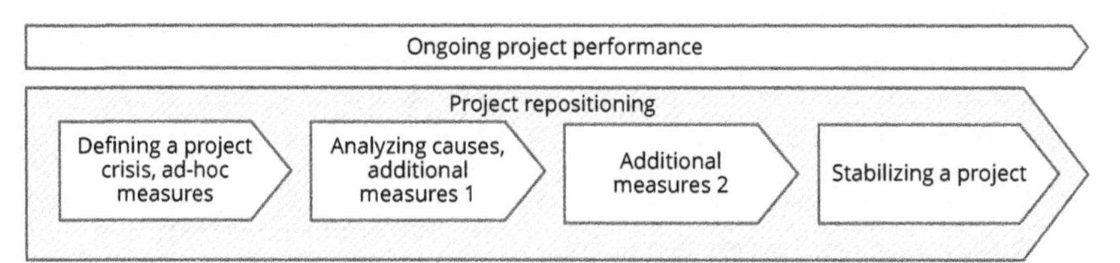

Fig. 11.8: Phases of the sub-process "Project repositioning".

Table 11.6: Sub-Process "Project Repositioning"—Responsibility Chart

Tool/document:
1 ... Situation analysis
2 ... Action plan
3 ... Communication plan
4 ... Project manual

Legend
P .... Performing
C...Contributing
I... Being informed
Co...Coordinating

Process tasks		Roles: Project owner	Project manager	Project team	Project team members	PM-Consultant	Expert pool managers	Representatives of project stakeholders	Tool/Document
1	Defining a project crisis, ad-hoc measures								
1.1	Interrupting the project routine, defining a project crisis	P	C	C			I		
1.2	Performing a rough situation analysis	C	Co	P		C			1
1.3	Planning ad-hoc measures	C	Co	P		C			
1.4	Deciding ad-hoc measures	P	Co	C			C		2
1.5	Performing ad-hoc measures		Co		P	C		C	
1.6	Controlling ad-hoc measures	P	Co	C					
1.7	Planning project crisis communication	C	Co	P		C			3
1.8	Performing first project crisis communication	P	Co	C		C	I	I	

(continues on next page)

Table 11.6: Sub-Process "Project Repositioning"—Responsibility Chart *(cont.)*

Tool/document: 1 ... Situation analysis 2 ... Action plan 3 ... Communication plan 4 ... Project manual Legend P .... Performing C...Contributing I... Being informed Co...Coordinating Process tasks		Roles							
		Project owner	Project manager	Project team	Project team members	PM-Consultant	Expert pool managers	Representatives of project stakeholders	Tool/Document
2	Analyzing causes, additional measures								
2.1	Analyzing causes of the crisis	C	Co	P		C		C	
2.2	Planning crisis management strategies		Co	P		C	C		
2.3	Deciding crisis management strategies	P	Co						
2.4	Planning additional measures 1-m	C	Co	P		C	C		2
2.5	Performing additional measures 1-m		Co		P	C		C	
2.6	Controlling additional measures 1-m	P	Co	C					
2.7	Communicating results of additional measures 1-m	C	Co		P	C	I	I	
3	Additional measures 2								
3.1	Planning additional measures n-x	C	Co	P		C			
3.2	Performing additional measures n-x		Co		P	C		C	
3.3	Controlling additional measures n-x	P	Co	C					
3.4	Communicating results of additional measures n-x	C	Co		P	C	I	I	
4	Project stabilizing, closing the project repositioning								
4.1	Further developing project structures, context relations	C	Co	P		C		C	
4.2	Documenting new project plans in the project manual		Co	P		C			4
4.3	Reflecting lessons learned	C	Co	P		C		C	
4.4	Communicating the closing of the project repositioning	C	Co		P	C		I	
5	Performing content work (parallel)				P			P	

The definition of a project crisis is a central task in the repositioning of a project. A project discontinuity is not defined because of objective criteria such as project ratios, but is the result of a communication process in the project. It is the conscious construction of a crisis as a project reality. It is only through the definition of a discontinuity that a specific meaning is given to this situation. The designation of a situation as a crisis or a chance acts as a "label", which ensures special organizational attention. This label legitimates radical measures to overcome the discontinuity.

In order to reposition a project, immediate measures must be planned and implemented, the causes of the crisis or the chance must be analyzed, possible coping strategies must be planned, and additional measures must be planned, implemented and controlled iteratively in several phases. The repositioning of a project can lead to a new project owner, project manager, or individual project team members and necessitates a redefinition of the project objectives and a redesign of the project stakeholder relations. A new project identity is developed. This should provide the basis for a successful continuation of the project.

In the course of stabilizing a project, it is to be agreed as to which new project rules and project values apply. Like the definition of a project crisis, its ending also represents an act of symbolic management. The ending of the project crisis should be performed as early as possible and as late as necessary.

Managing a project chance by repositioning is similar to managing a project crisis. The causes, on the one hand new potentials and on the other existential threat, and the objectives are different. However, the basic process of repositioning a project is similar.

11.5.2 Project Repositioning: Organization and Quality

Project Repositioning: Organization

As a rule, for repositioning a project, external experts (e.g., law experts and PR experts) are required, in addition to the members of the project organization, to provide know-how in the short term. If the project owner and the project manager are not themselves the cause of the project crisis, they should retain their roles in the management of the project discontinuity. However, managers of the permanent organization can be involved in the management of the project crisis and be given decision-making powers.

Decision-making powers are needed to be able to react in the short term, and social relations are necessary to ensure the acceptance of the management measures.

The communication structures of the project are to be redesigned. The extent and the intensity of project communication increase during the management of the project discontinuity. The objectives and participants of project meetings have to be redefined. The basis for meetings are project controlling reports, which are to be produced in the short term.

The communication of the causes of a project crisis, the management strategy, and the status of management measures to members of the project organization and representatives of project stakeholders is to be planned and professionally implemented. The success of the repositioning of a project is also dependent on the handling of project information. The basic communication strategy must be planned. The formats and frequency of dissemination of information and the target-group–specific information must be determined.

Project Repositioning: Quality

Crises and chances of projects often occur, but in practice they are hardly ever professionally managed. The repositioning of a project has only been formalized in a few project-oriented organizations. Relevant organizational competencies are usually lacking. The quality of the management of project discontinuities depends on the organizational competencies as well as on the individual competencies of the employees.

11.6 Project Transforming and Project Repositioning: Methods

Methods that can be used for project transforming and project repositioning are the analysis of the causes of a project crisis or project chance and the planning and controlling of immediate and additional measures. The methods of change management can also be used in project changes. The methods for managing a project crisis or a project chance are basically the same, but the reasons for applying them are different.

Analyzing the Causes of a Project Crisis or Project Chance

The basis for planning objectives, strategies, and measures for the transforming or repositioning of a project is an analysis of the causes which led to a project discontinuity. The strengths and weaknesses of the project structures and relations to project contexts are to be identified. A view of the project situation common to all members of the project organization is to be constructed. All project dimensions must be considered in the analysis.

Planning and Controlling Immediate and Additional Measures

When planning and controlling measures, a distinction must be made between immediate measures and additional measures for managing project discontinuities. Immediate measures effective in the short-term and medium-term are to be planned and implemented on the basis of a brief analysis of the project. Additional measures can then be agreed upon in several iterations, based on a more comprehensive analysis of causes and planning of management strategies.

The measures for managing a project discontinuity can be planned and controlled in a To-Do list. It is not recommended to include the measures in the project plans during the transforming or repositioning. On the contrary, the revision of the project plans is itself a measure to be implemented in a later phase of the project change.

Applying Change Management Methods

The perception of the management of project discontinuities as project changes promotes the application of change management methods (see Chapter 14) in transforming or repositioning projects.

For example, to formulate a project change vision, it is advisable to provide project stakeholders with an understanding of the urgency of the project change, to plan quick wins for the project change, and to develop an appropriate change communication plan for the project.

11.7 Project Transforming: Case Study "Establishing a Hospital"

In the case study "Establishing a hospital", the transforming of a project into a chain of two projects and a program for establishing a hospital is described.

Solution "Hospital"

A hospital holding organization, which operates several hospitals in Vienna, Austria, decided in 2009 to set up a new hospital. The "solution", which was to be provided by a project, is briefly described below. The following facts convey the size of the hospital:

> Number of beds: 800
> Number of employees: 2,000
> Area: Hospital 160,000 m^2, parking area 45,000 m^2
> Initial investment: €900 million

New concepts for hospital management—for example, for bed management and operating room management—were to be implemented in the hospital. The latest technologies in medical technology, information, and communication and facility management were to be employed. Two thousand employees were to be recruited and trained so that they would be ready for action at the time of the planned commissioning in 2016.[5] Coordination with other hospitals was to take place, as existing departments of other hospitals were planned to move to the new hospital.

A specific feature of the proposed solution was consideration of the principles of sustainable development in the planning and construction of the hospital.

Originally Planned Project

Based on a concept developed in 2007 and 2008, the project "Building a hospital" was started in 2009. It included the project phases "preliminary design", "design", and "construction" (see Fig. 11.9).

[5] The commissioning date was shifted to 2018 during the implementation.

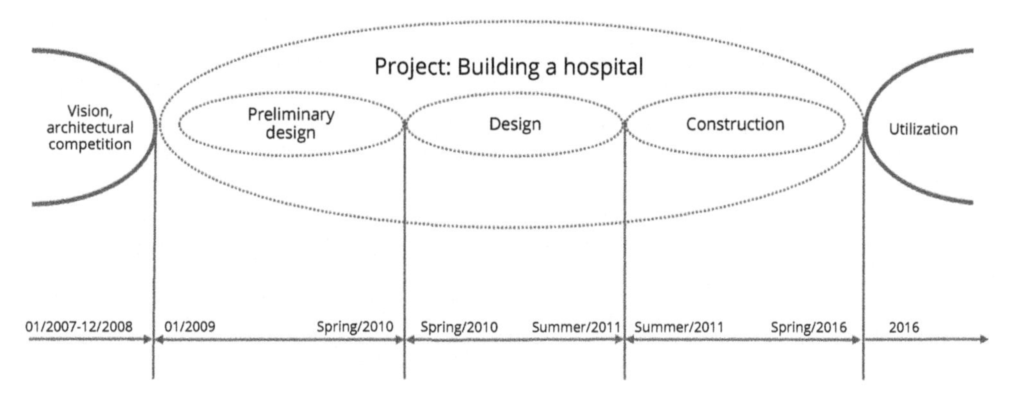

Fig. 11.9: Boundaries of the project "Building a hospital".

As can be seen from the terms chosen for the project phases, the project was perceived as a "construction project". This resulted in very narrow perception of the project boundaries, limited to the fulfillment of planning the object to be constructed and the construction work. No holistic definition of the solution "hospital" was provided. This would also have considered the hospital services, medical technology, information and communication technology, hospital organization, hospital personnel, facility management, and the stakeholder relations of the hospital.

Since these objects of consideration were not part of the defined solution, they would have had to be dealt with separately from the project, as soon as consciousness of them had arisen. This would have led to suboptimal results, since no integrative consideration of all solution components would have taken place. The excessively long project duration of eight years, without corresponding quality assurance by stage gates during this period, resulted in a high project risk.

Because of these facts, the general management of the hospital holding organization decided in 2011 to transform the project into a chain of projects and a program.

Transforming the Project "Building a Hospital"

In transforming the project "Building a hospital", two main objectives were pursued:

> The solution "hospital" should be viewed holistically and integratively in order to achieve an optimized outcome.
> Several projects and a program should be distinguished in order to reduce management complexity and to establish stage gates between projects for quality assurance.

The process "Project transforming" was performed by a team consisting of the project owner, the project manager, and his deputy, as well as the project team members. Consultants from RGC supported this transformation.

The change process started in the spring of 2011 and lasted about five months. On the basis of a project analysis, a chain of projects and a program were defined, consisting of a project "Conceptualizing the hospital", a project "Planning the hospital", and a program "Establishing the hospital" (see Fig. 11.10).

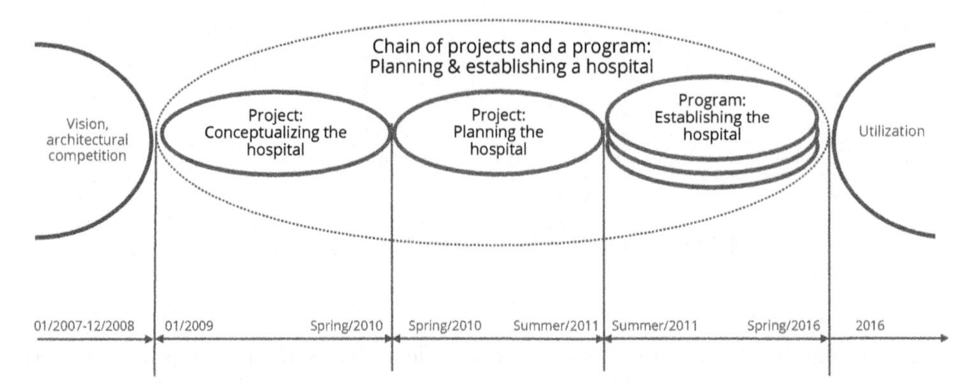

Fig. 11.10: Chain of projects and a program for "Planning and establishing the hospital".

Since the transformation took place during the project "Planning the hospital", the project "Conceptualizing the hospital" was only included in the project chain for the purpose of historiography. After defining the chain, the project "Planning the hospital" was restructured. In particular, its results were defined as specified requirements for the hospital and the initiated program. In September 2011, the program was formally launched.

Results of the Project Transformation

The program structure, the program organization, and a summary of the changed project dimensions are presented below as examples of the results of project transformation.

The structure of the program "Establishing the hospital" is shown in Figure 11.11. A chain consisting of a planning project and a preparation project was defined for each object of consideration—namely ICT, personnel, organization, and facility management.

This results in a stage gate at the end of each planning project. The project "Constructing the hospital" was planned to be performed in parallel. The subsequent commissioning was planned to take place in two overlapping projects.

Figure 11.12 shows the program organization chart. This shows the composition of the extensive program team. The projects active during the program starting are also shown.

The project dimensions changed as a result of the project transformation are summarized in Table 11.7.

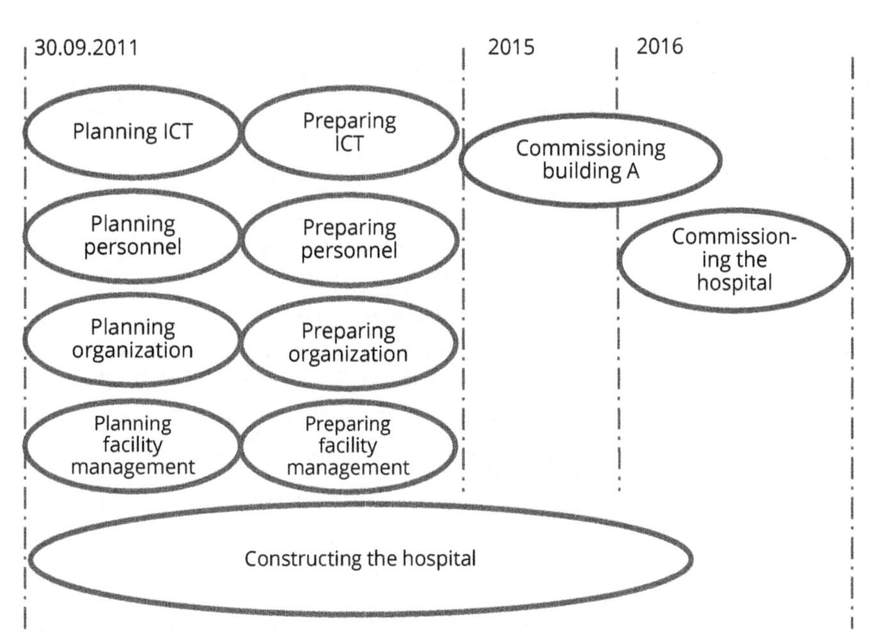

Fig. 11.11: Structure of the program "Establishing the hospital".

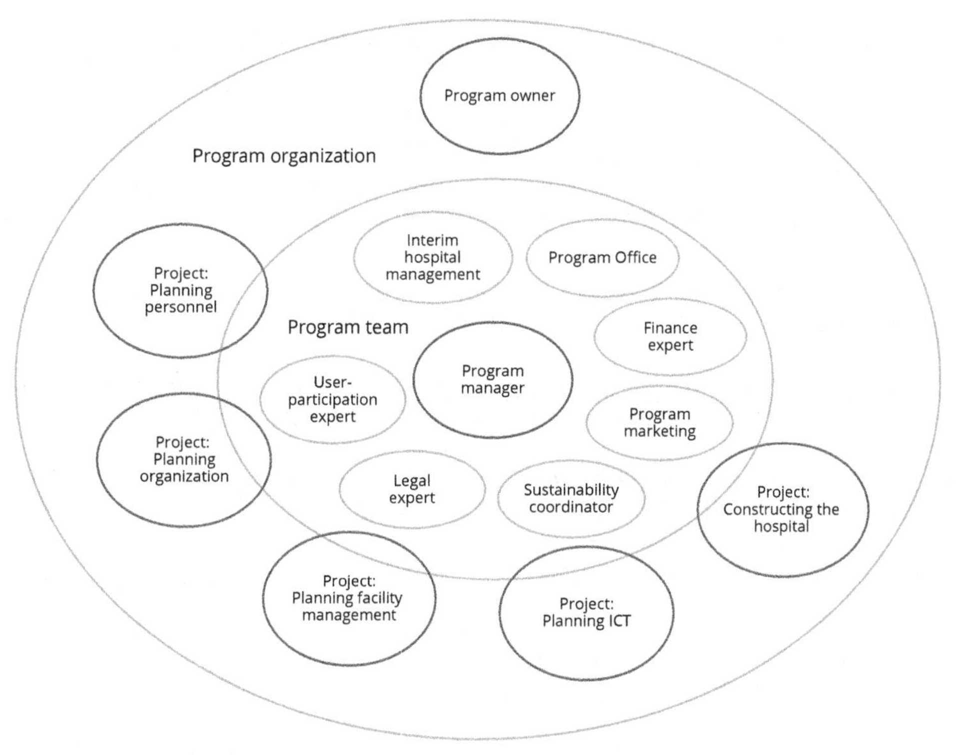

Fig. 11.12: Program organization chart "Establishing the hospital".

Table 11.7: Project Dimensions Changed as a Result of Project Transforming

Dimensions changed as results of project transforming	
Objectives	> Differentiation between investment objectives, program objectives and project objectives > Differentiation in economic, ecological and social objectives
Scope	> Construction of the hospital plus consideration of the hospital services, organization, personnel, medical technology, ICT, etc.
Plans	> Short-term project plans, mid-term program plans, long-term cost-benefit analyses for the investment > Consideration of opportunity costs > Operationalization of the principles of sustainability
Risk	> Risk minimization due to reducing the high complexity of a "big project"; defining several manageable projects within a chain of projects and a program > Different risk analyzes for the investment, the program and the projects
Culture	> Transparency, risk-orientation, empowerment (e.g. due to several projects in the program) > Clear distinction between the values of the hospital, of the program and of the single projects
Organization	> Establishment of a program management team plus several project organizations (with different project owners, project managers, project teams) > Integration responsibility of the program owners and the program managers > Differentiated communication formats for the program and the different projects
Personnel	> Additional personnel for the management of the program and the projects necessary, new tasks to be performed > Humane management approach applied, because manageable projects were defined
Stakeholder relations	> Identification of different stakeholders of the hospital, the program and the single projects > Different strategies for managing the relations to the stakeholders of the hospital, the program and the projects

Literature

Lewin, K.: Frontiers in Group Dynamics. Concept, Method and Reality in Social Science: Social Equilibria and Social Change, *Human Relations,* 1(1), 5–40, 1947.

Reibnitz, U. v.: *Szenario-Technik: Instrumente für die unternehmerische und persönliche Erfolgsplanung,* Gabler, Wiesbaden, 1992.

12 Sub-Process: Project Closing

When the content-related objectives of a project have been achieved, the project no longer has a right to exist as a social system. It is therefore to be closed. Project closing is a sub-process of the business process "Project managing" (see overview below).

The objective of project closing is the closing of a project from a managerial point of view. The know-how gained in the project is to be transferred to the project-oriented organization and other projects, the post-project phase is to be planned, the project performance is to be assessed, and symbolic closing actions are to be performed.

In addition to the members of the project organization, representatives of the project-oriented organization and of stakeholders are to be involved in project closing to ensure individual and organizational learning.

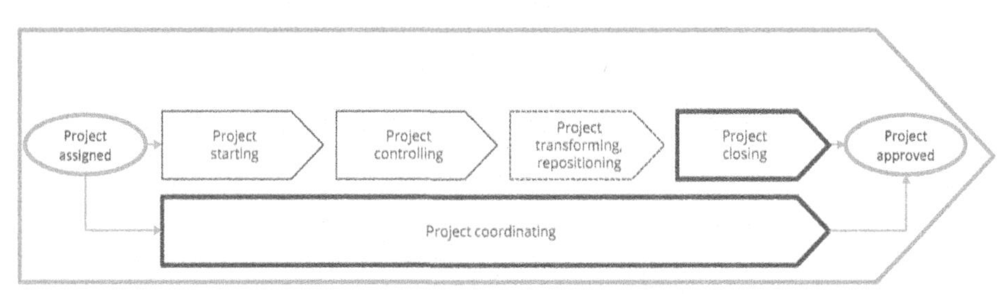

Overview: "Project closing" as a sub-process of the business process "Project managing".

12 Sub-Process: Project Closing

12.1 Project Closing: Objectives and Process

12.2 Project Closing: Organization and Quality

12.3 Methods: Planning the Post-Project Phase, Assessing Project Performance, Transferring Know-How, Performing Symbolic Closing Actions

12.1 Sub-Process: Project Closing

When the content-related objectives of a project have been achieved, the project no longer has a right to exist as a social system. This contradicts the basic objective of social systems: to ensure their survival. Project closing therefore requires a large emotional effort, just like project starting.

Challenges of project closing are that

> some often unattractive, residual tasks still have to be performed,
> some project stakeholders—for example, the users of the solution developed or some project team members—still have an interest in the continued existence of the project (see Excursus: Closing a project in context of starting new operations), and
> project team members have already been deployed in new projects and are therefore scarcely available for the tasks of project closing.

The processes "Project starting", "Project controlling", "Developing a solution", "Administrating a project", "Change managing", and "Benefits realization controlling" precede project closing. "Project coordinating" is still running in parallel. Ideally, the project closing process will have been planned and budgeted during project starting. Project closing begins with the decision to close a project and ends with the formal project acceptance by the project owner. Project closing takes one to two weeks, depending on project size.

Excursus: Closing a Project in the Context of Starting New Operations

Project closing can be delayed by the possible fear of a customer or the users of a developed solution of not having been delivered a qualitatively appropriate solution or of not being able to work with the solution.

Since the project organization is fundamentally interested in completing a project as quickly as possible, measures must be taken to provide the customer or the users with the necessary confidence in a solution and the competencies to apply the solution. At the same time, the closing of a project represents the start of new operations of the customer or of the users of a solution (see Fig. 12.1). Potentials for the planning of the project closing can be provided by a joint consideration of the closing situation and the starting situation. In this way, it is possible to plan processes for the post-project phase together with representatives of the customer or users and to provide relevant methods for this. Support functions for the application of a solution during operations can also be agreed upon.

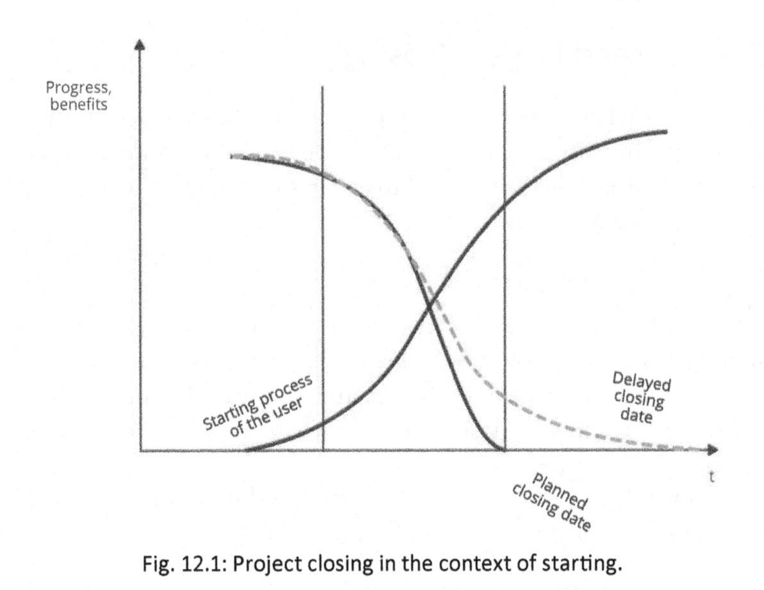

Fig. 12.1: Project closing in the context of starting.

12.2 Project Closing: Objectives and Process

Project Closing: Objectives

The objectives of project closing are the content-related and the emotional closing of a project. The closing should free up resources and energy for new projects. The know-how gained in the project is to be transferred to the project-oriented organization and other projects by means of project documentation and exchange of experience. A contribution to the knowledge management of the project-oriented company is to be made. In project closing, the content-related and time-related project boundaries have to be finally decided. Objectives and contents that will not be part of the project are either removed or are defined as objectives and tasks of the post-project phase. The project end is to be communicated both internally and externally. Possible symbolic closing actions are a final presentation and a "social" project closing event.

The objectives of the sub-process "Project closing", differentiated into economic, ecological, and social objectives, are described in Table 12.1.

Project Closing: Process

The tasks, roles, and responsibilities for fulfilling the task of project closing are shown in the responsibility chart of Table 12.2 (on page 324). A draft for the project closing should be produced during project starting, since the costs of project closing must also be taken into account in the project costs plan. This draft is to be concretized by the project team after the decision of the project owner to close a project.

Table 12.1: Objectives of the Sub-Process "Project Closing"

Objectives of the sub-process: Project closing
Economic objectives > Measures planned for the post-project phase > Updated version of the cost-benefit analysis or business case analysis provided > Know-how transferred to other projects and into the project-oriented organization > Project closing report prepared > Agreement made regarding the controlling of the benefits realization > Project stakeholder relations dissolved and new relations of the project-oriented organization established > Final project marketing performed > Project assessed and contributions of the members of the project organization to the project success evaluated > Project team dissolved > Project approved by the project owner > Project closing performed efficiently
Ecological objectives > Ecological consequences of project closing optimized
Social objectives > Project personnel led appropriately during project closing > Project stakeholder involved in project closing

Preparation for project closing includes planning the remaining work as well as planning the post-project phase, assessing the project success, dissolving existing stakeholder relations and establishing new ones, and preparing the project closing report. Various communication formats can be used during project closing. A project closing workshop with the project team, a final meeting with the project owner, and an exchange of experience workshop are possible formats. The project closing process is supported by symbolic actions such as the sending of thank-you letters or the giving of gifts, or the organization of a final "social" closing event.

The follow-up phase of the project closing process consists of completing and distributing the documentation, final project marketing, and closing the project cost center.

The closing of different project types involves different tasks (see Table 12.3 on page 325).

12.3 Project Closing: Organization and Quality

Project Closing: Organization

In addition to the members of the project organization, representatives of the project-oriented organization must be included in the project closing to secure individual and organizational learning. Stakeholder representatives are also involved in the dissolution of existing stakeholder relations or the establishment of new ones.

Various communication formats must be used for project closing. It is advisable to hold a project closing workshop. This makes it possible to use teamwork to assess the

Table 12.2: Sub-Process "Project Closing"—Responsibility Chart

Tool/document:
1 ... Structures for project closing
2 ... Invitation of participants for the project closing workshop
3 ... Info-material for the project closing workshop
4 ... Project closing reports

Legend
P .... Performing
C...Contributing
I... Being informed
Co...Coordinating

Process tasks		Roles						
		Project owner	Project manager	Project team	Project team members	Expert pool managers	Project stakeholders	Tool/Document
1	Planning and agreeing project closing							
1.1	Planning the project closing process		P	C				
1.2	Agreeing on the project closing process	C	P					1
2	Preparing the project closing							
2.1	Planning measures for the post-project phase		P		C			
2.2	Preparing the assessment of the project sucess		P		C		C	
2.3	Preparing the dissolution and establishment of the stakeholder relations		P		C			
2.4	Creating a draft of the project closing report		P		C		C	
3	Performing project closing communication							
3.1	Distributing info-material to participants of the closing communication	I	P	I			I	2, 3
3.2	Performing project closing workshop	C	Co	P			C	
3.3	Performing a final project owner meeting	D	C		C			
3.4	Performing an exchange of experience workshop		P			C	C	
3.5	Performing a "social" closing event	C	Co	P		C	C	
4	Following up project closing							
4.1	Dissolving and establishing stakeholder relations		P		C		C	
4.2	Finalizing project closing reports		P		C		C	4
4.3	Providing final data for the portfolio data base		P					
4.4	Distributing the project closing reports	I	P		I	I	I	
4.5	Closing the cost center	I	P					
4.6	Performing a final project marketing	I	P					
5	Performing content work (parallel)				P		P	

Table 12.3: Challenges in Closing Different Project Types

Closing different project types	
Successful project or unsuccessful project	> The assessment of the project success and the assessment of the contributions of members of the project organization to the project success are easier when closing a successful project. > To have to deal with failure during project closing produces a high level of social complexity.
Conception project or implementation project	> The basis for transferring know-how gained in a conception project into the following implementation project is to be created in project closing. > This is enabled by appropriate project documentation and the involvement of persons who will later work in the implementation project.
Repetitive project or pilot project	> In a repetitive project, learning for similar projects must be organized in project closing. > Knowledge management is particularly important in pilot projects.

project's success and the achievements of the members of the project organization. The workshop also provides an opportunity to inform team members about the allocation of personnel after the project end.

Project Closing: Quality

Just as with project controlling, the quality of project closing is ensured by the consistency of the project management methods used. For example, the project objectives plan and project stakeholder analysis are also methods of project closing. The project performance can be assessed during project closing on the basis of the project objectives agreed upon during project starting and adapted during project controlling. A "retrograde definition of meaning" by means of a redefinition of project objectives during project closing is often necessary. A holistic definition of project boundaries during project starting is the basis for successful project closing. There can only be "happy stakeholders" if all the components of a solution have been considered when defining project objectives.

The project stakeholder analysis enables the dissolution of existing stakeholder relations and the establishment of new relations of an appropriate quality of the project-oriented organization for the post-project phase. It is also the basis for final project marketing.

Appropriate social competencies are required to enable social project closing. To give and receive feedback as an individual and as an organization and to discuss positive and negative outcomes requires appropriate competencies.

Formal project closing is rarely carried out in practice. The necessary culture is often lacking. The right of the members of the project organization to receive appropriate feedback and to achieve emotional closure are often in conflict with the dynamics of daily business and an underestimation of the benefits of professional project management (see Table 12.4).

Table 12.4: NO-NOs for Project Closing

NO-NOs for project closing!
> Not defining a clear project end event. This leads to "never-ending stories".
> Not planning measures for the post-project phase. This might reduce the benefits of the solution developed.
> Seamless transition to new objectives, no appreciation of the results achieved.
> Not preparing a project closing report.
> Lack of strategic orientation regarding the dissolving of project relations with customers, suppliers, etc.
> No reflection on the learning experience of the project.
> Giving no feedback to members of project organization; no organization of individual learning.
> Individual rather than collective leave-taking of project team members.
> Not closing the project cost center.

12.4 Methods: Planning the Post-Project Phase, Assessing Project Performance, Transferring Know-How, Performing Symbolic Closing Actions

Specific methods and techniques, such as developing project closing reports and assessing project performance, can be used to fulfill the tasks of project closing. The project plans used during project starting and project controlling can be used again. For example, the project performance can be assessed by comparing the planned project objectives in the project objectives plan with the realized project objectives, or the project stakeholder analysis can be used to dissolve project stakeholder relations. Professional project management is characterized by a consistent project management approach of this kind.

Planning the Post-Project Phase

The post-project phase must be planned during the project closing process. Techniques such as a project stakeholder analysis, an investment analysis, and a To-Do list can be used.

The basis for planning the post-project phase is the planning of project stakeholder relations. As the project as a temporary organization is dissolved during project closing, the project's relations with project stakeholders are also to be dissolved. However, new relations between former project stakeholders and the organizational units of the project-oriented organization can be established. For example, it can be in the interest of the project-oriented organization to establish a new long-term supplier relation with a supplier whose cooperation was secured in the project for the first time. The responsibility for managing the relationship with the supplier in the post-project phase is, for example, to be assigned to the purchasing department.

An example of using the project stakeholder analysis during project closing is shown in Figure 12.2 for the project "Developing Values4Business Value".

Case Study: Developing Values4Business Value—Project Stakeholder Relations During Project Closing

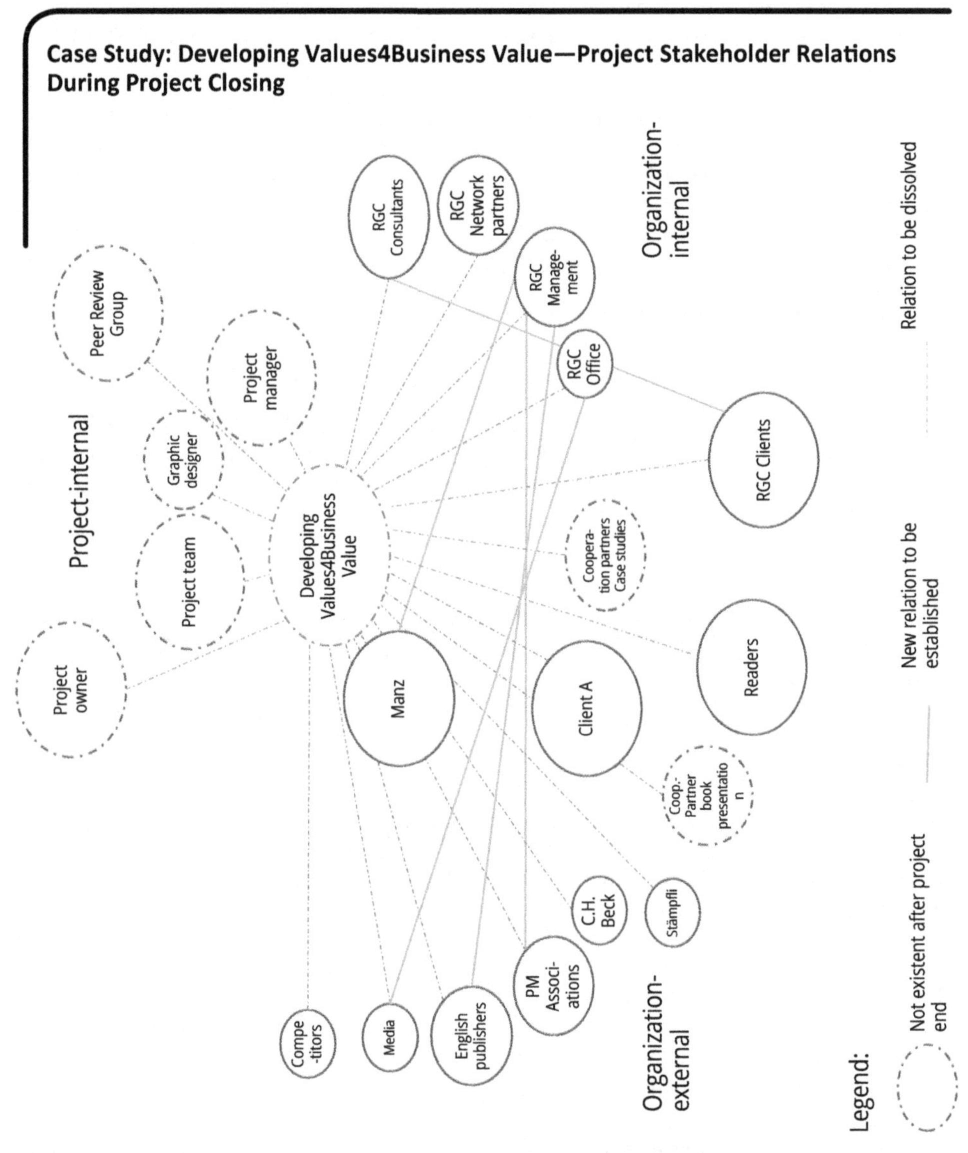

Fig. 12.2: Managing stakeholder relations during the project closing of the project "Developing Values4Business Value".

The closing of the project "Developing Values4Business Value" took place three months after the text of this book was completed. Project closing had already been planned at the beginning of March 2017.

On the one hand, measures were planned for the dissolution of stakeholder relations. This dissolution is symbolized by dotted lines between the project and the project stakeholders. A project closing workshop, including an assessment of the project performance and mutual feedback, was planned for closing the project-internal relations. It was planned to invite representatives of internal and external project stakeholders to the book presentation. VIPs and members of the peer review group were to be "made happy" with a copy of *PROJEKT. PROGRAMM.CHANGE.*

On the other hand, it was planned to establish new relations between representatives of the permanent organization of RGC and selected "former" project stakeholders. For example, the RGC management was to establish relations with Manz Verlag, the publishing house Taylor & Francis, and the project management associations. The consultants were, of course, to manage customer relations in the new context of "Values4Business Value"; the RGC Office employees were to manage the new media contacts.

Measures for managing stakeholder relations are to be planned and implemented. Methods for the dissolution of existing project stakeholder relations are letters of thanks, gifts, and a "social" project closing event. In this way, appreciation for good cooperation with customers, partners, and suppliers in the project can be expressed. A final "social" event enables emotional closure of a project and supports the dissolution of the project team.

The tasks to be performed after the project end and the responsibilities for these tasks are to be planned in a To-Do list "post-project phase". It must be ensured that, even after the project organization has been dissolved, the performance quality of tasks required by the solution implemented by the project is assured. Typical tasks of the post-project phase are, for example, maintenance tasks and further training of users of a new solution. In the To-Do list "post-project phase", it should also be specified whether benefits realization controlling after the project end is planned.

As an example, Table 12.5 shows the To-Do list "post-project phase" for the project "Developing Values4Business Value".

Assessing the Project Performance

In the project closing process, the project success and the performance of the members of the project organization must be assessed.

The project objectives adapted during the implementation of the project are the basis for the assessment of the project performance. It is important to provide the solution that the project owner and project stakeholders expect at the end of the project, rather

Case Study: Developing Values4Business Value—To-Do in the Post-Project Phase

Table 12.5: To-Do List "Post-Project Phase" for the Project "Developing Values4Business Value"

To-Do List "Post-Project Phase"
Developing Values4Business Value

Process	Date of arrangement	Responsible	Completion date
Additional information of RGC networking partners about "Values4Business Values"	3/1/2017	All consultants	July 2017
Reflecting on the application of "Values4Business Values" in consulting contracts	3/1/2017	Management and consultants	after July 2017
Adapting additional seminars	3/1/2017	Consultants	after June 2017
Adapting the designs of consultancy services	3/1/2017	Consultants	after June 2017
Applying the further developed approaches for the RGC management	3/1/2017	All RGC employees	after June 2017
Reflecting on the application of the further developed approaches for the RGC management	3/1/2017	All RGC employees	20.12.2017
...			

A draft of the To-Do list "post-project phase" was developed in March 2017.

The To-Do list shows only a few measures, since a follow-up project "Stabilizing Values4-Business Value" was planned to further implement the change. The development of the English version of the book, further marketing of the management paradigm "Values4Business Value", and further development of the RGC process management approach, etc., were planned as contents of this follow-up project, as results of the work package "Planning a follow-up project".

than the one defined during project starting. Sometimes only when the project is completed it is clear which objectives have been achieved in the project. This clarification corresponds to a "retrograde definition of meaning" of the project.

The evaluation of the project performance can be done by the members of the project organization and by project stakeholders. In the case of contracting projects, the customer's opinion is of particular interest. Getting feedback from the customer's representatives is part of the quality management system in many companies. Various reflection techniques can be used to assess the performance of the project. Figure 12.3 shows a project performance matrix. In this matrix, the relation between the project results achieved and the process of cooperation in the project can be considered.

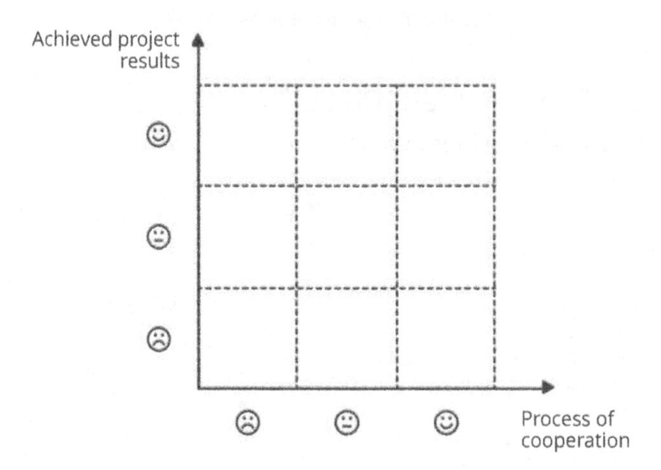

Fig. 12.3: Project performance matrix.

A more formal technique for assessing the performance of a project is a questionnaire. This must be completed by the members of the project organization. The evaluations are to be discussed together.

In order to assess the success of the project, a final adaptation of the cost–benefit analysis or business case analysis of the investment realized by a project is also required. The aim of this analysis is to take into account the final actual data of the project and current assumptions about the post-project phase. On the basis of these updated results of the investment analysis, it can be assessed whether a basis for "sustainable business value" has been created by the project.

Performance of Members of the Project Organization

During project closing, an explicit assessment of the performance of the members of the project organization can take place. The objective of these performance assessments is to appreciate the performance and to organize learning for the individual members of the project organization. This learning chance should be exploited by the project manager and the individual project team members, but also by the project owner.

The criteria for assessing the project performance of the individual members of the project organization are their contributions to the realization of the project. This creates a direct link between the project success and the project-related performance of the members of the project organization.

Assessment of the project manager of the project					
by ☐ the project owner ☐ the project team ☐ Others					
Criterion	Very poor	Poor	Avg.	Good	Very good
Designing the process "Project managing"					
Business value-orientation					
Appropriate use of project management methods					
Managing the project context relations					
Promoting Project team working					
Product and industry competences					

Fig. 12.4: Criteria for assessing a project manager.

Individual feedback and questionnaires can be used as techniques for assessing the performance of individual members of the project organization. Individual feedback can be given in individual meetings, but also in the project team as part of a project closing workshop. Prerequisites for the assessment of the project performance of individuals are appropriate social competencies of those who give and receive feedback.

A questionnaire for a 360° feedback to the project manager and the criteria for this assessment are shown in Figures 12.4 and 12.5.

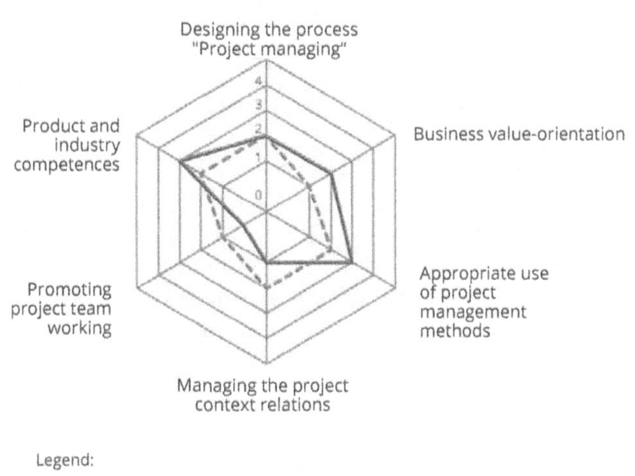

Fig. 12.5: Assessment of the project manager by different project stakeholders.

Performance assessments of the members of the project organization can be the basis for project-related bonuses. The results of the performance assessments are not only relevant to the project, but can also be used as information for planning personnel development and measures for the disposition of personnel in the project-oriented organization.

Transferring Know-How

The most important methods for transferring the know-how acquired in a project to other projects and to the project-oriented organizations involved are the project closing report and exchange of experience workshops.

The project closing report summarizes the main project results and the know-how acquired in the project. It provides final project information for all members of the project organization and selected representatives of project stakeholders. In principle, only one project closing report has to be prepared. However, target-group–specific adjustments are possible. If specific topics are of interest only for specific target groups, the project closing report must be supplemented by special reports for these target groups.

The project closing report can be structured according to project phases and project stakeholders. It should be formulated as briefly as possible and in a keyword-like manner. The final project plans are to be enclosed as a supplement to the final project report. Project plans are also an instrument of knowledge management for project-oriented organizations. They can serve as a basis for planning similar projects.

The know-how acquired in a project can be conveyed in exchange-of-experience workshops. However, the cost of such a workshop is no longer to be borne by the project, but is a general cost of the project-oriented organization. Final project marketing can be performed in the form of a project presentation and possibly through articles about the project in a newsletter and on the homepage of the project-oriented organization.

Symbolic Actions During Project Closing

Important symbolic actions for closing projects are the closing of the project cost center, the organization of a final "social" project event, and formal project approval by the project owner. The formal approval of a project by the project owner can be performed either by means of minutes or by an approval note on the project assignment. Project approval is the last event in the project. Upon project approval, the project manager and other project team members are formally relieved of responsibility for the project.

13 Program Initiating and Program Managing

If a program is needed to implement an investment, it must be initiated. As in project initiating, initial program plans have to be developed and program strategies have to be defined; the relations to projects of the project portfolio of the project-oriented organization have to be analyzed and a program proposal has to be created. In addition, the first projects of the program to be performed are to be initiated.

The business process "Program managing" includes the sub-processes program starting, program coordinating, program marketing, program controlling, program transforming, program repositioning, and program closing.

The objective of program managing is the integration of the projects of a program in order to realize the program objectives. The methods to be employed in program managing are similar to those in project managing. Specific challenges in program managing are the structuring of programs and the designing of program organizations.

The business processes "Program initiating" and "Program managing" are performed in accordance with the values underlying these processes. The use of program management methods is illustrated by a case study of a power supply company (on page 359).

Chapter 3	Strategic managing and investing	Strategic managing (period 1)	Strategic managing (period 2)
	Project	Conceptualizing an investment	Implementing an investment object
Chapter 3	Strategic managing and investing	Planning an investment object	Developing a solution Controlling the benefits realization
Chapter 4	Managing requirements & projects	Defining initial requirements	Controlling requirements
Chapter 5 / Chapter 6	Project initiating / Project managing	Project initiating	Project managing
Chapter 13	Program initiating and program managing	Program initiating	Programm managing
Chapter 14	Change initiating and change managing	Change initiating	Change managing

Overview: "Program initiating" and "Program managing" in context.

13 Program Initiating and Program Managing

13.1 Program Initiating

Program Initiating: Objectives

Program initiating is a business process of project-oriented organizations. The objectives of program initiating are:

> the decision to perform a comprehensive business process as a program is made,
> the program is assigned to the program manager,
> a basis for program starting is existing,
> a basis for starting the first projects of the program is existing, and
> selected stakeholders were involved in the initiating process.

A non-objective of program initiating is that detailed program plans were developed. These are developed in program starting.

Definition: Program initiating

Program initiating is a business process of project-oriented organizations. Its objectives are to decide to perform a comprehensive business process as a program, to assign the program to the program manager, and to ensure the existence of a basis for starting a program and starting the first projects of the program.

The reason for initiating a program is usually a "top-down" decision based on strategic management (see Chapter 3). A "bottom-up" need is an exception for initiating a program. However, a case that is often seen in practice is the transforming of an ongoing project into a program. In this case, the complexity and scope of the business processes to be performed were obviously underestimated at the beginning. A case study involving this situation is given in Chapter 11.

Program Initiating: Process

The business process "Program initiating" is shown in Figure 13.1. Responsibilities for fulfilling the process tasks are shown in the responsibility chart in Table 13.1.

The start event of program initiating is the decision to initiate a program. This requires the decision to implement an investment to have been made. The end event is that the program is assigned or not. Initiating a program involves the following tasks:

> constructing appropriate boundaries for the program and developing initial program plans,
> analyzing the relations of the program with other projects and programs in the project portfolio of the organization,
> developing program strategies,

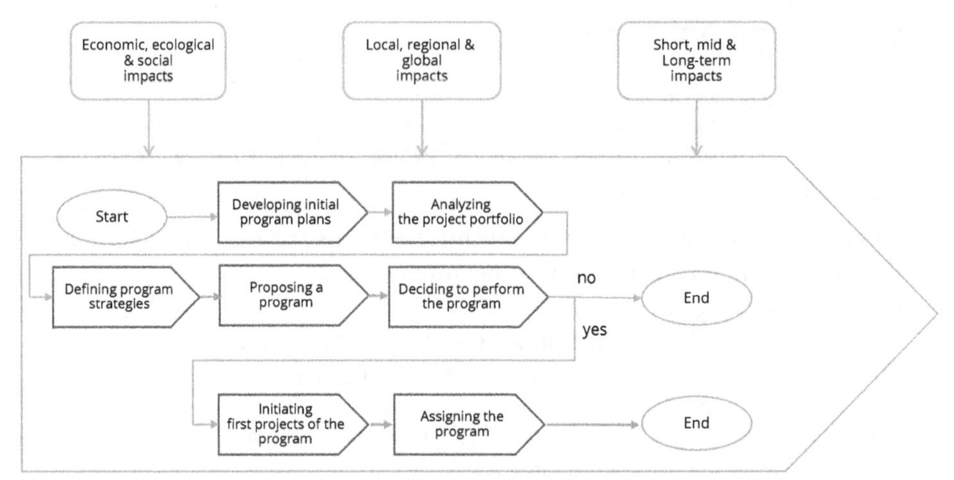

Fig. 13.1: Program initiating—Flow chart.

Table 13.1: Program Initiating—Responsibility Chart

Sub-process: Program initiating									
Legend: P ... Performing C ... Contributing I ... Being informed Co ... Coordinating	Roles								
Process tasks	Initiator	Initiating team	PM Office	Project Portfolio Group	Expert pool managers	Program owner	Program manager	Program stakeholders	Tool/Document
Preparing initial program plans		P	C		C			C	1
Analyzing the project portfolio		C	P						2
Defining program strategies	I	P	C						3
Proposing a program	I	P	C						4
Deciding to perform the program	I	C	C	P	I	I		I	5
Assigning the program			I		I	P	C	I	6
Initiating first projects of the program	I	P	C	C	C	I	C	C	7

Tool/Document:
1 ... Initial program plans
2 ... Project portfolio data base
3 ... Documentation of the program strategies
4 ... Program proposal
5 ... Minutes of the meeting
6 ... Program assignment
7 ... Initial project plans

> deciding that a program is the appropriate form of organization for performing the business process under consideration,
> initiating the first projects of the program, and
> assigning the program manager and the program team to the program.

The basis for developing initial program plans is the construction of the program boundaries. As with projects (see Chapter 6), program boundaries are appropriately set if they consider the necessary program objectives and program scope to meet the requirements of the investor and essential stakeholders. On the basis of the initial program plans, program strategies for the realization of the program objectives can be agreed upon, and the program owner, program manager, and program team can be selected.

Since the first projects of the program can be started immediately after program starting, these projects are also to be initiated within the framework of program initiating.

Program Initiating: Methods

The following methods can be used in program initiating:

> planning the objects of consideration, planning the program objectives, establishing the program strategies,
> developing the program breakdown structure, planning the program schedule,
> planning the program budget, analyzing the program risks, and
> analyzing the program stakeholders and creating the program organization chart.

Other methods for program initiating are analyzing the project portfolio and developing a program proposal. The structures of the program proposal and the program assignment should correspond to those of the project proposal and the project assignment. The program proposal shows the structures and contexts of the proposed program. In addition to the initial program plans, the attachments to a program proposal also include the results of the initiating of the first projects of the program. This information allows the project portfolio group to decide whether the program and the first projects of the program will be performed.

Program Initiating: Organization

The necessary roles for program initiating can be seen in the responsibility chart in Table 13.1. These are the initiator of the investment, the initiation team, the PM Office, the project portfolio group, the expert pool managers, the program owners to be selected, the program managers to be assigned, and selected program stakeholders. Ideally, the program proposal is prepared in cooperation with the designated program manager and in coordination with the selected program owner.

Because of the complexity of programs, it is advisable to implement the conceptualization of an investment as a project. The objective of such a conception project is the planning of the investment and the initiation of the program. The sequence of the conception project and the program results in a chain of a project and a program (see Fig. 13.2).

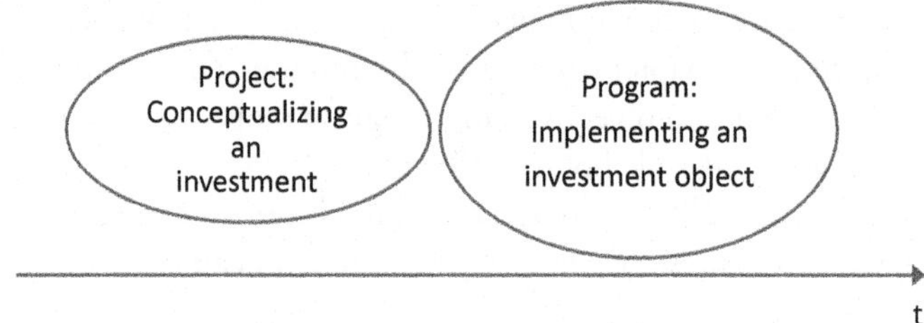

Fig. 13.2: Chain: Conception project and implementation program.

In the conception project, the investment object is planned and the initial plans of the implementation program are developed. In program starting, these plans are detailed and concretized. In order to ensure continuity, members of the organization of the conception project will also hold roles in the program organization.

13.2 Program Managing: Overview and Design of the Business Process

Program Managing: Objectives

The basic objective of the business process "Program managing" is to contribute to the successful performance of a program by professional management. Operationalization of this objective is made possible by differentiating the economic, ecological, and social objectives of program managing. These objectives are shown in Table 13.2.

Program managing is to be performed in addition to managing the individual projects of a program. The development of a "big program picture", the application of methods for program controlling, and use of program communication formats are tools for integrating the projects of a program. Compliance with content-related standards and project management standards in performing the projects of a program also has an integrating effect. Program managing creates an added value that cannot be achieved by just managing the individual projects of the program.

Dimensions of Programs to Be Considered

The dimensions of programs to be considered in program management are the same as in project management. The following structural dimensions of programs are to be managed:

> the solution requirements, the objects of consideration of a program, the program objectives, and the program strategies,

Table 13.2: Objectives of the Business Process "Program Managing"

Objectives of the process: Program managing
Economic objectives > Program complexity, program dynamics, and relations to program contexts managed > Program starting, program coordinating, program marketing, program controlling, and program closing performed professionally; possibly also program transforming or program repositioning > Economic impacts optimized for the change and for the implemented investment
Ecological objectives > Local, regional, and global ecological impacts of the program considered > Ecological impacts optimized for the change and for the implemented investment
Social objectives > Program personnel recruited and allocated, incentive systems applied, development of project personnel realized > Local, regional, and global social impacts of the project considered > Social impacts optimized for the change and for the implemented investment > Stakeholders involved in program management

> the program scope, the program schedule, the program personnel, the program resources, the program budget, and the program risk, and
> the program organization, the program culture, and the program infrastructure.

The following contextual dimensions of programs are to be managed:

> the pre-program and post-program phases,
> the program stakeholders,
> other projects and programs implemented at the same time,
> the objectives and strategies of the organization performing a program, and
> the investment implemented by a program.

Program Managing: Process

An overview of the process of program managing is shown in Figure 13.3. The start event of a program is the assignment of the program to the program manager and the program team by the program owner. The formal end event is the approval of the program by the program owner. The business process "Program managing" includes the sub-processes program starting, program coordinating, program marketing, program controlling, program transforming or repositioning, and program closing. The overview also shows how the principles of sustainable development are considered.

Because of the great importance of program marketing, it is a separate sub-process, unlike project managing. Professional communication of program objectives and program structures is a success factor for programs. Through program marketing, stakeholders can gain an understanding of the significance of a program, and

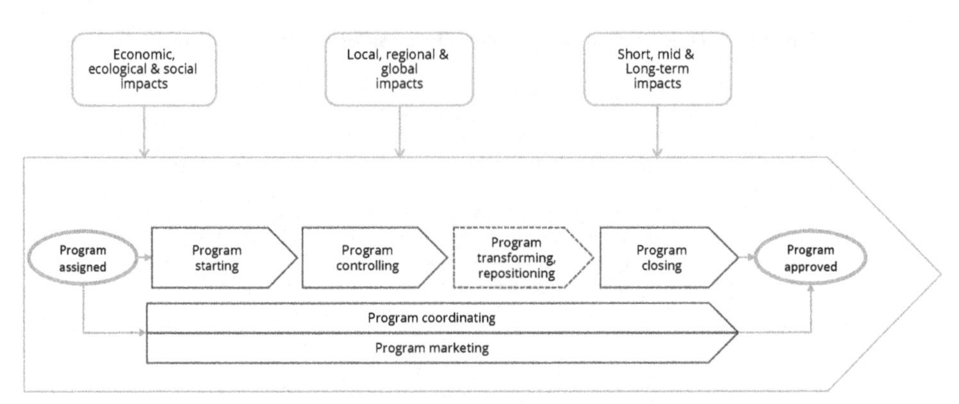

Fig. 13.3: Business process "Program managing"—flow chart.

management attention, know-how, and resources for the performance of the program can be secured. The marketing of the individual projects of programs has to be coordinated with the strategies and measures of program marketing.

Program starting and program closing are performed only once in a program. Program controlling is performed periodically. Program coordinating and program marketing must be carried out continuously during the program. Detailed descriptions of the objectives, tasks, responsibilities, and results of the sub-processes of program managing can be found in Section 13.3. The sub-processes of program managing are related. These relations are similar to those for project managing (see Chapter 6) and are therefore not described again.

Program Managing: Designing the Business Process

Program managing is to be designed according to the specific needs of a program. The use of methods for program managing, checklists, program communication formats, an appropriate program infrastructure, and possibly program consultants must be planned accordingly.

The definition of general rules regarding the program management methods to be used and of general structures of the program organization is a corporate governance task (see Chapter 16). These general rules are to be defined in conjunction with the general rules for projects.

The methods for program managing are to be distinguished according to the sub-processes of program managing. They correspond to the methods for project managing—that is, in analogy with projects, there is a program objectives plan, an objects of consideration plan, a program breakdown structure, a program bar chart, a program stakeholder analysis, etc. The recommendations given in Table 6.2 (on page 106) regarding the use of methods for project managing can therefore be applied to program managing. There is a difference, however, regarding the need to use the methods. Because of the high complexity and dynamics of programs, a "must" should apply to all methods.

The methods listed in Table 6.2 (page 106) and described in Chapters 7 to 11 are also relevant for programs. Section 13.4 discusses specifics regarding the use of methods for program managing.

Repetitive projects are performed often in programs. For example, in an ERP program of an international concern, projects for the implementation of the same solution in individual group companies are repetitive projects. Another example of a program for establishing a hospital is a set of repetitive organization projects performed for different hospital departments.

It is advisable to develop standards for carrying out repetitive projects in a program. Standard project plans (for example, standard work breakdown structures, standard work package specifications, or standard milestone plans) can be developed and used for the repetitive projects. For the fulfillment of the content-related processes of repetitive projects, standardized structures, methods, rules, and templates can be defined and prescribed (see Excursus).

Excursus: Standards for the Fulfillment of Content-Related Processes in Repetitive Projects

In a real estate program, which aimed at the renovation of a number of similar facilities, the following standards were developed and implemented for the fulfillment of content-related processes:

Standard process for performing a frequency analysis for a facility:

- A frequency analysis had to be carried out for each facility in order to be able to define the appropriate renovation objectives for each facility.
- The objectives, the processes, the responsibilities, and the results of the frequency analysis were specified in a process description.

Design manual as a standard for the design of the facilities:

- To support the investor's corporate identity, design standards were specified.
- Examples of this are the use of fonts and colors or the installation of vending machines in the facilities.

Standards regarding the rental of units of the facilities:

- Standards were developed to ensure good cooperation with the tenants of units.
- Examples of this are standard rental agreements and defined quality standards regarding possible tenants for the rental of units.

The objective of the use of program standards is to assure common procedures in the projects of a program. The projects of a program are linked by the application of standards. The relative autonomy of the projects of a program is therefore lower than those of projects which do not belong to programs. The development and use of program standards ensures quality and promotes organizational learning in programs.

The need for the use of information and technology (ICT) tools and consultants in programs is not significantly different from that for projects.

13.3 Program Managing: Sub-Processes

The objectives of the sub-processes of program managing are described in Table 13.3. These can be understood as key performance indicators for assessing the quality of these sub-processes. As an example, the tasks, results and responsibilities of the sub-process "program starting" are shown in a responsibility chart in Table 13.4 (on page 345).

13.4 Program Managing: Specifics in Applying Methods

Planning Program Objectives

Because of the large scope and long duration of programs, their objectives are more open and dynamic than those of projects. In the course of program initiating, initial program objectives are defined, which are detailed in program starting and periodically adapted in program controlling.

In program starting, the objective of the above-mentioned real estate program, for example, was the renovation of some 40 facilities. The number of facilities that could be renovated was, however, dependent on the financing possibilities. These were not yet clear during program starting. Securing financing for individual facilities was a task to be carried out as part of the program. Furthermore, the program later differentiated between large and small facilities, as these had very different financing requirements. In the end, about 20 small and seven large facilities were renovated within the scope of the program.

Programs are often implemented to pursue savings objectives. However, concrete savings opportunities are only developed by conceptual planning of possible solutions by projects during the performance of the program. Such conception projects are instruments for clarifying the program objectives during the program performance.

The medium-term nature and the dynamics of programs may necessitate the planning of alternative program objectives with different end-events and deadlines. Clear boundaries to the content and timing of programs should, however, be set. Programs should not be "open ended".

Table 13.3: Objectives of the Sub-Processes of Program Managing

Program managing: Sub-processes	Objectives
Program starting	> Program personnel recruited and allocated, incentive systems developed, development of project personnel planned > Information from the pre-project phase transferred into the program > Expectations about the post-project phase defined > Appropriate program plans developed > Measures planned for program risk management as well as for program discontinuities > Program organization designed, program culture developed > Objectives and structures defined for the following sub-processes of the program management > Communication relations with program stakeholders established > Initial program marketing performed > Documentation "Program starting" developed, distributed and filed > Program starting performed efficiently > Program stakeholders involved in program starting
Program coordinating	> Members of the program organization coordinated > Representatives of program stakeholders coordinated > Program resources ensured > Program progress ensured through coordinating the current projects and work packages > Finalized projects and work packages approved > Relations between projects of the program coordinated > Program coordinating performed efficiently > Program stakeholders involved in program coordinating
Program marketing	> Communication with members of the program organization and with stakeholders through appropriate media performed > Interest in the program ensured > Program communication plan updated constantly and basis for program marketing developed further (slogans, figures, etc.) > Program marketing performed efficiently > Program stakeholders involved in program marketing

(continues on next page)

Table 13.3: Objectives of the Sub-Processes of Program Managing ***(cont')***

Program controlling	> Program and project management personnel assessed and allocated, development controlled > Program status constructed by the members of the program organization > Measures for directing the program agreed and arranged > Program plans further developed > Program stakeholder relations analyzed and measures for further development agreed > Program controlling reports developed > Organizational learning of the program promoted > Program controlling performed efficiently > Program stakeholder involved in program controlling
Program transforming and program repositioning	> Program discontinuity managed successfully > Possible damages for the program minimalized and possible benefits for the program maximized > Basis for a successful continuation of the program established > Program transforming or program repositioning performed efficiently > Representatives of program stakeholders involved in program transforming or program repositioning
Program closing	> Program and project management personnel assessed and released > Content related remaining work finalized > As-is program documentation developed > Final program marketing performed > Agreement for the post-program phase reached (incl. controlling of the benefits realization) > Program success assessed, members of the program organization assessed > Program organization dissolved > Program closing report developed > Know-how transferred into the permanent organization and into other programs and projects > Program closing performed efficiently > Relations to program stakeholders dissolved

Table 13.4: Program Starting—Responsibility Chart

Sub-process: Program starting

Tool/Document:
1 ... List of applied methods for program managing
2 ... Invitation to participants to the program starting formats
3 ... Information material for the program starting formats
4 ... Documentation „Program starting"

Legend
P .... Performing
C...Contributing
I... Being informed
Co...Coordinating

Process tasks	Roles							
	Program owner	Program manager	Program Office	Program team (incl. Project managers)	Program team members	Expert pool managers	External stakeholders	Tool/Document
Planning program starting								
Checking program assignment and results of the pre-program phase		P						
Selecting start communication formats	I	P	C					
Selecting program team members	I	C				P	C	
Selecting methods to be applied		P	C					
Agreeing on the starting process with program owner	D	P						1
Preparing program start communication								
Preparing the start communication I, II, . . .		P	C					
Inviting participants to different communication formats	I	P	C			C	C	2
Documenting results of the pre-program phase		C	P		C		C	
Developing drafts of program plans		C	P		C		C	
Developing info material for the start communication formats		C	P		C		C	3
Performing program start communication								
Distributing info material to participants	I	C	P					
Performing the start communication I	C	Co		P		C	C	
Developing a draft of the documentation "Program starting"		C	P					
Performing start communication II, ...	C	Co		P				
Following up program start communication								
Finalizing documentation "Program starting"		C	P				C	
Agreeing with program owner on starting documentation	D	P	C					4
Performing inital program marketing	C	P	C		C		C	
Distribution documentation "Program starting"	I	C	P			I	I	
Filing documentation "Program starting"	C	Co	C		P	C	C	
Performing initial content work (parallel)		Co			P		C	

Developing the Program Breakdown Structure

In a program, projects but also work packages independent of projects are performed. A program breakdown structure must thus include both the projects of the program and work packages not belonging to projects, such as program managing. For clear differentiation, different symbols should be used in the program breakdown structure for the representation of projects and work packages (e.g., ellipses and boxes; see Fig. 13.4).

The example of the real estate program in Figure 13.4 shows a number of facilities that were renovated. Repetitive projects and chains of projects for conceptualizing, architecturally planning, and renovating of facilities can be seen.

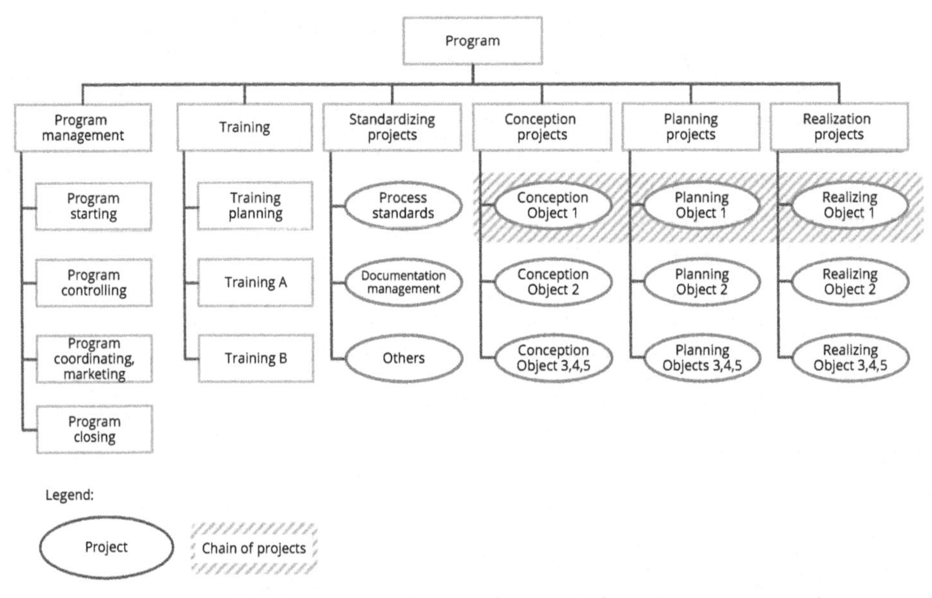

Fig. 13.4: Program breakdown structure (example of a real estate program).

In a program breakdown structure, either projects can be shown as a whole or the phases of individual projects can also be presented. The work packages of projects are not to be shown. These are presented in the individual work breakdown structures. As a result, the level of detail of the program breakdown structure is kept relatively low. In programs, projects can also be implemented sequentially. This results in chains of projects in programs (see Fig. 13.4).

As with the development of a work breakdown structure of a project, it is advisable to use the objects of consideration of a program as a basis for developing the structure. As in the case of projects, objects of consideration can be the structural and contextual dimensions of organizations—namely, their services and products, organizational structures, personnel structures, infrastructures, budget and financing, and stakeholder relations.

The breakdown of the program should be process oriented. Because of the large scope of programs and the simultaneous implementation of several projects, however, there will be a stronger object orientation than in the structuring of projects.

Programs must be structured so as to establish project boundaries that are as holistic as possible, so that individual projects quickly lead to results in the sense of minimal viable products. As a result, relatively autonomous projects should be defined which are loosely or moderately connected. The aim of this "cutting" of programs into projects is to reduce the complexity of programs by creating understandable and manageable projects.

The projects of a program are executed both in parallel and sequentially. The sequential execution of projects in programs produces chains of projects. The construction of a chain of projects promotes the clear differentiation of sequential projects and the explicit managing of the relations between the projects in the chain. The differentiation is performed by defining project-specific objectives, stakeholders, and roles. The integration of the projects in a chain is performed by planning each follow-up project in the previous project and by ensuring continuity of personnel in the chain of projects. The differentiation of projects in a chain promotes the agility of programs, since flexible planning of objectives is possible.

Organizational learning in programs can be secured by defining pilot projects in programs. In pilot projects, time and space must be provided to reflect upon and document experience gained. Experience gained is to be made available to follow-up projects in the program. This organizational learning creates significant opportunities for increasing the efficiency of programs.

The dynamics of programs can make it necessary to "split" or "merge" projects in program controlling. This dividing of a project into several projects or merging of two or more projects represents a discontinuity for the corresponding projects, since their structures are fundamentally altered. The underlying programs should be sufficiently robust that such changes of projects do not lead to program discontinuities.

Managing Program Stakeholders

In addition to managing project-specific stakeholder relations, programs must also manage relations with program stakeholders. The individual projects of a program and the program as a whole can have common stakeholders but also different ones. It is necessary to differentiate between these in order to be able to manage the stakeholder relations appropriately in the corresponding context.

Stakeholders can have different expectations for individual projects of a program and the program as a whole. This accounts for the complexity of programs. Unified strategies for managing stakeholder relations and coordination between those responsible for stakeholder relations in a program and in individual projects are necessary.

Managing Program Risks

Because of the size, complexity, and uniqueness of programs, they are subject to high risks. The risk of failing or stopping is large in programs. In addition to the individual projects of a program, the program as a whole must also be subjected to risk managing. Risk analysis and risk policy measures to avoid or promote risks, to provide for risks, and to control risks are risk management tasks in programs. Program risk managing can be performed either globally or in a detailed manner.

In global risk managing, the risks of the program are analyzed without considering the risks of the individual projects of the program and the relations between these risks. This approach is the same as for risk managing in projects. The program is treated like a single project.

In detailed program risk managing, the risks of the individual projects of a program and the relations between these risks are considered. Because of these relations, the program risks are not equal to the sum of the project risks. The risks of projects can be positively or negatively correlated, or they can be uncorrelated. If the risks of the projects of a program are assumed to be uncorrelated, the program risk is the sum of the project risks. If the project risks are positively correlated, the program risk is higher than the sum of the project risks; if they are negatively correlated, the program risk is lower than the sum of the project risks.

The basis for a detailed program risk analysis is an analysis of the risks of the individual projects of a program. Correlations between project risks can then be analyzed. There are positive correlations between project risks

- in project chains between pilot and follow-up projects or conception project and implementation project,
- in the case of cooperation with the same suppliers or partners in several projects of the program,
- when using the same technologies in multiple projects of the program, and
- when implementing several projects for the same customer(s).

By using the same supplier or the same technology for several projects, "economies of scale" and learning potentials can be exploited, and this is reflected in reduced project costs. On the other hand, there is a higher level of dependence on a selected supplier or technology. If the supplier fails or if the technology is insufficiently developed, not one but several projects suffer damage at the same time. One has to weigh lower program costs against the risk of a larger program loss.

Negative correlations between projects result when a risk which becomes acute in one project excludes a risk in another project. Such relations between projects will rarely exist. By managing the correlations of project risks, the program risk can be influenced. The objective of risk management in programs is primarily the reduction of positive correlations between project risks.

On the basis of a risk analysis, the decision makers of a program can implement a variety of risk policy measures according to their risk appetite (risk friendly, risk neutral, or risk averse).

Although in practice basic conditions for a detailed risk analysis in programs, such as the performing of risk analyses for projects, are often not met, it is advisable to consider correlations for strategic thinking, rather than as element of a quantitative calculation model.

13.5 Program Organizing

Program Organization Chart and Program Roles

A key difference between projects and programs is in their organizational design. Elements of the organizational structure of programs are the projects of the program and program-specific roles—namely, the program owner team, the program manager, the program team, the program office, and possibly a product team. The program roles and their relations can be visualized in a program organization chart. Figure 13.5 shows a standard program organization chart.

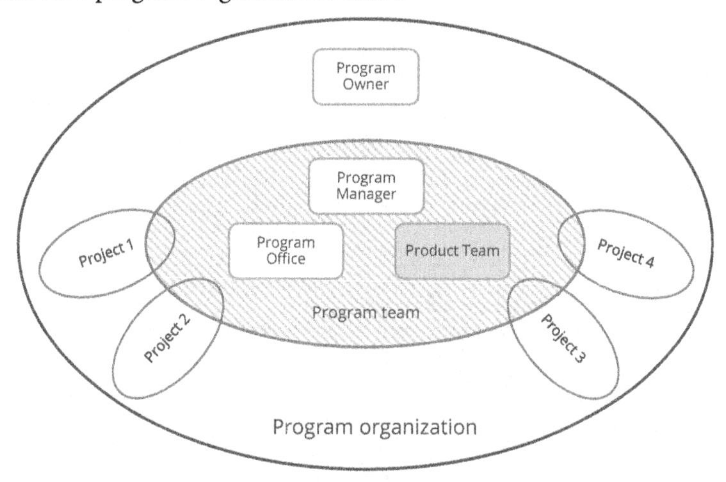

Fig. 13.5: Standard program organization chart.

The program owner assigns the program manager with the performance of a program. Strategic program decisions, such as the selection of projects to be started, the alteration of program priorities, and the specification of program strategies, are made by the program owner team.

Furthermore, the program owner team decides about project-related topics which are beyond the decision-making authority of the program team, project owner, or project manager. The planning of relations between the program owner team and the project owners is challenging. The program owner team is responsible for the program as a whole and is thus more strategically oriented than the project owners. The project owners are more oriented towards the content and are also qualified to make fundamental decisions regarding the content. A project owner of a project of a program has less authority than a project owner of a project performed independently. The potential for conflict between the interests of the program owner team and those of the project owners can be reduced by means of multiple role assignments. This means that a person might hold the roles of program owner team member and project owner at the same time.

The program owner team should be composed of managers of major organizational units affected by the program. It should not include more than three persons. A spokesperson for the program owner team is to be nominated as the primary contact person for the program manager. An extended program owner team may also include representatives of project partners and major suppliers as members. In this case, one frequently refers in practice to a steering committee. A general description of the role of "program owner team" is given in Table 13.5.

Table 13.5: Role Description: Program Owner Team

Role: Program Owner Team	
Objectives	> Realization of program objectives ensured strategically > Program objectives aligned with the strategic objectives of the project-oriented organization > Program manager assigned with the performance of the program > Strategic program controlling performed > Project manager directed
Organizational position	> Part of the program organization > Program manager reports the program owner team
Tasks in project starting	> Informing about the program contexts > Ensuring the allocation of program resources > Contributing to initial program marketing > Participating in the program starting workshop > Performing program owner meetings
Tasks in project controlling	> Program agreeing with other programs, projects and organizational objectives > Analyzing program progress reports > Selecting project owners and project managers for the projects of the program > Taking strategic program decisions > Selecting projects to be performed within the program > Ensuring the allocation of additional program resources > Contributing to program marketing > Performing program owner meetings
Tasks in project transforming and project repositioning	> Defining program discontinuities > Contributing to cause analysis of the program discontinuity > Taking decisions about strategies for program transforming or program repositioning > Contributing to the implementation of measures for program transforming or program repositioning > Closing program transformations or repositionings
Tasks in project closing	> Performing program owner meetings > Participating in the program closing workshop > Approving the program
Formal decision-making authority	> Assigning the program manager > Providing the program budget > Changing program objectives > Defining program discontinuities > Stopping the program > Approving the program

The program manager is responsible for the realization of the program objectives. He or she ensures the professional fulfillment of the business process "Program managing". Because of the extensive management and marketing requirements of programs, it is advisable to establish a program office to support the program manager.

In the program office, program managing is institutionalized, creating a "home base" for the program. The program office serves as a contact for the members of the project organizations and representatives of stakeholders. The program office also organizes program marketing. It develops a program marketing plan and plans the budget for program marketing. A small program office can be staffed by a program management assistant and a program administrator. A larger program office can also include further program management experts, process management experts, and controllers.

The roles of the program manager and the program office are described in Tables 13.6 and 13.7 (on next pages).

The program team consists of the program manager, representatives of the program office, the product team and the project managers of the projects which are performed at a given point in time and projects which are shortly to be started. The composition of the program team therefore changes over the course of time since different projects are active at different times. The tasks of the program team are to provide information to the members of the program team, to secure synergies, to resolve conflicts, to set priorities in the program and to implement program controlling. The role of the program team is described in Table 13.8 (page 353).

Formats for Program Communication

In addition to the communication formats of the projects of a program, such as project owner meetings or project team meetings, special communication formats are required in programs. One can make a basic distinction between communication formats for program managing and those for communicating with program stakeholders. An overview of the communication formats for program managing can be found in Table 13.9 (page 354). This also shows that the frequent communication in the program office is distinguished from the periodic, but relatively rare, communication in the program team or with the program owner.

As an example of internal program communication, the internal communication plan of the program "Establishing a hospital" is shown in Figure 13.6 (on page 355). This shows the high communication demands on the program manager and the members of the program team.

Program Managing Competencies

Explicit individual and organizational competencies for program managing are required in the project-oriented organization. The holders of program roles handle

Table 13.6: Role Description: Program Manager

Role: Program Manager	
Objectives	> Realization of program objectives ensured > Program controlled together with the project managers > Networking of the projects of the program promoted, synergies assured > Program team directed > Program represented towards program stakeholders
Organizational position	> Part of the program team > Reports to the program owner team > Is supported organizationally by the program office > Contact person for project owners and project managers of the projects of the program
Tasks in project starting	> Managing the program starting together with the program office > Performing program team meetings and a program starting workshop > Developing appropriate program plans, designing the program organization together with the program office > Analyzing and planning the program context relations together with the program office > Performing an initial program marketing together with the program office > Developing the documentation "program starting" together with the program office
Tasks in project controlling	> Managing program controlling together with the program office > Performing program team meetings and attending program owner meetings > Controlling the external program resources > Planning directive measures together with the program team > Determining priorities within the program > Adapting program plans together with the program office > Developing program progress reports together with the program office > Performing program marketing together with the program office > Starting projects of the program together with the project owners and project managers > Defining project discontinuities together withe the respective project owners > Contributing to the closing of projects within the program > Transferring know-how into other projects within the program
Tasks in project transforming and project repositioning	> Developing ad-hoc measures together with the program team > Analyzing causes of program discontinuities together with the program team > Developing strategies for transforming or repositioning together with the program team > Implementing measures for transforming or repositioning together with the program team > Closing a program transformation or repositioning together with the program owner
Tasks in project closing	> Managing program closing together with the program office > Performing the program closing meetings and a program closing workshop together with the program team > Developing program closing reports together with the program office > Attending the final program owner meeting > Transferring know-how into the project-oriented organization and into other programs and projects > Reaching agreements for the post-program phase > Performing a final program marketing
Formal decision-making authority	> Determining project priorities within the program > Taking responsibility for the program budget > Scheduling program owner team meetings > Deciding about managing relation to stakeholders > Starting projects together with the respective project owners and project managers > Changing project objectives, defining project discontinuities and closing projects together with the respective project owners

Table 13.7: Role Description: Program Office

Role: Program Office	
Objectives	> Program manager and projects within the program supported > Compliance of the management standards in the program ensured > Program administration performed
Organizational position	> Reports to the program manger > Directed by the program manager > Head of the program office is a member of the program team
Tasks	> Preparing program owner meetings, program workshops and program team meetings > Ensuring the infrastructure for program managing (rooms, ICT, etc.) > Supporting the program manager in the program managing sub-processes > Supporting the program manager in developing and adapting the program plans > Supporting the program manager in developing the program progress reports > Supporting the program manager in performing the program marketing > Supporting individual projects within the program in project managing and in project marketing > Performing contract administration, personnel administration, filing within the program > Ensuring compliance of program management standards in projects within the program
Formal decision-making authority	> Deciding about the processes of the program office

Table 13.8: Role Description: Program Team

Role: Program Team	
Objectives	> Interests of the program represented > Common "Big Project Picture" developed > Ensuring synergies in the program and solving conflicts > Co-responsibility for the program success understood
Organizational position	> Directed by the program manager > Members: Program manager, head of program office, project managers of the current projects of the program
Tasks	> Performing program team meetings > Providing information about the program an its contexts > Contributing to program controlling > Reaching agreements in the program team
Formal decision-making authority	> Reaching agreements in the program team > Reaching agreements with the program owner

program complexity and fulfill challenging integrative functions. Program managers therefore need several years of project management experience and high social competencies. Appropriate organizational competencies are an essential dimension of the maturity of a project-oriented organization.[1]

[1] The RGC model for analyzing the maturity of a project-oriented organization is dealt with in Chapter 15.

Table 13.9: Communication Formats for Program Managing

Communication formats for program managing				
Format	Objectives	Participants	Frequency	Duration
Program start workshop	Starting the program in the program team	Program team, guests, program owner team, representatives of stakeholders	1x in the program start process	2-3 days
Program owner meeting	Strategic program controlling	Program owner, program manager, guests (if needed)	1x per month or every 2nd month	2-3 hours
Program team meeting	Program controlling	Program team, guests (if needed)	1x per month	2-3 hours
Program office meeting	Operational program managing, preparing program controlling	Program manager, program Office staff, guests (if needed)	1x per week	2 hours
Program Office stand-up meeting	Coordinating the tasks of the Program Office	Program Office staff	1x or 2x per week	30 minutes
Program closing workshop	Closing the program in the program team	Program team, guests, program owner team, representatives of stakeholders	1x in the program closing process	2 days

13.6 Program Initiating and Program Managing: Values

The values of the RGC management paradigm can be interpreted for project initiating and project managing (see Fig. 13.7, page 356).

Holistic Definition of Solutions and Boundaries

The development of a holistic solution by a program requires the definition of holistic program boundaries. Since programs have a longer duration than projects, as well as a larger scope and higher risks, program boundaries have to be more flexible than project boundaries. Program boundaries are more open and dynamic. But they must still be clearly defined—for example, by program objectives and a program end date. Only then can they provide structure and orientation for the projects of a program.

A holistic approach to program managing means that the dimensions of programs, such as program objectives, program scope, program organization, and program contexts, are comprehensively understood, and that all sub-processes of program managing and their relations are considered. Information from the individual projects of a program provides an essential basis for program managing.

Sustainable Business Value as a Success Criterion

Sustainable business value for the performing organization is striven for as the result of a program. Like projects, programs are performed in the contexts of investments

Communication format	Duration in hours	Frequency	Program manager	Program manager deputy	Program manager deputy	Project managers	Medical director	Personnel	Organization	Facility management	Controlling	IT	Construction	City of Vienna	Consultant construction	Consultant organization	Consultant statics	Monitoring	Construction supervision	Accompanying control
Jour fixe medical coordinator	2	Monthly	x				x		x							x				
Jour fixe program manager	1,5	weekly	x	x	x	x		x	x	x	x		x							
Core team & program team	1,5	Monthly	x	x	x		x	x	x	x			x			x				
Core team	1,5	Monthly					x		x							x				
Jour fixe program management & controlling	1	weekly	x	x	x															x
Jour fixe controlling & site monitoring	1	weekly	x	x	x					x			x					x		x
Program coordination	2	weekly	x	x	x				x	x	x		x		x		x	x	X	x
Liquidity management	1	Monthly																		
Week preparation meeting	1	weekly																		
Program team quarterly	4,5	Once per quarter	x	x	x	x	x	x	x	x	x	x	x	x						
Contracting	1,5	14-day																x	x	x
Jour fixe ombudswoman	1	Monthly	x	x	x	x														
Program advisory board	2	Twice a year	x	x	x	x	x		x	x			x							
Change coordination requests	1	weekly							x						x		x	x		x

Fig. 13.6: Internal communication plan for the program "Establishing a hospital".

and medium-term objectives and strategies of the implementing organization. The contributions made by a program to the sustainable development of the organization can be analyzed in a cost–benefit analysis. The economic as well as the ecological and social costs and benefits of different stakeholders can be considered. As the quality of the information and assumptions improve during the performance of a program, the cost–benefit analysis has to be adapted periodically.

The principles of sustainable development also influence the business process "Program managing" and its methods. The principles can be considered, for example, in the program objectives plan or in the program stakeholder analysis.

Holistic definition of solutions and boundaries	Sustainable business value as success criterion	Quick realization of results and efficiency	Iterative approach
> Defining a problem solution holistically in a program to avoid sub-optimizations > Defining the program boundaries (content, time, and social-related) holistically	> Contributing to the business value by a program > Controlling the benefits realization during and after the program > Considering principles of sustainable development in program management	> Ensuring quick wins in the program > Stopping a program as early as possible in case of failure > Managing a program efficiently and flexibly	> "Cutting" a program to allow an iterative approach > Creating chains of projects in a program > Using agile methods in projects of programs > Iteratively managing by Program initiating, starting, controlling)

Context orientation	Continuous and discontinuous learning	Frequent, visually supported communication	Empowerment and resilience
> Designing an integrated program organization by involving stakeholders > Management of and for stakeholders > Managing relations of a program to other projects and programs	> Using feedbacks and reflections for organizational and individual learning in a program > Promoting program changes	> Performing stand-up meetings, reflections by the Program Office and the program team > Using scrum boards, program plans, score cards, etc. for communicating	> Empowering the program > Representing the product owner interests in the program organization > Empowering the projects within the program

Fig. 13.7: Values of the RGC management paradigm—program initiating and program managing.

Quick Realization of Results and Efficiency

Programs last several years, but should not take longer than three years because of the dynamics in the organization's environment. Because of this longer duration, the achievement of interim results is particularly important for maintaining the motivation to continue a program. The interim results of a program are achieved in the projects of the program. Quick wins of a program—that is, quick and easily achieved results—must therefore be planned and controlled beyond project boundaries.

Quick realization of results of program managing is also possible. Although programs require a high degree of integration, the planning and controlling effort for programs is relatively low. Though all management dimensions of programs are to be considered, managing takes place at a high level of abstraction. The objects of program managing are the individual projects and their relations. Therefore, only individual projects (or possibly also their phases) are shown in a program breakdown structure. It will also suffice to use the phases of the projects as the most detailed level in the program bar chart. More detailed planning is decentralized and used for empowerment at the project level.

In programs, however, not only should program managing be efficient, but the business processes for developing solutions should be performed efficiently. For programs with repetitive projects, program standards can significantly increase efficiency.

Iterative Approach

Defining programs with multiple parallel and sequential projects instead of performing "large projects" represents an iterative approach. Prioritizing the projects of a program based on their contributions to business value, on content-related dependencies, and on scarce resources also corresponds to an iterative approach.

The high complexity and dynamics of programs necessitates an iterative approach to managing the requirements to be met by programs. A program strategy can be the use of agile methods in some projects of a program. An iterative approach is also used in program managing: Initial program plans are developed in program initiating, the program plans are detailed in program starting, and the plans are adapted in program controlling.

Context Orientation

As in a project, context orientation in a program means that contextual dimensions of a program such as the pre-program and post-program phases and relations with other projects and programs are considered to ensure the program's success.

Stakeholder orientation in the program means that program stakeholders are identified, and the relations with them are explicitly managed and differentiated from relations with the stakeholders of the projects of a program. "Management for stakeholders" is also relevant in program managing.

There may be conflicts of interest between stakeholders, which also contributes to the complexity of a program. Overall interests taken care of by the product team on the program level might be in conflict with specific interests of product teams on the project level. An explicit stakeholder management strategy is needed. The early involvement and integration of selected stakeholders into the program organization reduces the program risks and creates opportunities for optimization.

Continuous and Discontinuous Learning

The process of performing a program is more dynamic than that of a project. The high complexity and dynamics of programs necessitate continuous learning, but sometimes also discontinuous learning. As tools for individual and organizational learning, feedback and reflections promote the cultural development of programs.

In programs, there are often repetitive projects. The learning of such programs can be organized by means of pilot projects and by the development of program-specific standards.

A need for formal changes of programs is likely because of their high complexity and dynamics. The transformation or repositioning of a program is therefore to be considered as a possible sub-process of program managing.

Frequent, Visually-Supported Communication

The complexity and dynamics of programs necessitate frequent interaction between members of the program organization as well as communications with representatives of program stakeholders. Management of programs is therefore characterized by a particularly large need for communication. Integration of the projects of programs takes place mainly through program communication. Frequent, short-term communications with program owners or in program teams are hardly possible, but such communications are possible and necessary for the continuous program coordinating in the program office.

An important function of the communication is the contribution to program culture development. Through various meetings, trust is built, networking between the projects of the program is promoted, and stress is reduced by collective problem solving.

Visual aids such as program plans and program scorecards can be used for program communication. Appropriate room and ICT infrastructure must be provided to support program communication.

Empowerment and Resilience

Empowerment in a program can occur on the program level; on the level of the projects of the program; on the levels of the program team, project teams, and sub-teams; but also on the level of individual members of the program and project organizations. It is assumed that empowerment increases the motivation of employees and also their loyalty. But this requires clear rules for cooperation. A specific area of conflict with regard to empowerment is the division of decision-making authority between the program and the project level. This question is particularly relevant to the role understandings of program owners, program managers, project owners, and project managers.

The resilience of programs must be ensured. That means, for example, that the stability of a program should not be jeopardized, even if individual projects of the program have problems or are in a state of instability. The resilience of programs can be promoted by agile and flexible structures, as well as by redundant program structures.

13.7 Program Managing: Benefits

Program managing enables programs to be implemented and ensures the quality of program results. Without appropriate managing, programs cannot be performed at all or only very inefficiently because of their complexity and dynamics.

The results of programs are the solutions which exist at the program end as a basis for achieving sustainable business value. The benefits of program managing are thus in ensuring efficiency and effectiveness. The efficiency of a program, for example, is ensured by coordinating the relations between projects, the use of resources, and the possible transfer of know-how between projects. Effectiveness is ensured by appropriate "business focus".

13.8 Program Managing: Case Study of a Power Supply Company

In the following case study, the structures actually used by a power supply company for implementing an ERP solution are described. An ideal-type structure for this program is also presented and interpreted.

Reason and Contexts for ERP Implementation

In 2013, the power supply company merged with a power production company. Until then, the two companies had operated two independent ERP systems.[2] These systems were to be merged into a common ERP system. The following business tasks were to be supported by the new ERP system:[3]

> financial management (FI): credit accounting, debit accounting, general accounting, cash management, auditing, . . .
> facility management (FI-AA): facility operations, facility simulation, insurance management, . . .
> controlling (CO/IM/PS/PPM): overhead cost controlling, product cost controlling, profit center accounting, investment management, project portfolio management, project management, . . .
> purchasing (SRM/MM): operational and strategic purchasing, supplier management, contract management, . . .
> materials management (MM): inventory management, materials planning, . . .
> maintenance (PM): problem analysis, repair, . . .
> technical operation (PM): work clearance management, shift book documentation,
> sales (SD/CS): customer order processing, excluding billing.

[2] Enterprise resource planning (ERP) supports the corporate tasks of planning and controlling resources such as capital, personnel, equipment, materials, and services in accordance with the corporate purpose in a timely and appropriate manner.

[3] The abbreviations refer to the various modules of the system.

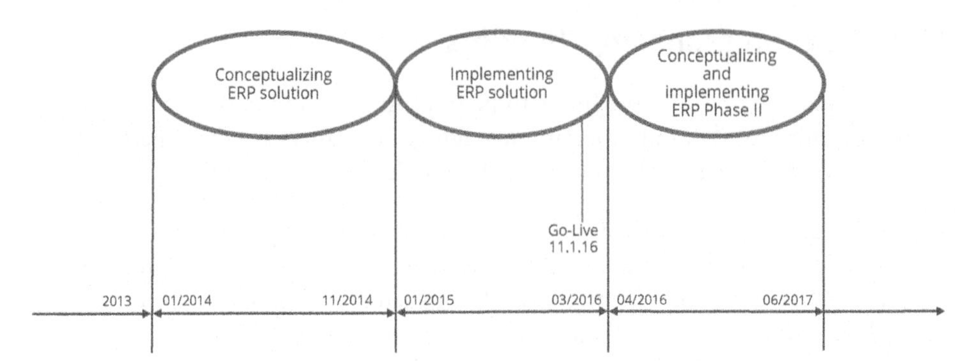

Fig. 13.8: Chain of projects for ERP conceptualizing and ERP implementing in a power supply company.

After the merger, a project for conceptualizing the ERP solution and initiating the project "Implementing ERP solution" was started in January 2014. The "go live" milestone for using the core modules of the ERP system was set for 11 January 2016. After the completion of the implementation project in March 2016, a follow-up project for conceptualizing and implementing additional ERP solutions (Phase II) was started. These relations are shown in the representation of the resulting chain of projects in Figure 13.8.

A program for the reorganization of the company was implemented in parallel with the chain of projects for the conception and implementation of the ERP system. The objective of this program was to develop the basic organizational structures for the company as a whole and for its individual organizational units. In contrast to the ERP implementation, the program concept was used for this reorganization.

Project "ERP Implementing"

The following objectives were defined for the project "ERP implementing":

- implemented and accepted core modules,[4]
- harmonized IT processes and data of the organizational units,
- integration with external systems performed,
- data migration performed,
- financial reports of all organizational units integrated into the starting balance for 2016,
- the implementation of the additional modules conceptualized, and
- users of the core modules trained.

Non-objectives were reorganization measures, consideration of the requirements of earlier projects not taken care of, and consideration of non-ERP processes.

[4] The implemented core modules are shown in the lines of Figure 13.11 on page 363.

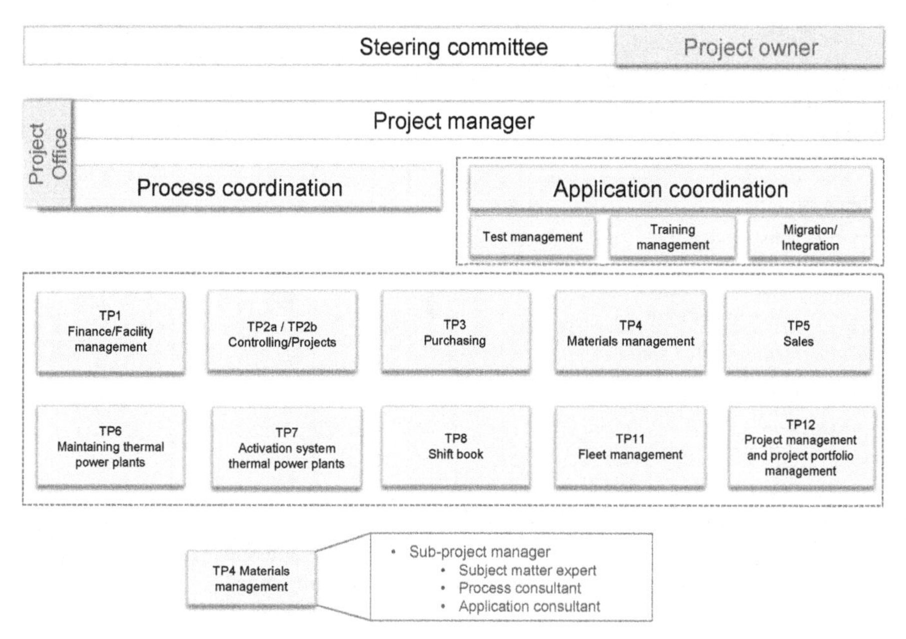

Fig. 13.9: Organization chart of the project "Implementing ERP solution".

The organizational structure of the project "ERP implementing" is shown in Figure 13.9. It can be seen that a sub-project was defined for each module, and that integrating structures existed with a steering committee/project owner, project manager, and two coordination teams. Each sub-project had a sub-project manager, technical experts, process consultants, and application consultants. The terms "project" and "sub-project" were used—that is, it was not perceived as a program.

As shown in the columns in Figure 13.10, the following phases were performed for each sub-project:

> customizing, developing, forms,
> work flow, permissions, migrating,
> testing, training, and
> integrating/interfaces/ESB.BI/BW/DWH and IT security.

Interpretation of the Structures for the Implementation of ERP

The following can be seen from the plans of the project "ERP implementing":

> The chain of projects started with a conception project. To provide a basis for the implementation, clearly important analyses were performed, alternative solutions were defined, a solution to be implemented was selected, and the implementation project was initiated. It was necessary to perform the conception as a project because of the high strategic importance of the decisions to be made.

Project management											
	Application coordination										
Process coordination	Customizing	Development	Forms	Workflow	Authorizations	Migration	Test	Training	Integration / interfaces / ESB	BI / BW / DWH	IT Security
Finance/facility management											
Controlling/projects											
Purchasing											
Materials management											
Sales											
Maintaining thermal power plants											
Activation system thermal power plants											
Shift book KWK											
Shift book MVA											
Energy services											
Fleet management											

Fig. 13.10: Objects of the process coordination and application coordination in the project "ERP implementing".

- A second implementation phase for additional modules can also be seen. This differentiation between the implementations of core modules and additional modules was useful, because it enabled clear prioritization and an iterative approach. The chain of projects, however, allowed for little integration between the two sequential projects.
- The implementation of individual modules were referred to as sub-projects. The perception of sub-projects has the disadvantage that relatively rigid boundaries are created. This strong focus on the individual sub-projects often makes it impossible to look at relations to other sub-projects. Sub-projects also have no (sub-) project owners. As a result, sub-projects receive little (top) management attention.

  Because of their relative autonomy, sub-projects can actually have project characteristics. But then instead of a project with sub-projects one should call it a "program with projects".
- The reorganization program was formally decoupled from the implementation of ERP and planned as a parallel activity. The program concept was therefore used in different ways. It was not sufficiently oriented towards the actual needs for differentiation through projects and integration through programs.

Ideal-Type Structures—Program "Implementing Reorganization"

Ideal-type structures for the case described above are presented below. The objects of consideration plan, the program objectives plan, the program breakdown structure, the program organization chart, and the program assignment are used as methods. The main results of program initiating would be in this overview form.

In order to provide an appropriate integrative view of essential activities related to the reorganization of the company, it is recommended to consider the reorganization and ERP implementation together as a program. The ERP solutions are strongly linked to the organizational solutions and should therefore be managed together, integratively. Since the reorganization of the company was the central objective of the program and the ERP solutions only represented infrastructure support, the ideal-type program could be called " Implementing reorganization".

A basis for the program "Implementing reorganization" is provided by the project "Conceptualizing reorganization" and the project "Creating a basis for reorganization" (see Fig. 13.11). Just as in the actual implementation, the objective of the project "Conceptualizing reorganization" would be to plan the changed company necessitated by the merger. The project "Creating a basis for reorganization" could be based on these decisions. The objectives of this project would be basic planning of the organization of the power supply company and the related ERP concept. After the program "Implementing reorganization", it might be useful to implement a project to stabilize the company's new structures.

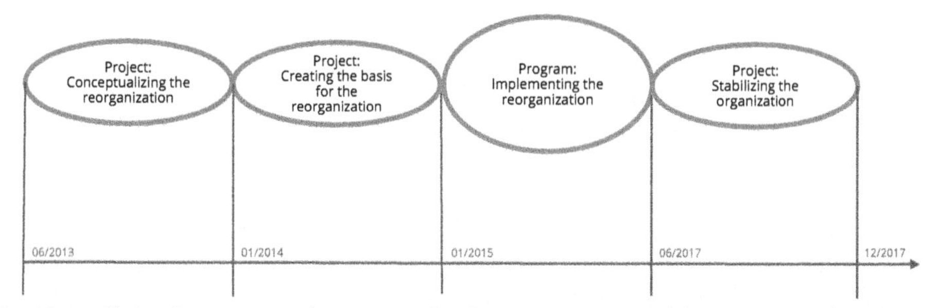

Fig. 13.11: Chain of a project and a program for the reorganization of the power supply company.

An overview of the objects to be considered by the program is shown in an objects of consideration plan in Figure 13.12. The objects of consideration form a basis for formulating the program objectives. A rough description of the program objectives can be found in Figure 13.13 (page 365).

The program breakdown structure (see Fig. 13.14, page 366) and the program organization chart (see Fig. 13.15, page 367) are based on the following assumptions:

- The originally separate topics of reorganization, implementation of ERP core modules, and implementation of additional ERP modules are combined into a program. This produces holistic program boundaries, which enable integrated managing of these topics by the program organization.
- An iterative approach is implemented. On the one hand, a chain is formed from two sequential projects and a program, and on the other hand, pilot projects are used for learning, in order to be able to implement follow-up projects subsequently. The additional modules are implemented after the core modules.

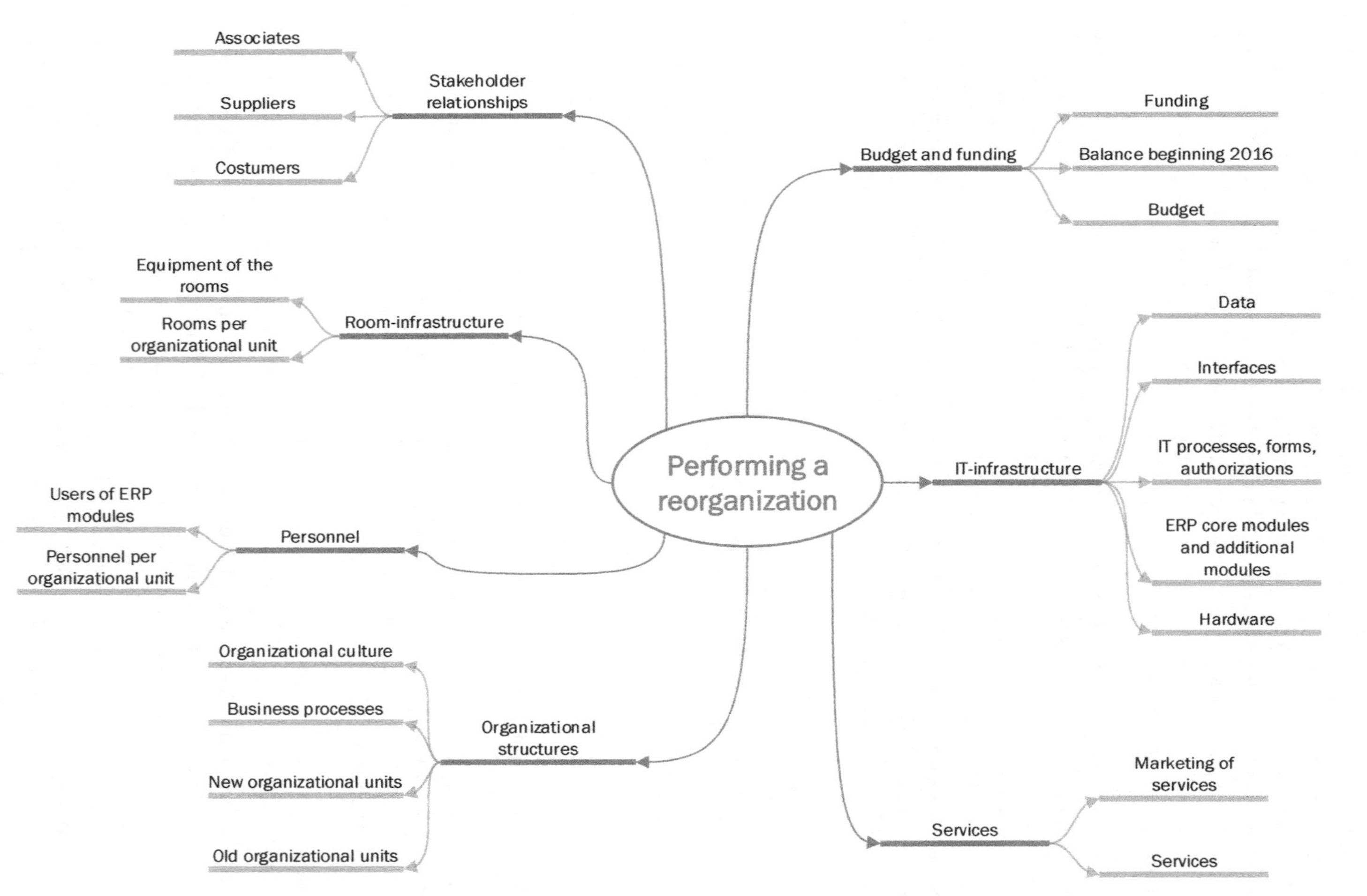

Fig. 13.12: Objects of consideration of the program “Implementing reorganization”.

Objectives of the Program: Implementing Reorganization
Economic program objectives
Service-related program objectives
Services of merged companies integrated
Marketing tools for extended services developed
Inital marketing performed
Organization-related program objectives
Old organizational units of the company dissolved
New organizational units established
Business processes agreed and documented
Organizational culture further developed
Personnel-related program objectives
Personnel for each organizational unit defined, vacancies filled and persons qualified
User of ERP moduls trained for the applying the modules
Room infrastructure
Rooms for each organizational unit provided
Rooms equipped appropriately
IT infrastructure-related program objectives
Hardware available for using the ERP modules
ERP core modules and additional modules implemented and accepted
IT processes and data of the organizational units harmonized
Integration with external systems performed, authorizations assured
Data migration performed
Budget and finance-related objectives
Both companies considered in one budget
Opening balance sheet 2016 in consideration of the merged company developed
Financing of the merged energy supplier optimized
Ecological program objectives
Mobility demand reduced for employees
Paperless office developed further
Social program objectives
Customers of energy supllier informed about the merger
Supplier relations of the merged company integrated
Repositioning within the group structures performed
Employees of both companies integrated
Non-objectives
Personnel reduced
Processes considered which are not relevant for the ERP application

Fig. 13.13: Objectives of the program "Implementing reorganization".

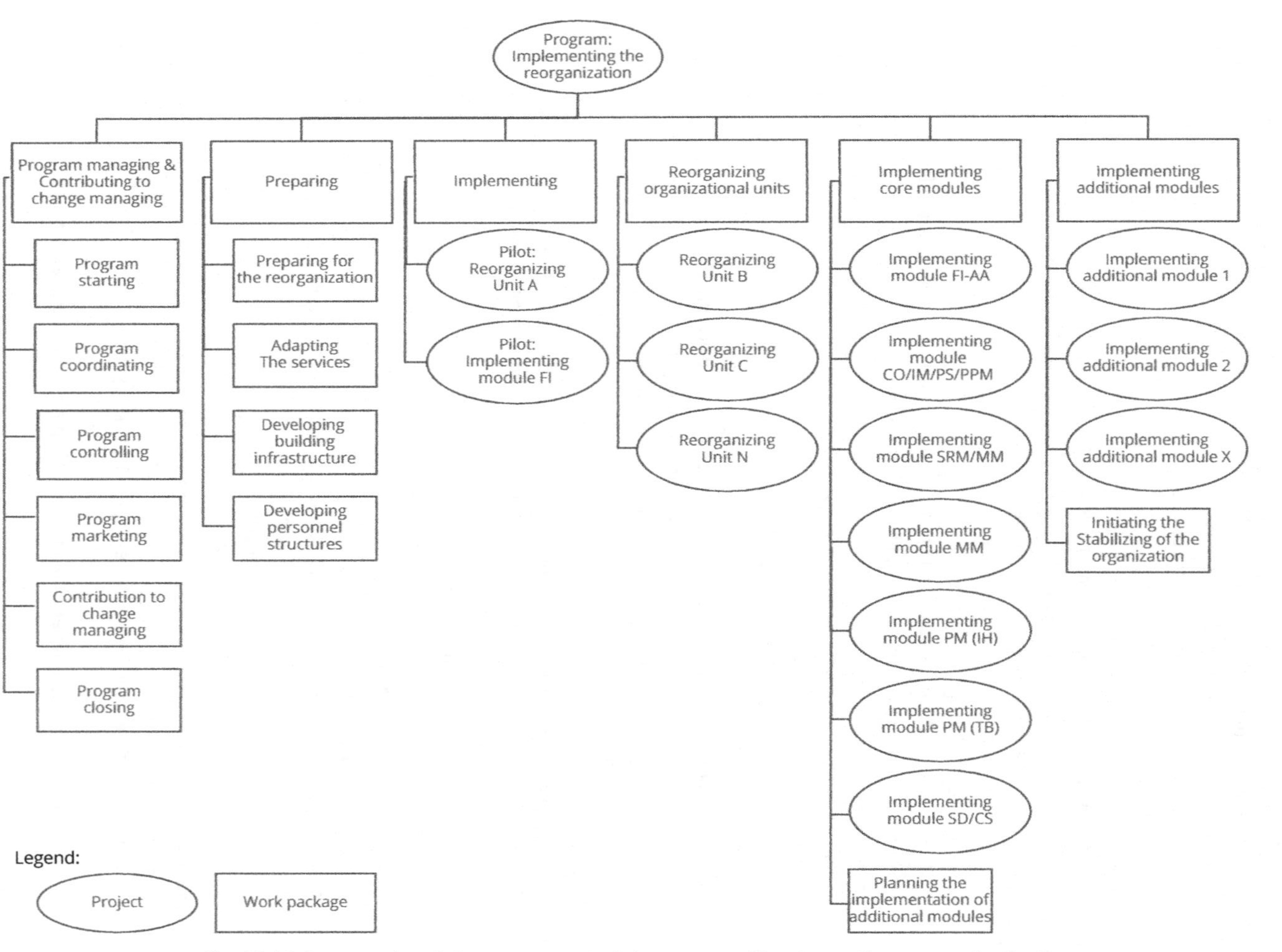

Fig. 13.14: Program breakdown structure of the program “Implementing reorganization”.

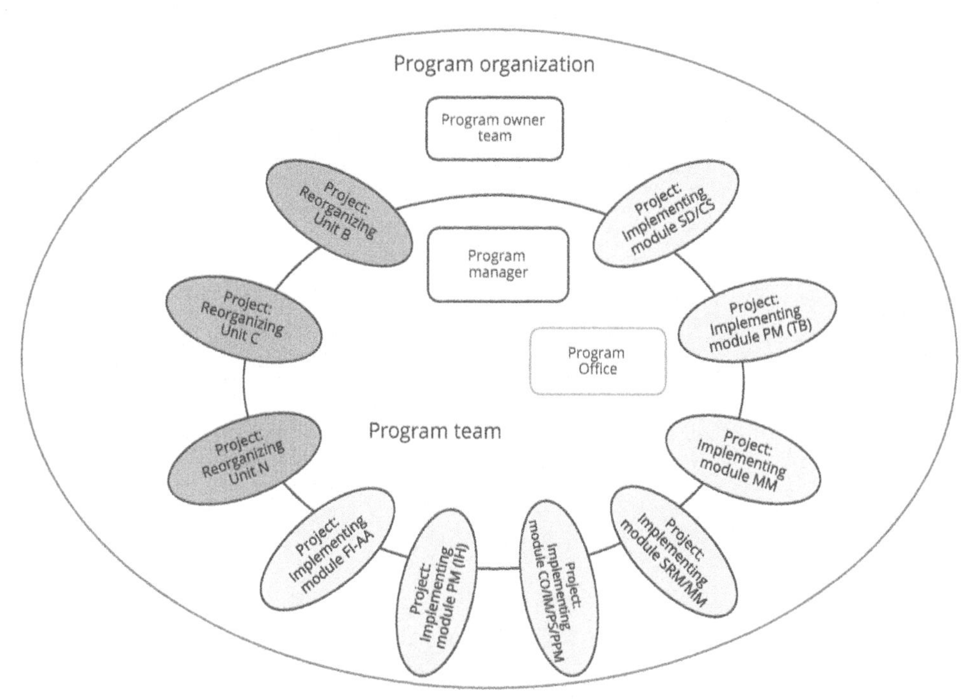

Fig. 13.15: Organization chart of the program "Implementing reorganization".

> Instead of a project with sub-projects, a program with several projects is established. This allows project owners to be assigned for each project, which increases management attention in the program. It also provides a basis for differentiated plans for the individual projects of the program, which reduces the complexity of the program plans.
> The organization chart of the program changes over time, since only the current projects of the program are to be shown. The program organization evolves.
> The program starts after the closing of the project "Creating a basis for reorganization". The program is also initiated as part of this project. The program ends when all organizational units of the power supply company are reorganized and the ERP solution is supported. The content of the program still includes initiating the following project, "Performing stabilization".

Program organization charts change over time. Only current projects are to be shown. The first period, for example, would be the period of performing the project "ERP conceptualizing" and the two pilot projects. A subsequent period would be the period of reorganization of the organizational units and the implementation of the ERP core modules. The organization chart for these periods is shown as an example in Figure 13.14. The already completed pilot projects and the projects for the implementation of the additional ERP modules are not included in this representation.

The essential information of these rough program plans provides a basis for the program assignment.

14 Change Initiating and Change Managing

Changes are achieved by projects or by chains of projects and/or programs. The structural relation between changes and projects can be seen in the change architectures—that is, in the structuring of change process "by projects".

"Change managing" is a business process of organizations that contributes to the successful implementation of changes. In practice, change managing is either not performed explicitly or is often reduced to change communication.

Similar to project and program managing, change managing includes the sub-processes "Change starting", "Change coordinating", "Change controlling", "Change communicating", and "Change closing". In change managing, the focus is on managing a change and not on the fulfillment of content-related tasks. Change management roles—namely, change owner, change manager, and change agents—must be defined for change managing.

To enable successful changes, the change managing process has to be appropriately designed, and its relations with project or program managing have to be managed. Project-oriented organizations can use their project management competencies for change managing.

An initial definition of the change dimensions to be considered, the necessary change processes, the change stakeholders, etc. is achieved as results of the business process "Change initiating". Examples of change initiating and change managing are presented in Section 14.9 (starting on page 395) for the case study "Values4Business Value".

The business processes "Change initiating" and "Change managing" are shown on next page in the context of other relevant business processes.

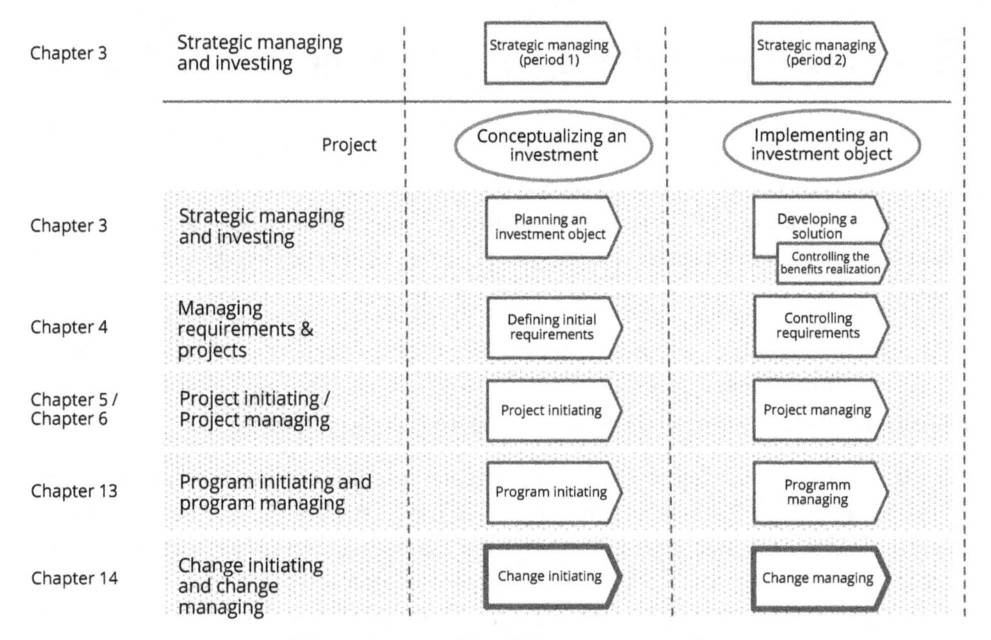

Overview: "Change initiating" and "Change managing" in context.

14 Change Initiating and Change Managing

14.1 Objects of Change, Change Definition, and Change Types

Objects of Change

The objects of changes are social systems. The social systems that are considered here are permanent organizations (see Fig. 14.1).[1]

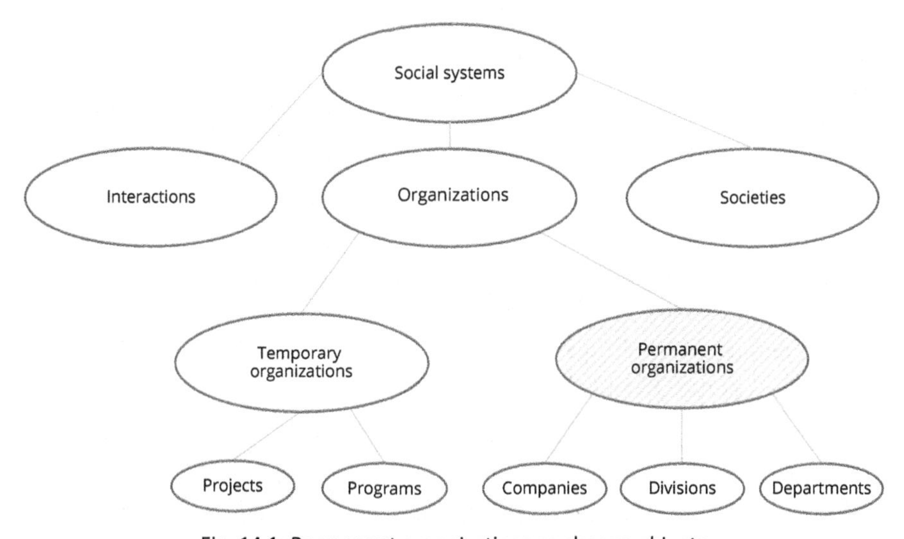

Fig. 14.1: Permanent organizations as change objects.

Dimensions of a permanent organization that can be changed are their services and products, markets, organizational structures, cultures, personnel structures and infrastructures, the budget, the financing, and their stakeholder relations.

Change Definition

A change in an organizational context is a continuous or discontinuous development of an organization in which one or more dimensions of the organization are affected.[2,3] It is assumed that in a change the social structures and stakeholder relations of the organization are affected. It is therefore a "social" change.

Change Types

In considering the demand of an organization for change and the number of change dimensions affected, one can differentiate between the first-order changes "Organizational learning" and "Further developing" and the second-order changes "Transforming" and "Radical repositioning" (see Fig. 14.2). First-order changes contribute to the

[1] In Chapter 11 changes of temporary organizations are discussed.
[2] Lewin defines change as the transformation of an existing dynamic balance of an organization into a new one.
[3] See Lewin, K., 1947, p. 5–40.

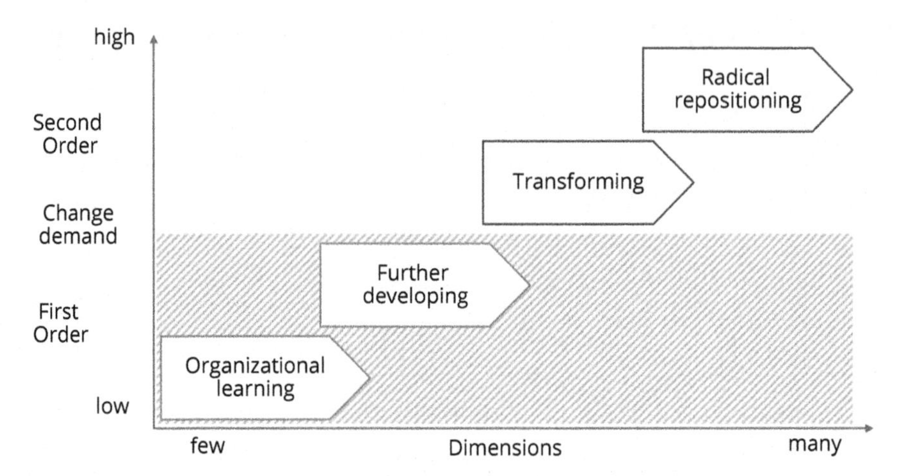

Fig. 14.2: Change types of permanent organizations.

continuous development of organizations. Second-order changes represent discontinuities and lead to new identities of organizations.[4]

A relatively small demand for change, which necessitates "Organizational learning", is a result of the ongoing business activities of an organization. It can, for example, be a customer's wish to adapt a business process. This demand can be met by a small change. In organizational learning, usually only one or a few dimensions of an organization are affected—in this case, a business process and some personnel. Continuous improvements of this kind are integrated into ongoing business activities and require no projects for their implementation.

A moderate demand for change, which requires "Further developing" of an organization, may, for example, be caused by an increase in market dynamics. To ensure continued business success, innovations such as the development of a new product or the entering into a new market may be necessary. Changes of this kind affect several dimensions of an organization. They require appropriate management attention—for example, by implementation as a project or a chain of projects.

Signs of future threats, but also of new opportunities for an organization, can lead to a medium to high demand for change. Changes that can result from threats or new opportunities are, for example, the acquisition of another organization, a merger with another organization, or the development of a new business model. It can be assumed that this produces a demand for strategic and cultural reorientation. Many or all dimensions of organizations are to be considered in these changes. "Transforming" is the appropriate change type for this. A chain of projects, or possibly a chain formed by a project and a program, is demanded for transforming.

[4] See Gareis, R., 2010, p. 318.

The greatest demand for change in an organization is when the organization is in a crisis. Financial losses over a prolonged period, the withdrawal of customers, bad relations with stakeholders, etc. are indicators of this. Here, all dimensions of an organization must be considered in the change; "Radical repositioning" is necessary.

The basic objective in such a case is to ensure the survival of the organization. To achieve this, liquidity must be secured, positive financial ratios must be achieved, opportunities for future development are to be reassured, and so on. Radical repositioning usually requires several projects or a program for its performance.

The most relevant types of change in practice are "Organizational learning", "Further developing", and "Transforming". "Radical repositioning" is only relevant in situations of organizational crisis, which do not occur frequently in an organization.

Although "Organizational learning" is relevant, its implementation does not require any projects, and it is therefore not further discussed here.

Definition: Change

A change is a continuous or discontinuous development of an organization in which one or more dimensions of the organization are affected. One can differentiate between the change types "Organizational learning", "Further developing", "Transforming", and "Radical repositioning".

14.2 Change Architecture and Change Roles

Chains of Change Processes and Change Projects

Different chains of change processes are needed for realizing different change types. All change types include the business processes of planning, implementing, and stabilizing, but to different extents. As a result, different organizations—namely, working groups, small projects, projects, or programs, are required to perform the individual processes. This results in chains of projects or of projects and programs, which make up the "change architectures". As examples, the architectures for the change types "Further developing" and "Transforming" are discussed below.

Changes by projects!

Changes are realized by projects, chains of projects, and/or programs. The structural relation between changes and projects can be seen in their respective change architectures.

Figure 14.3 shows the architecture of the change "Further developing".

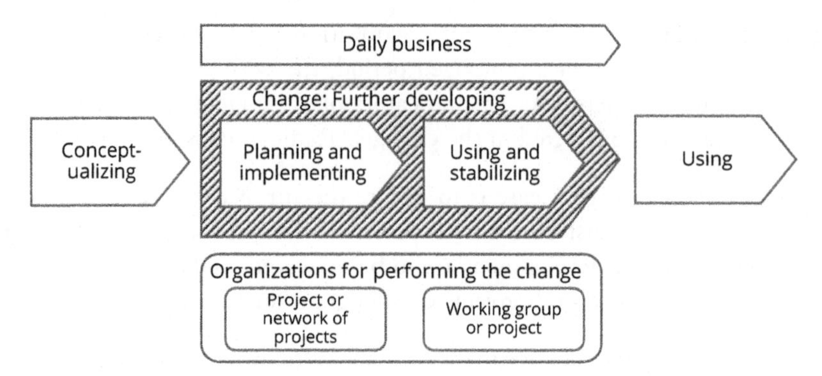

Fig. 14.3: Architecture of the change "Further developing".

To further develop an organization, the business processes "Planning and implementing" and "Using and stabilizing" must be performed. Stabilization of individual solution components can usually be performed as part of implementing. "Planning and implementing" can be performed by a project or a network of projects. In "Further developing", no programs are usually necessary. "Using and stabilizing" can be implemented by either a working group or a project.

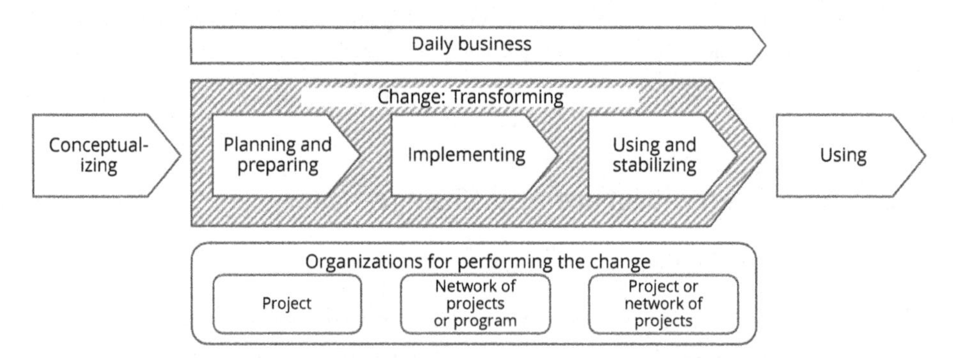

Fig. 14.4: Architecture of the change "Transforming".

Figure 14.4 shows the chain of change processes of the change "Transforming". For transforming an organization, the business processes "Planning and preparing," "Implementing", and "Using and stabilizing" must be implemented. Again, the stabilization of individual solution components can usually take place as part of implementing. The stabilization of the overall solution can only take place after the end of the implementation of the solution.

The fulfillment of the individual business processes in the change "Transforming" can be performed by different temporary organizations. The process "Planning and preparing" can be performed by a project. The process "Implementing" can be performed by a network of projects or a program, depending on the scope of the process.

"Using and stabilizing" can be performed by either a project or a network of projects. This results in a chain of projects, networks of projects, and programs for transforming an organization. Change management ensures the integrative consideration of these chains of projects or programs and, thereby, the optimization of results.

Figures 14.3 and 14.4 show that changes always take place in parallel with ongoing business activities of an organization. This contributes to the complexity of changes and the associated challenges for change managing. To ensure focus on a change, it is sometimes attempted to reduce the scope of the ongoing business activities or to optimize some of these processes before stating the change. The measures which are implemented in a change are to be clearly differentiated from the organization's ongoing business activities.

In the Excursus "Projects for performing second-order changes" the use of projects for performing the changes "Transforming" and "Radical repositioning" is described and special challenges in managing these changes are presented.

Excursus: Projects for Performing Second-Order Changes

Flexibility and Dynamics in Realizing Changes by Projects

The management of second-order changes, "Transforming" and "Radical repositioning", is often performed by the permanent organizations. However, the structures and the processes designed for the management of continuous routine processes is not suitable for carrying out these socially complex business processes. Hierarchy prevents unbureaucratic and quick problem solving in critical situations. Projects, on the other hand, promote organizational flexibility and dynamism. Flat project organizations can satisfy the extensive communication requirements for managing a discontinuity of the organization.

Characteristics of the Changes "Transforming" and "Radical Repositioning"

The changes "Transforming" and "Radical repositioning" are characterized by high uncertainty. The existing structures of the organization are questioned. New solutions are not obvious. The consequences of the measures which are implemented and new facts can alter the change status and the approach on an almost daily basis. There is a need for creativity, and there is a lot of time pressure and pressure to make decisions.

The high social complexity of second-order changes and of associated projects is attributable to the affected stakeholders. Professional communication with change stakeholders is needed.

Managing Projects for Implementing the Changes "Transforming" and "Radical Repositioning"

The challenges for managing projects for performing second-order changes are greater than for "normal" projects and require good project management competencies. Because projects for transforming or repositioning are (relatively) unique projects, experiences with similar projects and standards cannot be relied upon.

In project starting, a project organization must be designed that is adequate to meet these challenges, and project plans must be drawn up which are as flexible as possible. It is often advisable to create alternative project plans for different scenarios. The staffing of the roles of project owner and project manager significantly influences the project success.

In designing the project organization, it is important to ensure that expertise is available for the analyses, the strategic planning, and the implementation measures, in addition to decision-making competencies and the required "relationship capital". Decision-making competencies are necessary to react quickly and to be able to decide relatively autonomously on strategies and measures. Dependence on hierarchical decision structures must be minimized. Key decision makers of the social system to be changed must therefore hold roles in the project organization. The "relationship capital" in the project must ensure acceptance by affected parties of the decisions made.

In designing the project organization, special attention must be paid to communication structures. The project organization is responsible for the intensive exchange of information, the construction of current "change realities", and regular reflections on the successes and failure of measures undertaken. Representatives of project stakeholders are to be included as appropriate.

Symbolic management is important in second-order changes. The significance of a change can be highlighted by events and rituals. Project plans can be used to ensure that the members of the project organization have a common view of the project. Confidence in a professional change approach can be assured by performing project management.

Change Roles

As shown in Table 14.1, both change management roles and project and program roles are necessary for fulfilling the tasks in changes. Change owners, change managers and change agents, a change team, and possibly change management consultants are used to fulfill the tasks of change management. These change management roles are described in Table 14.2.

Change Roles and Communication Formats for Transforming and Repositioning

For transforming and radical repositioning, change management roles are to be held by specific change personnel. The people who are to fulfill change management roles must be assigned for the whole duration of the change. Because the change management roles must be fulfilled in addition to the project and program roles, a clear definition of these roles and the cooperation between them must be respected. For example, it must be ensured that the change manager informs the members of the project organization about the change vision, so that this is understood and accepted. This enables project role holders to contribute to change communication. Selected project results must be jointly planned as quick wins of the change by the change manager and members of the project organization.

Table 14.1: Roles in Changes

Change management roles
> Change owner > Change manager > Change agent > Change management consultant > Change team
Project roles
> Project owner > Project manager > Project team member > Project contributor > Project team > Sub-team
Program roles
> Program owner > Program manager > Program team member > Program contributor > Program team

Table 14.2: Description of Change Management Roles

Change management roles	Tasks
Change owner	> Taking responsibility for the results of the change > Making strategic decisions in the change > Allowing processes in the change process chain to be performed > Possibly stopping the change > Marketing the change > Ensuring resources for the change > Contributing to change projects and change programs (as owner, manager)
Change manager	> Change managing together with the change management team > Directing the change management team > Ensuring transitions between processes in the change process chain > Ensuring the compliance of change management standards > Marketing the change > Contributing to change projects and change programs (as owner, manager, expert)
Change agent	> Contributing to the change management team > Informing employees of the own organizational unit about the change > Implementing change measures in the own organizational unit > Contributing to change projects
Change management consultant	> Contributing to developing competences of the change organization > Collecting information, developing hypotheses, planning change interventions > Contributing to using change management methods > Providing analyses, observing the change organization > Moderating meetings, workshops, preparing documentations
Change team	> Managing the change processes (led by the change manager) > Developing a common „Big Change Pictures" > Ensuring synergies in the changes > Solving conflicts in the change > Organizing learning in the change

There are special communication formats for transforming and repositioning, such as change owner meetings and change team meetings. A distinction must be made between the formats of change management communication and the formats of change communication. Besides the change manager and change agents, the managers of current projects or programs also participate in change meetings.

Change Roles and Communication Formats for Further Developing

For the less complex change "Further developing", change management roles can be fulfilled by the holders of project roles. This means, for example, that the project manager of a project "Planning and implementing" can also fulfill the role of change manager.

If the further developing is organized as a chain of projects, the owner roles in the chain should be fulfilled by the same person in order to provide continuity in the change. Since different project managers can be used in a chain of projects, the additional fulfillment of the change manager role for the duration of each project must be agreed upon with these managers. This results in a transfer of the change manager role in chains of projects.

The management communication formats for further developing and the project "Planning and implementing" can also be combined to keep the process of change managing as lean as possible. This means that change managing issues can also be dealt with in project owner meetings and project team meetings. In addition, the communication formats of the permanent organization, such as profit center meetings, can be used for change management communication.

In principle, this means that in the change "Further developing", change managing and project managing can be performed together to create synergies and keep the management effort as low as possible.

In "Transforming" and "Radical repositioning", change management roles are to be assigned to change personnel for the duration of the change. For the less complex change "Further developing", change management roles can be combined with project roles for the duration of each project. In chains of projects, however, continuity of the roles "project owner/change owner" is to be ensured.

14.3 Change Initiating

Change Initiating: Objectives

The basis for a change can be either the results of top-down strategic planning in an organization or bottom-up needs identified by a stakeholder intervention. To be able to start a change, it must first be initiated.

The objective of initiating a change is to make decisions about the managing of a change. These decisions relate to

> the change type to be applied,
> the boundaries of the organization to be changed,
> the change dimensions to be considered,
> the necessary change roles and their staffing, and
> the change vision.

Change Initiating: Process

Initiating a change is like initiating a project or a program. The business process "Change initiating" includes the following tasks:

> developing initial change plans,
> coordinating initial change plans with initial project plans,
> defining change strategies,
> developing a change proposal and deciding whether to implement the change, and
> (possibly) assigning the change.

The process of change initiating is shown in Figure 14.5.

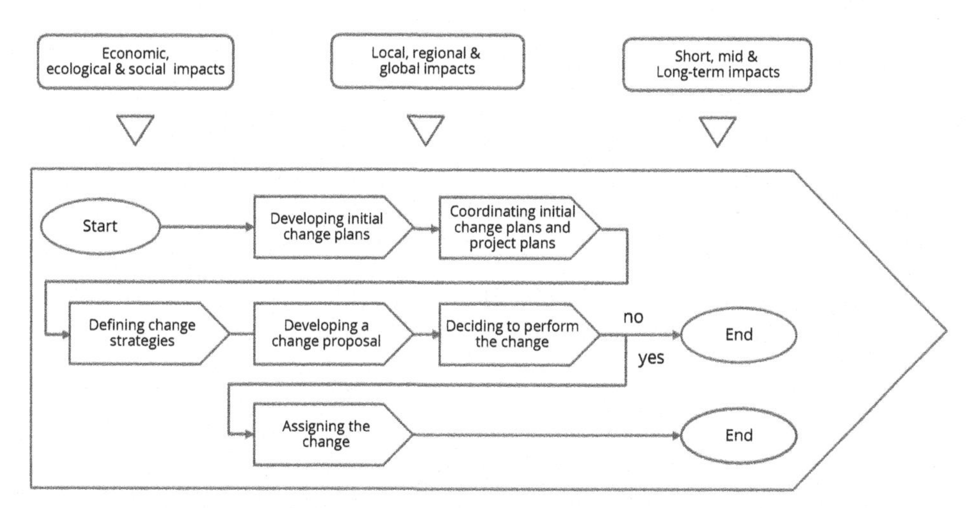

Fig. 14.5: The business process "Change initiating"—Flow chart.

The results of change initiating provide a basis for strategic managing. They feed into managing the strategic objectives of an organization (see Fig. 14.6).

For the change "Further developing", it is advisable to implement the process "Change initiating" at the same time as the process "Project initiating". Table 14.3 (page 381) shows a responsibility chart which illustrates this integrated initiating process.

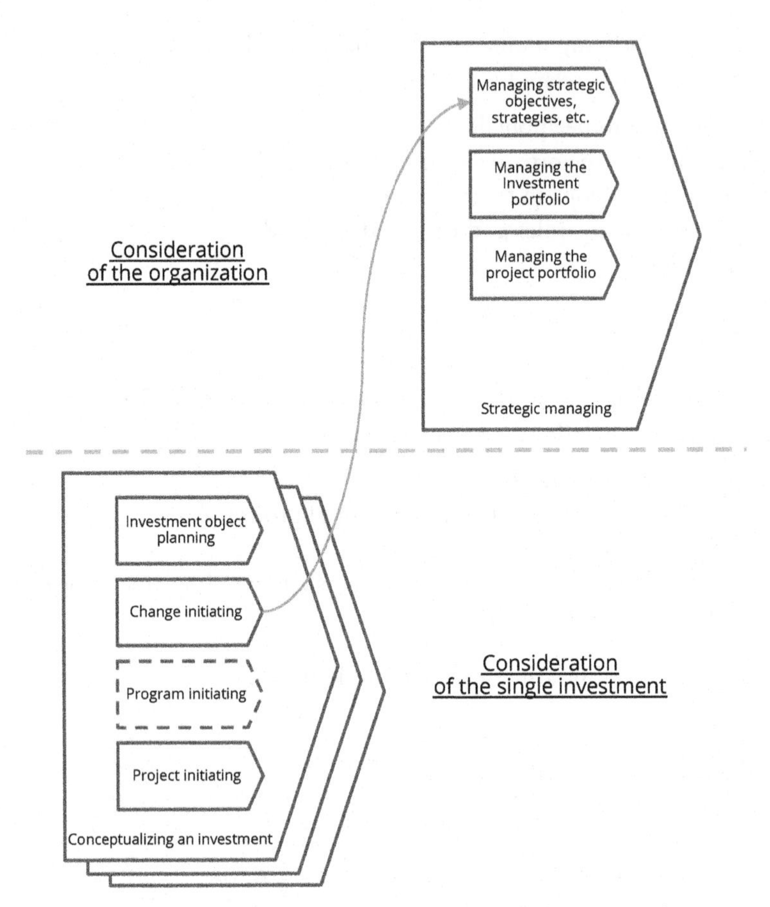

Fig. 14.6: Relation between change initiating and managing the strategic objectives of an organization.

Decisions regarding the implementation of a change and a project are to be made by a decision-making committee and a project portfolio group (see Chapter 15).

Change Initiating: Methods

Developing initial change plans involves analyzing the change contexts, specifying the boundaries of the organization to be changed, planning the dimensions to be changed and the change type, identifying key change stakeholders, formulating the change vision, developing the change architecture, and planning the change roles.

Kotter sees the formulating of the change vision as the most important change management method.[5] In the change vision, the change objectives are defined, and a future target state is described. A change vision should be clear, concise, and easily understandable. It should excite, inspire, and challenge but be achievable at the same

[5] See Kotter J.P., 1996, p. 72.

Table 14.3: Integrative Initiation of a Change and a Project

Change and project initiating								
	Roles							
Legend P .... Performing C...Contributing I... Being informed Co...Coordinating Process tasks	Initiator	Initiation team	Decision committee	Expert pool managers	Change owner/ Project owners	Change manager/ Project managers	Change stakeholder/Project stakeholders	Tool/Document
Developing initial change plans	I	P, Co		C			C	1
Developing initial project plans	I	P, Co		C			C	2
Analyzing the project portfolio		P, Co		C				3
Coordinating initial change and initial project plans	C	P, Co		C			C	
Developing change strategies	C	P, Co		C			C	
Developing a change proposal and project proposals	P	C, Co						4
Deciding to perform the change and the projects	C	C, Co	P					5
Assigning the change and the projects		Co		I	P	C	I	6

Tool/Document:
1 ... Initial change plans
2 ... Initial project plans
3 ... Project portfolio data base
4 ... Change proposal, project proposals
5 ... Minutes of the meeting
6 ... Change assignment, project assignments

time. The change vision must be distinguished from the overall vision of the organization. Change visions for a number of changes that may be implemented in parallel can contribute to the vision of the organization (see Fig. 14.7).

As with projects, change strategies for the actualization of a change vision can relate to procurement, the use of technology, the use of methods, the planning of stakeholder

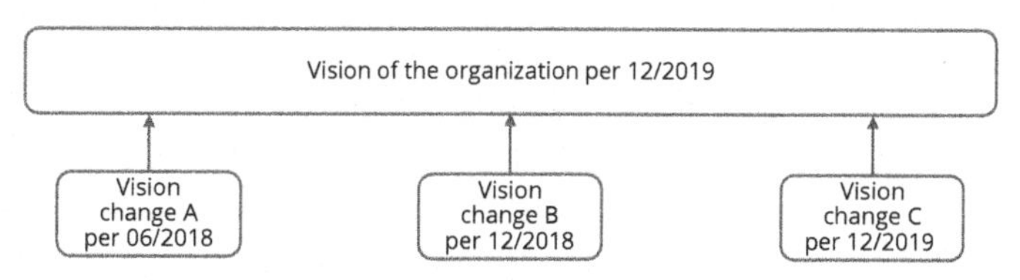

Fig. 14.7: Relation between a change vision and the organization's vision.

relations, and change communication. The difference between change strategies and project strategies is that change strategies apply to the change as a whole and not only to a project of the change, and that they are established for the duration of the change and not only for the duration of a project. Initial change plans resulting from the application of methods for change initiating are the basis for the development of the change proposal and change assignment.

Examples of initial change plans and a change proposal for the case study "Values4-Business Value" are presented in Section 14.9 (beginning on page 395). The development of the change proposal and the initial change plans was supported by using templates for change initiating. These templates can also be used for change managing.

14.4 Change Managing: Business Process

Change Managing: Objectives

"Change managing" is a business process of project-oriented organizations that should contribute to the successful realization of changes. It includes the sub-processes "Change starting", "Change coordinating", "Change controlling", "Change communicating", and "Change closing". Change managing focuses on managing a change and not on the performance of content-related tasks involved in a change, such as organizational design or personnel development.

The objectives of change managing are described in Table 14.4.

The start event for change managing is the change assignment. The end event is the approval of the change. Change managing is preceded by "Change initiating".

Table 14.4: Objectives of the Business Process "Change Managing"

Objectives of the process: Change managing	
Economic objectives	> Complexity and dynamics of the change managed > Change starting, change coordinating, change controlling, change communicating, and change closing performed professionally > Economic impacts optimized for the implemented investment > Quick wins ensured
Ecological objectives	> Local, regional and global ecological impacts of the change considered > Ecological impacts optimized for the implemented investment
Social objectives	> Change management personnel recruited, allocated, incentive system developed and applied > Change management personnel assessed, developed, released > Local, regional and global social impacts of the change considered > Social impacts optimized for the implemented investment > Acceptance of the change ensured > Stakeholder involved in change managing

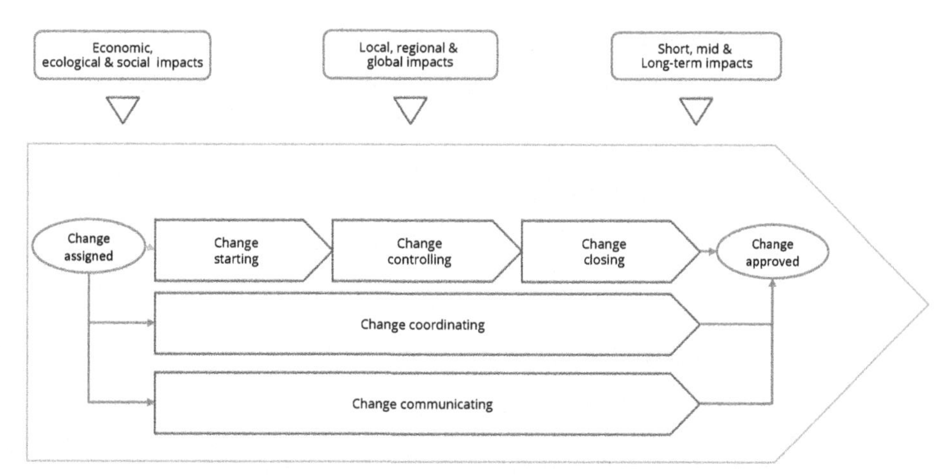

Fig. 14.8: Business process change managing—Flow chart.

The business processes "Project managing", and possibly "Program managing" and "Developing content-related solutions", are performed in parallel with change managing. Subsequent business processes are "Using the solutions" and possibly "Benefits realization controlling".

Change Managing: Process

The process "Change managing" is shown in Figure 14.8. This also makes visible how the principles of sustainable development are considered. Detailed descriptions of the objectives and tasks of the sub-processes of change managing can be found in Table 14.5. The scope of the change management tasks to be fulfilled is larger for the transforming and repositioning than for further developing.

Change managing should achieve sustainable change results, taking into account the economic, environmental, and social impact of the change and the investment realized by a change. If change managing is reduced to the performance of change communication or is confused with the performance of content-related tasks, these change objectives cannot be realized.

14.5 Change Managing: Designing the Business Process

Establishing general rules for designing change organizations and using change management methods is a corporate governance task. Within the framework of these rules, the business process "Change managing" is to be designed in accordance with specific requirements in a particular situation.

The use of methods and communication formats for change managing, the use of a change management infrastructure, and possibly the use of change management consultants must be planned.

Table 14.5: Objectives and Tasks of the Sub-Processes of Change Managing

Change starting	
Objectives	> Change management personnel recruited and allocated, incentive system developed, change management personnel developed > Structures developed for realizing the change objectives and for managing the complexity and dynamics of the change
Tasks	> Analyzing initial change plans, developed during change initiating > Verifying the boundaries and the change dimensions of the organization to be changed > Designing the organization for performing the change > Verifying change management roles and staffing > Developing organizational and individual competences as well as an incentive system for the change > Developing the change vision and change strategies for realizing the vision further > Convincing selected change stakeholders about the change urgency > Planning quick wins > Planning the relation to the process „benefits realization controlling"
Change coordinating	
Objectives	> The change continuously coordinated with members of the change organization and representatives of change stakeholders > Progress and acceptance of the change ensured
Tasks	> Continuous coordinating of the members of the change organization > Continuous coordinating of the change team with representatives of change stakeholders > Frequently organizing program meetings, program office meetings, and project meetings
Change controlling	
Objectives	> Change controlled and directive measures planned > Change management personnel assessed and further development planned > Structures for realizing the change objectives and for managing the complexity and dynamics of the change controlled and optimized
Tasks	> Analyzing the change status by considering information from the change projects and programs > Analyzing change structures and relations to change stakeholders > If needed, defining directive measures for redesigning the relations to change stakeholders > If needed, developing the change organization further > If needed, adapting the change vision and change strategies > Controlling the realization of quick wins > Managing the relation to benefits realization controlling > Developing ideas for stabilizing the organization
Change communicating	
Objectives	> Information about the change vision, the change organization and the change process provided > Orientation to the change stakeholder provided > Commitment of employees for the change increased > Employees prepared for positive and negative consequences of the change > Dialogues held with change stakeholders by change agents
Tasks	> Planning objectives and strategies of the change communication > Developing a change communication plan > Preparing the change communication periodically > Performing change communication > Reflecting the results of the change communication
Change closing	
Objectives	> Change management personnel assessed and released > Change organization dissolved > Additional measures for benefits realization defined
Tasks	> Planning measures for the post-change phase > Organizing for change lessons learned > Giving feedback to the change organization and its members

Change Managing: Methods

Methods for change managing are shown in Table 14.6. Recommendations are provided for "must" and "can" use of these methods. The objectives of the change management

Table 14.6: Methods for Change Managing

Methods for change managing	Further developing	Transforming, Repositioning
Developing a change proposal and a change assignment	Must	Must
Defining the organization to be changed, the change dimensions, change type and change contexts	Must	Must
Designing the chain of change processes and the change organization	Must	Must
Performing a change stakeholder analysis	Must	Must
Formulating a change vision	Must	Must
Planning the change communication	Must	Must
Developing a sense of change urgency	Can	Must
Performing a change management training	Can	Must
Defining a change incentive system	Can	Can
Planning quick wins	Can	Must
Developing a change score card	Can	Must
Planning the stabilizing of the organization	Can	Must

methods are listed in Table 14.7. Change plans resulting from the use of methods for further developing are usually less detailed than those for transforming or repositioning.

Forms to support the application of the methods are shown in examples for the case study "Values4Business Value" in Section 14.9 (beginning on page 395).

Methods for change managing are to be distinguished from methods for content-related work in changes. Methods for content-related work can be differentiated according to change dimensions, as listed in Table 14.8.

Changes are implemented in the form of projects, networks of project, or programs. Methods for project and program initiating, project and program managing, and managing of networks of projects are therefore to be used in changes (see Chapters 5, 6, and 16).

Change Managing: Communication Formats of the Change Organization

In changes, communication formats for change managing are required in addition to those for project or program managing. The communication formats for change managing used by the change organization are:

- > workshops for change starting and change closing,
- > change owner meetings for change starting, change controlling, and change closing,
- > change team meetings for change starting, change controlling, and change closing.

In the case of further developing, the communication formats for change managing can be combined with those for project managing.

Table 14.7: Objectives of the Methods for Change Managing

Methods for change managing	Objectives of applying the method
Defining the organization to be changed, the change dimensions, change type and change contexts	> Organization units identified which are affected by the change > Relevant change dimensions identified > Change type defined based of the number of change dimensions and the change demand > Change contexts described (objectives and strategies of the organization, other ongoing changes, etc.)
Designing the chain of change processes and the change organization	> Change processes identified > Objectives, starting and ending event, pre-process and post-process phase for each process defined > Organization for performing each process defined and designed > Change management roles listed and role players assigned
Formulating a change vision	> Change objectives to be realized in the future defined > Change vision described for each change dimension > Change vision described by a slogan or metaphor
Performing a change stakeholder analysis	> Change stakeholders differentiated by their influence and buy-in to the change > Expectations of change stakeholders analyzed > Strategies, measures and responsibilities for managing relations with change stakeholders planned and controlled
Developing a sense of change urgency	> Change stakeholders identified who are lacking a sense of urgency for the change > Data and facts prepared for each stakeholder, to prove weaknesses and to show potentials
Planning the change communication	> Change communication strategies defined > Change communication planned periodically, differentiated for each change stakeholder > Change communication controlled periodically
Planning quick wins	> Quick wins, differentiated by economic, ecological and social quick wins, defined > Realization dates, responsibilities, realization effort and benefit of quick wins planned > Usage of quick wins in the change communication planned
Developing a change score card	> Change status visualized and interpreted in a score card > Overall evaluation and evaluation by different criteria (meeting change objectives, relations to change stakeholders, etc.)

Table 14.8: Methods for Content-Related Change Work

Change dimension	Method
Strategic managing	> Performing a SWOT analysis > Performing a product-market portfolio analysis > Using the scenario technique > Defining an organization vision, etc.
Organizational designing	> Developing an organization chart > Developing job descriptions > Performing process management > Defining values > Establishing rules, etc.
Personnel managing	> Performing management by objectives > Planning personnel development > Defining incentive systems, etc.
Marketing	> Branding > Performing events > Performing campaigns, etc.
Infrastructure managing	> Managing the ICT > Managing the rooms infrastructure, etc.
Controlling und financing	> Planning the budget > Developing a finance plan, etc.
Stakeholder relations managing	> Performing a stakeholder analysis > Establishing frame contracts with suppliers, etc.

Change Managing: Communication Formats with Change Stakeholders

The communication formats of the change organization are to be distinguished from those for communication with change stakeholders. These are to be planned in a change-specific manner.

Communications with change stakeholders can be performed by means of digital media, print media, and personal communication. The intranet, emails, electronic newsletters, postings on social media, etc. can be used. Relevant print media are, for example, change folders or an employee newspaper. Personal communication formats are, for example, workshops, presentations, openings, open space events, world cafés, or roadshows. Change stakeholders will also receive information about changes in the communication formats of the permanent organization, such as board meetings or departmental meetings.

Infrastructure for Change Managing

Professional change managing requires the use of an appropriate ICT infrastructure and an appropriate room infrastructure. Templates for change managing can be provided electronically. Collaboration portals, telephone conferences, and video conferences can be used. A room infrastructure must be provided for change management meetings and possibly for a change office.

Use of Change Management Consultants

Consultants can be used in changes both to support content-related work and for change managing. The use of change management consultants is particularly advisable for change initiating and change starting, which is when the basic change structures are created. The decision about whether to use consultants should be made by the change organization. Consulting roles can be performed either by qualified employees of the project-oriented organization or by external consultants. The client systems for change management consultants are the organizations to be changed.

14.6 Change Initiating and Change Managing: Values

The values of the RGC management paradigm (see Fig. 14.9) are interpreted as follows for change managing.

Holistic Definition of Solutions and Boundaries

The boundaries of the organization that is to be changed must be defined holistically. This means that all organizational units that are affected by a change are considered in the change process.

Holistic definition of solutions and boundaries	Sustainable business value as success criterion	Quick realization of results and efficiency	Iterative approach
> Defining a change vision holistically > Defining the boundaries of an organization to be changed holistically > Considering all relevant change dimensions in change management	> Contributing to the business value by a change > Optimizing the business value by controlling the benefits realization > Considering principles of sustainable development in change management	> Ensuring quick wins in a change > Stopping unsuccessful changes in time > Performing the change efficiently	> Structuring a change process-oriented > Using a chain of projects and programs for performing a change > Performing change managing iteratively (change initiating, starting, controlling)

Context-orientation	Continuous and discontinuous change	Frequent, visually supported communication	Empowerment and resilience
> Considering relations to parallel ongoing changes > Designing an integrated change organization > Managing of and for change stakeholders > Involving stakeholders in change managing	> Promoting organizational and individual learning in the change > Promoting changes of a change	> Communicating with change plans, score cards etc. > Performing stand up-meetings, reflections in the change organization	> Empowering the change organization > Empowering change projects and change programs > Developing resilient structures by a change

Fig. 14.9: RGC management values and change managing.

To enable holistic solutions, all relevant change dimensions must be considered. These can be the services of the organization, its markets, business processes, organizational structure, personnel, infrastructure, budget and financing, as well as its stakeholder relations. The definition of the relevant change dimensions provides a basis for a holistic definition of the change vision.

A holistic approach to change managing means that all sub-processes of change managing and their relations are considered and that methods for change managing are used in relation to each other.

Sustainable Business Value as Success Criterion

"Agility is the ability to both create and respond to change in order to profit in a turbulent business environment".[6]

To be agile, therefore, means not only to react, but also to act in order to create benefits.

To achieve a sustainable business value for the changed organization is the success criterion of a change. The economic, ecological, and social consequences of a change, but also its short, medium, or long-term nature, as well as its local, regional, and global consequences, are to be considered. The benefits of a change can be optimized by benefits realization controlling.

The principles of sustainable development also influence the business process "Change managing". So, for example, the objectives of change managing can be differentiated into economic, ecological, and social objectives, and stakeholders can be involved in change managing. The principles of sustainability can be applied in the methods of change managing—for example, in the development of the change vision.

Quick Realization of Results and Efficiency

In change managing, quickly and relatively easily achievable results are identified as "quick wins". Quick wins arise in the projects performed for a change. The planning and controlling of quick wins therefore requires coordination between the change manager and the corresponding project managers.

The timing of quick wins can be seen in Figure 14.10. The benefits of quick wins include not only motivating employees of the organization to be changed, but also optimizing the cost–benefit ratio of the investment implemented by a change (see Fig. 14.11).

The change process must be structured so that the probability of change success can be assessed at an early stage on the basis of interim results. An unsuccessful change should be discontinued at an early stage.

[6] See Highsmith J., 2010, p. XXIII.

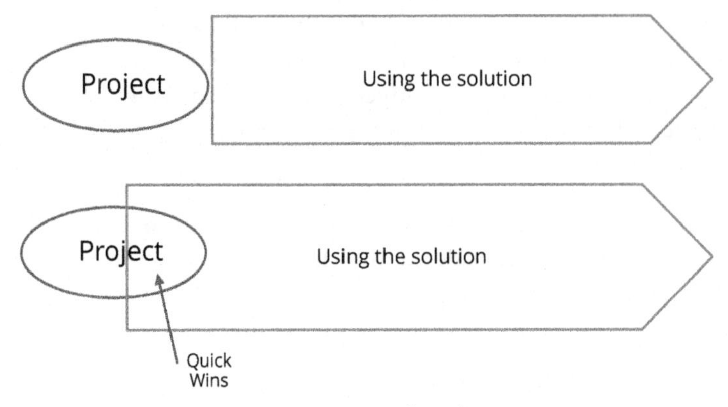

Fig. 14.10: Timing of quick wins.

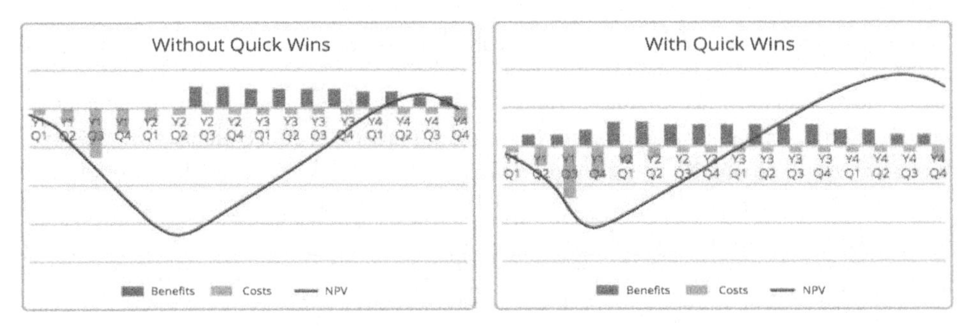

Fig. 14.11: Optimizing the cost–benefit ratio of an investment by quick wins.

Changes and change managing must both be efficient and lean. Only measures that contribute to realizing the change vision are to be implemented in a change. Communication with change stakeholders is appropriate if not too much information is provided and information is not provided at the wrong time. Standardized processes can often significantly improve efficiency in change managing. The institutionalization of change managing in the organization is a precondition for this.

Iterative Approach

In changes, an iterative approach can contribute to achieving sustainable business value. In an iterative approach, the definition of requirements and the development of the content-related solution are performed iteratively. The solution requirements to be fulfilled by the change are concretized and prioritized on an ongoing basis.

Quick wins in the change are made possible by creating "minimum viable products". Different components of the solution are stabilized during development in multiple cycles of designing, developing, testing, and applying (see Fig. 14.12).

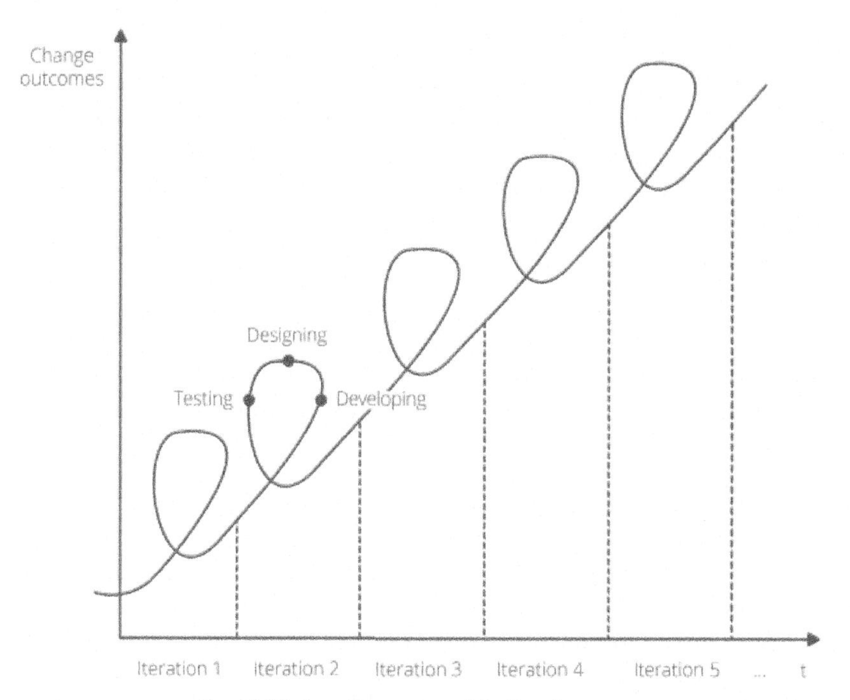

Fig. 14.12: Iterative approach in the change process.

Change managing can also be performed iteratively: At a meta-level, an iterative plan is the result of a process-oriented change architecture. This creates chains of processes as well as chains of projects or programs. The cyclical development of change plans such as the change vision, the change stakeholder analysis, the change communication plan, quick wins plan, etc. in change initiating, change starting, and change controlling also corresponds to an iterative approach.

Context Orientation

Context orientation in change managing means that change contexts such as the investment implemented by the change, the objectives and strategies of the implementing organization, and relations with change stakeholders are all considered.

Change stakeholders are to be identified, and relations with them are to be explicitly managed to ensure sustainable results. A distinction must be made between change and project stakeholders, as not all project stakeholders are "affected" by the change, and stakeholders have different expectations for a project and the associated change.

As in project managing, "management for stakeholders" can also be practiced in change managing, stakeholders can be involved in managing, and an integrated change organization can be designed.

Continuous and Discontinuous Learning

The complexity and dynamics of a change necessitate continuous learning, but sometimes also discontinuous learning. Learning by members of the change organization and by the change projects must be promoted by frequent reflections and feedback. This ensures efficiency in the change process and innovative change results. A "change of a change" may be necessary—for instance, as a result of new conditions in the implementing organization.

Frequent, Visually Supported Communications

In changes, a distinction must be made between the change management communication in the change organization and the change communication with change stakeholders.

Appropriate communication with change stakeholders is an important change managing task. The objectives and strategies of the communication, its intensity, the formats used, etc. must be explicitly planned. The visualization tools of the projects, of a possible program, and of the change itself can be used. According to Kotter, the change vision is the most important tool for change communication with stakeholders.[7]

Appropriate spatial and ICT infrastructure is to be provided to support communications within the change organization and with change stakeholders.

Empowerment and Resilience

Empowerment and resilience can be interpreted as relating to the change objectives and to the change process.

The objective of a change is to create structures that are empowered and resilient. To ensure the sustainability of the changed organization, it should promote the individual responsibility of employees and of the permanent and temporary organizational units as much as possible. The structures created should also ensure a high degree of stability by means of an appropriate "flow equilibrium".

Empowerment in the change process means that both the change organization and the projects and/or programs performed are given responsibility for implementing the change. The transfer of responsibility to different temporary organizations requires clear structures and rules. The strong differentiation of the change architecture is to be balanced by extensive integration measures.

The stability of the change process must not be jeopardized, even if some of the change projects have problems. The resilience of a change is to be ensured in a similar way to that of a program. To this end, agile and redundant structures are to be created and quick wins secured, as these contribute to the stability of a change.

[7] Kotter, J.P., 1996, p. 85ff.

Values determine the objectives pursued in change managing, the tasks to be performed, the methods to be used, and the roles to be fulfilled.

14.7 Change Managing: Benefits

The benefits of change managing are contribution to sustainable change results and the assurance of an efficient and transparent change process. Effective results are made possible by giving a change vision a strong business focus. In detail, the following benefits of professional change managing can be defined:

- Competitive advantages such as high quality of results, low change costs, and short change duration,
- Transparency through providing realistic change plans and consistent change reports,
- Clarity about the change status,
- Appropriate bases for change-related management decisions,
- Individual and organizational learning through reflections,
- Fulfilled expectations of change stakeholders through appropriate stakeholder communication and involvement,
- Optimized relations between the change and its projects and programs,
- Optimized cost–benefit ratio of the investment implemented, and
- Necessary information for project portfolio management.

Without change managing, there is a high risk that change results will not be accepted by stakeholders. If change managing is reduced to the implementation of change communication, or if it is confused with the performance of content-related tasks, this benefit cannot be achieved.

14.8 Relations Between Change Managing and Project or Program Managing

"Changes by projects!"—This strategy requires an analysis of both general and change type-specific relations between change managing and project or program managing.

General Relations Between Change Managing and Project or Program Managing

The following general relations between change managing and project or program managing exist:

- The managing of projects or programs usually takes place within the scope of changes. For this reason, change managing defines objectives, rules, and standards to be considered in managing projects or programs of a change.

- Project or program communication is an integral part of change communication. Project or program communication must be coordinated with change communication.
- Change stakeholders are to be distinguished from project or program stakeholders. On the one hand, there are project stakeholders who have only project-related expectations. On the other hand, project stakeholders can also be change stakeholders. In this case, they have different expectations for the project and the change.
- "External" contracting projects contribute to changes in customer organizations. A shared change managing and project managing understanding of the contractor and of the customer organization creates win-win situations.
- Quick wins of changes are achieved in projects. The planning of quick wins must therefore be performed jointly by members of the change organization and the respective project organizations.
- Program and project owners, program and project managers, and program and project team members contribute to changes.
- There are differences between program and project managers and change managers: Program and project managers are responsible for the program or project results and perform integrative functions in programs or projects. Change managers are responsible for change results and the integration in changes.
- The role of project manager also differs from the role of change manager in its short-term orientation (for a project) compared to the medium-term orientation (for a change). To make cooperation more efficient, there is a need to ensure appropriate role understandings of the change managers and program or project managers.
- Projects and programs as temporary organizations may also require changes. The members of the program and project organizations may therefore also require change managing competencies. Second-order changes of projects are project transforming and project repositioning.
- Change management methods can be used in projects, even if no changes are formally defined. For example, planning the stabilization of a solution in the post-project phase should be part of a project.

Change Type-Specific Relations Between Change Managing and Project or Program Managing

Different approaches are needed to manage the change types "Further developing", "Transforming", and "Radical repositioning". In further developing, change management roles are to be fulfilled by project role holders, and no separate formats for change management communication are required. Fewer change managing methods are used than in transforming, and the change plans are less detailed. This keeps the managing of further developments "lean".

In further developing, change management roles are fulfilled by those responsible for fulfilling project roles. This means that the project owner is also the change owner, the project manager is also the change manager, and the project team is also the change team. To ensure continuity in a chain of projects, the project owner of an implementation project should also be the project owner of a possible subsequent stabilization project. The project manager and the project team are to be newly defined for the stabilization project. The change manager and change team are thereby also newly established.

Although change managing in further developing is less formalized in an organizational sense, it must nevertheless be ensured that change managing tasks are fulfilled professionally. So, for example, change managing tasks must be shown in the work breakdown structure in addition to project managing tasks as "(contribution to) change managing".

In transforming, persons are explicitly assigned to the change management roles. Only in exceptional cases will the same persons perform change roles and program or project roles. Separate formats are used for change management communication. The change owner must coordinate with the program owner and the project owners, and the change manager must coordinate with the program manager and the project managers. The contributions of the program and the projects to the change are thereby coordinated.

The owners, managers, and team members of programs and projects require appropriate change managing competencies in order to be able to make the following contributions to the change "Transforming":

> contribution to defining strategies and measures for managing change stakeholder relations,
> contribution to change communication,
> contribution to the planning of quick wins,
> contribution to the planning of stabilization measures,
> contribution to determining the change status as part of change controlling, and
> contributions to adapting the change vision and the change structures.

The efficiency of a change as well as of its projects and programs and the quality of the results can be ensured by considering these relations between change managing and project or program managing.

14.9 Change Initiating and Change Managing: Case Study "Values4Business Value"

Perception of "Values4Business Value" as a Change

In conceptualizing "Values4Business Value", the further development of RGC management approaches, services, and products was identified as a change for RGC. The change was perceived as "Further developing". The change and the associated investment were given the name "Values4Business Value", because value orientation in management and the securing of sustainable business value for customers were key objectives.

The change was initiated together with the initiation of the project "Developing Values-4Business Value". The following initial change plans which were developed are shown in Tables 14.9 to 14.14 and above in Figures 14.10 and 14.11 (page 390):

- change vision,
- change architecture,
- change management organization,
- change stakeholder analysis,
- relations with change stakeholders,
- change management communication plan, and
- change communication plan.

The initial change plans were attached to the change proposal (see Table 14.15 on page 402). The change assignment was issued together with the project assignment. The initial change plans were adapted during change starting, and in addition, a quick-wins plan (see Table 14.16 on page 404) was developed.

Initial Change Plans

The vision of the change "Values4Business Value" was defined for December 2019 (see Table 14.9).

"Values4Business Value" was chosen as a slogan, and for each change dimension vision statements were formulated. The objectives defined in the change vision were to contribute to the realization of the strategic objectives of RGC:

- contributions made to sustainable project, program, and change managing in the society,
- sustainable business values secured for RGC customers,
- evolutionary growth of RGC in the DACH region,
- Theme leadership in process, project, program, and change managing in the scientific and consulting community,
- management of RGC further professionalized, and
- employee satisfaction ensured by value-oriented management.

The architecture of the change "Values4Business Value" and its context in time can be seen in Figure 14.13. The change included the process "Developing Values 4Business Value", which was performed by a project, and the process "Stabilizing Values4Business Value", which was performed by a small project. Before these processes, "Values4-Business Value" was conceptualized by a working group. "Values4Business Value"

Table 14.9: Case Study "Values4Business Value"—Change Vision

Change Vision

Values4Business Value

V. 1.001 v. S. Füreder per 5.10.2015

Change dimension	Vision statement
Services and products	• Services and products for sustainable project, program, and change managing developed and communicated in the society • Book PROJECT.PROGRAM.CHANGE published in German and in English
Businesses	• Contributed to an evolutionary growth of the RGC in the DACH region by management innovations
Organization	• Management of the RGC professionalized further by values orientation and application of further devolped management approaches • Stakeholder-orientation improved
Personnel	• Personnel bound through attractive jobs • Contributed to an international recognition of the RGC consultants as management experts and management innovators
Stakeholder realtions	• Basis developed for assuring RGC customers sustainable business values, empowerment, resilience and efficiency, context orientation, and stakeholder orientation • Contributed to the thematic leadership in the scientifc & consulting community by publishing PROJECT.PROGRAM.CHANGE • Additional cooperation with publisher realized
Change slogan	
Values4Business Value	

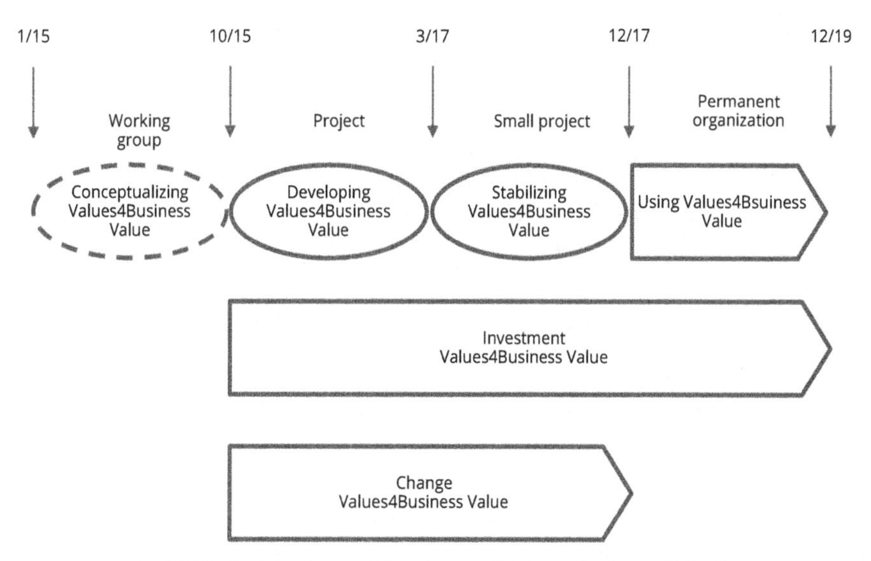

Fig. 14.13: Architecture of the change "Values4Business Value".

was supposed to be used by the permanent organization of RGC. Table 14.10 briefly describes the two change processes.

The change owner, the change manager, and the change team were defined as change management roles. These change management roles were performed by RGC employees, together with the roles project owner, project manager, and project team of the project "Developing Values4Business Value".

It was planned to retain the project owner in the follow-up project "Using and Stabilizing Values4Business Value". It was expected that the project manager and project

Table 14.10: Processes of the Change "Values4Business Value"

Processes of the Change "Values4Business Value"

Values4Business Value

V. 1.001 v. S. Füreder per 5.10.2015

Process: Developing Values4Business Value	
Process objectives	• RGC Management approaches and services further developed and documented in a book • Customers bound and won by further developed management approaches • Basis for increased revenues and contribution margins developed • Cooperation with publisher developed further • Peer Review Group established for quality assurance • RGC management professionalized and personnel developed further
Process start date	10/1/2015
Process end date	3/31/2017
Pre-process phase	Conceptualizing Values4Business Value
Post-process phase	Using and stabilizing Values4Business Value
Organization for implementing the processes	Project
Owner	R. Gareis, L. Gareis
Manager	S. Füreder
Process: Stabilizing Values4Business Value	
Process objectives	• RGC services optimized • RGC approach for porcess managment developed further • Additional personnel development performed • Additional marketing for Values4Business Values performed
Process start date	4/3/2017
Process end date	12/22/2017
Pre-process phase	Developing Values4Business Value
Post-process phase	Using Values4Business Value
Organization for implementing the processes	Small project
Owner	R. Gareis, L. Gareis
Manager	S. Füreder

Table 14.11: Case Study "Values4Business Value"—Change Management Roles

Change Management Roles

Values4Business Value

V. 1.001 v. S. Füreder per 5.10.2015

Change owner	Roland Gareis, Lorenz Gareis
Change manager	Susanne Füreder
Change team	Roland Gareis, Lorenz Gareis, Susanne Füreder
Change agents	Members of the Peer Review Group

team roles, and thus also the change manager and the change team, would be held by different people than in the development project. The members of the peer review group were identified as "informal" change agents, who contributed to communicating "Values4Business Value" to the community (see Table 14.11).

The participants, contents, dates, and responsibilities for the different formats for change management communication were specified in a change management communication plan (see Table 14.12).

The results of a change stakeholder analysis developed in October 2015 are shown in Figure 14.14. Change stakeholders were differentiated according to their power to

Table 14.12: Case Study "Values4Business Value"—Change Management Communication Plan

Change Management Communication Plan

Values4Business Value

V. 1.001 v. S. Füreder per 5.10.2015

Communication format	Participants	Content	Date	Responsible
Change start workshop	Change team and change owner	Starting the change, detailing the change plans	10/2015	Change manager
Change team controlling meeting	Change team	Controlling the change	Once per month	Change manager
Change owner meeting	Change owner, Change manager	Strategic change management decisions	Once per month	Change manager
Peer Review Group workshop	Change owner, Change manager, Peer Review Group members	Communicating the change	Twice	Change manager
Change closing workshop	Change team and Change owner	Closing the change	03/2017	Change manager

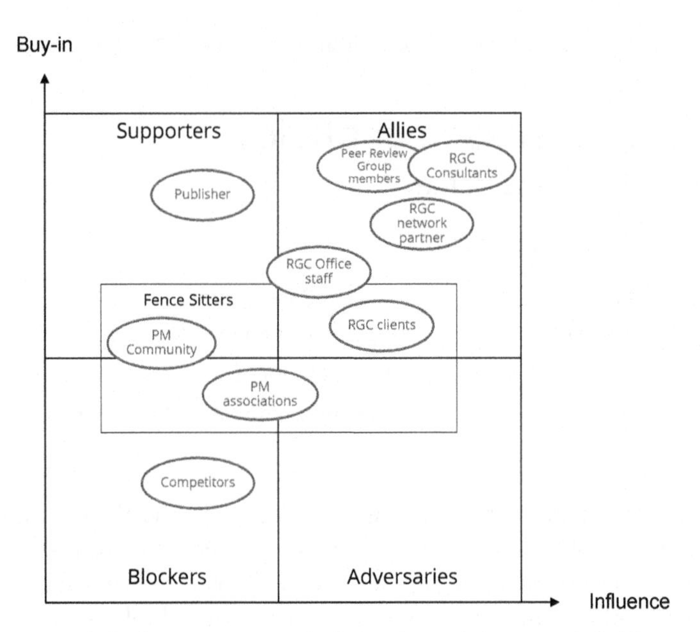

Fig. 14.14: Case study "Values4Business Value"—Analysis of change stakeholders.

influence the change and their buy-in. This analysis was periodically adapted during the change. This also provided a basis for adapting the strategies and measures for managing relations with change stakeholders. The strategies and measures for managing the relations as of October 2015 are described in Table 14.13.

Assumptions based on the stakeholder analysis were that stakeholders close to RGC, such as the consultants and network partners, would have both the opportunity to influence the change and the willingness to support the change. RGC customers would have to be informed and convinced about the benefits of the change results.

The project management associations such as the IPMA (International Project Management Association) or the PMI (Project Management Institute) were considered as "undecided" and had to be appropriately informed to make them "allies".

The analysis of change stakeholders and the planning of relations with them formed the basis for the development of an initial change communication plan for the period October to December 2015 (see Table 14.14 on page 402).

Change proposal

In the change proposal, the key information about the change "Values4Business Value" was summarized and supplemented by the initial change plans that had been developed. This information provided the basis for the RGC management decision to assign the change as planned.

Table 14.13: Case Study "Values4Business Value"—Relations with Change Stakeholders

Relations with Change Stakeholders
Values4Business Value

V. 1.001 v. S. Füreder per 5.10.2015

Stakeholder	Interpretation of the position (Influence, Buy in)	Strategy for managing the relation	Measures for managing the relation	Responsible
Allies				
Peer Review Group members	Have the possibility to represent "Values4Business Value" in the management community; identify with the change	Cooperating	Planning communication objectives and content	Change owner
RGC consultants	Inform customers about the change and the management innovations; inform the RGC office staff and the networking partners	Cooperating closely	Content-related communication, contribution in the Peer Review Group, implementation of innovations in customer contracts	Change owner
RGC networking partners	Are interested in the results, but are not involved intensively	Cooperating	Contribution in the Peer Review Group; implementation of innovations in customer contracts	Change owner
RGC office staff	Are interested in the results; have also additional tasks to fulfill in the change	Cooperating closely	Contribution to the change; continuous information about the change in the "RGC Grenzchen"	Change owner
Supporters				
Publishers	Can support the information of the management community; are interested in the success of the publication	Cooperating	Common marketing of the publication	Change manager
Blockers				
Competitors	Are interested in the management innovations	Not cooperating		
Adversaries				
"Fence Sitters"				
RGC customers	Are interested in the management innovations	Informing and cooperating		RGC consultants
Management community	Are interested in the management innovations	Informing	Informing about essential innovations of the publication, about new services through different media	Change owner, Change manager
Project management associations	Are interested in the management innovations; possibly expect content-related conflicts of interest	Informing	Information about management innovations, about connections to other shared approaches	Change owner

Additional Change Plans

In change starting, the initial change plans were adapted and a quick wins plan was developed (see Table 14.16 on page 404). Because the project "Developing Values-4Business Value" and consequently the associated change took some time, it seemed important to identify quick wins to secure benefits for RGC and motivation of the members of the change organization.

Essential quick wins were to be secured by "prototyping". Contributions to various conferences and the design of seminars on the topics "Agility & Processes", "Sustainable Development", "Benefits Realization Management", "Agility & Projects", etc. were identified as quick wins.

This prototyping was to be a first step in collecting essential insights for the further development of management approaches.

Table 14.14: Case Study "Values4Business Value"—Change Communication Plan: October to Decemter 2015

Change Communication Plan: October to December 2015

Values4Business Value

V. 1.001 v. S. Füreder per 5.10.2015

Change stakeholder	Communication format	Content	Date	Responsible
RGC internal change stakeholders				
RGC staff	RGC Grenzchen	Information aber the change, about roles in the change	October and December	Change owner
RGC consultants	Project team meetings	Information aber the change, about roles in the change; decisions about content-related cooperation	Once per month	Change owner
RGC networking partners	Mail	Information aber the change, about roles in the change	November	Change owner
RGC external change stakeholders				
Publisher Manz	Meeting	Information about the change, about content and form of the publication; planning the book marketing	December	Change manager

Table 14.15: Case Study "Values4Business Value"—Change Proposal

Change Proposal

Values4Business Value

V. 1.001 v. S. Füreder per 5.10.2015

Change type	
Further developing	
Change owner	**Change manager**
Roland Gareis, Lorenz Gareis	Susanne Füreder
Change team members	
Roland Gareis, Lorenz Gareis, Susanne Füreder	
Change vision	
Services and products for sustainable project, program, and change managing developed and communicated in the society	
Book PROJECT.PROGRAM.CHANGE published in German and in English	
Contributed to an evolutionary growth of the RGC in the DACH region by management innovations	
Management of the RGC professionalized further by values orientation and application of further devolped management approaches	
Stakeholder orientation improved	
Personnel bound through attractive jobs	
Contributed to an international recognition of the RGC consultants as management experts and management innovators	
Basis developed for assuring RGC customers sustainable business values, empowerment, resilience and efficiency, context-orientation, and stakeholder-orientation	
Contributed to the thematic leadership in the scientifc and consulting community by publishing PROJECT.PROGRAM.CHANGE	
Additional cooperation with publisher realized	

(continues on next page)

Table 14.15: Case Study "Values4Business Value"—Change Proposal ***(cont.)***

Organization to be changed	
RGC	
Change dimensions	
Services and products, markets	
Organization, personnel, stakeholder realtions	
Relations to strategic objectives of the organization	
Contribution to a sustainable project managment, program management, and change management developed in the society	
Sustainable business values assured for RGC customers	
Thematic leadership for process, project, program, and change managing represented in the scientifc and consulting community	
Evolutionary growth of the RGC realized	
Professional management of the RGC developed further	
Change stakeholders	
RGC consultants, RGC network partners, RGC office, publisher	
Peer Review Group	
RGC customers	
Management community, project management associations	
Change architecture	**Change management costs**
Project: Developing Values4Business Value	Communication with RGC consultants, networking partners, management community, and RGC customers
Small project: Stabilizing Values4Business Value	Stakeholder specific analysis of the sense of urgency for the change
Change start date	**Change end date**
05.10.2015	20.12.2017
Annex	
Initial change plans	

Change initiator

Change initiating team

Table 14.16: Case Study "Values4Business Value"—Quick Wins Plan

Quick Wins Plan
Values4Business Value

V. 1.001 v. S. Füreder per 5.10.2015

WBS-Code	Work Package	Categorization					Planned date of realization	Responsible
		Impact			Realization effort L / M / H	Benefits L / M / H		
		economic	ecological	social				
1.2.2	Prototyping by the conference "Agility & Processes"	x		x	M	M	1/16	R. Gareis
1.2.3	Prototyping by the forum "Developing sustainable"	x			L	M	2/16	R. Gareis
1.2.4	Prototyping by the seminar "Agility & Projects"	x		x	M	H	3/16	L. Gareis
1.2.6	Prototyping by consulting a digital transformation	x		x	M	M	3/16–6/16	M. Stummer
1.2.7	Prototyping by further developing sPROJECT"	x	x	x	M	L	3/16–6/16	L. Gareis
1.2.9	Prototyping by HP 16 "Benefits Realization Controlling"	x		x	M	H	5/16	R. Gareis

Literature

Gareis, R.: Changes of Organizations by Projects, *International Journal of Project Management,* 28(4), 314–327, 2010.

Highsmith, J.: *Agile Project Management,* 2nd edition, Addison-Wesley, Boston, MA, 2010.

Highsmith, J.: *Agile Software Development Ecosystems,* Addison-Wesley, Boston, MA, 2002.

Kotter, J. P.: Leading Change, *Harvard Business Review Press,* Boston, MA, 1996.

Lewin, K.: Frontiers in Group Dynamics. Concept, Method and Reality in Social Science: Social Equilibria and Social Change, *Human Relations,* 1(1), 5–40, 1947.

15 Strategies, Structures, and Cultures of the Project-Oriented Organization

An organization that regularly defines projects and programs to perform relatively unique and comprehensive business processes can be perceived as a "project-oriented organization". A project-oriented organization defines "managing by projects" as an organizational strategy. It has specific structures and cultures for managing projects, programs and project portfolios. There are specific permanent organizational units of a project-oriented organization—namely, a project portfolio group, a PMO (Project Management Office), and expert pools.

Because of the specific strategies, structures, and cultures, a project-oriented organization also has specific guidelines and rules for corporate governance. These are created when developing as a project-oriented organization.

An example of an organization chart of a project-oriented organization is shown below.

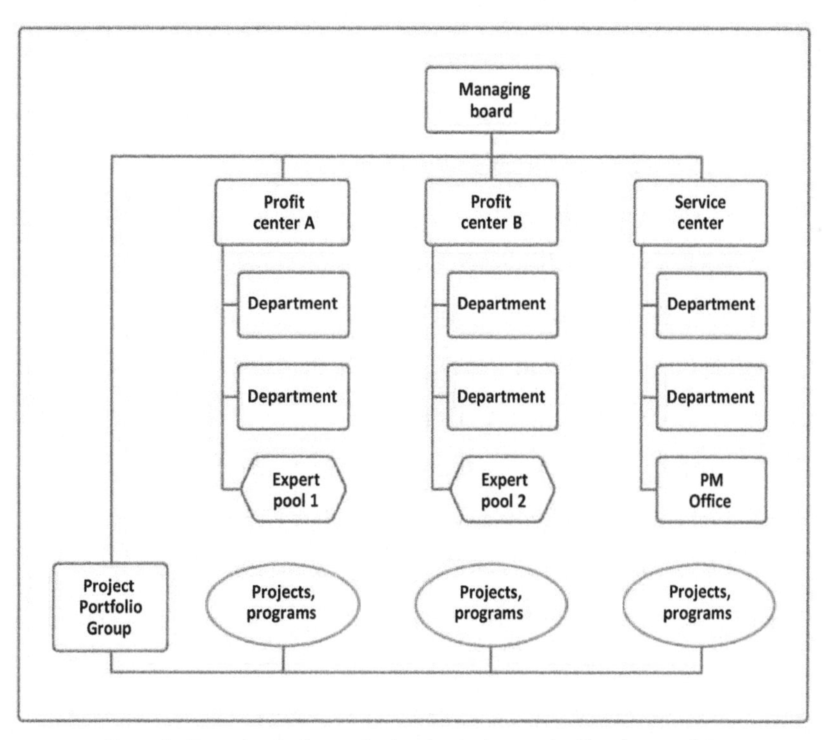

Organization chart of a project-oriented organization (example).

15 Strategies, Structures, and Cultures of the Project-Oriented Organization

15.1 Project-Oriented Organization: Overview

15.1.1 Project-Oriented Organization: Definition

An organization that performs repetitive routine business processes "by the permanent organization" and performs relatively unique, risky, and comprehensive processes "by projects and programs" can be perceived as a "project-oriented organization". The temporary organizations are combined with the permanent line organization units.

A project-oriented organization is characterized by projects and programs. In addition to the repetitive, non-comprehensive processes of the line organization, several projects are started, managed, closed, or discontinued simultaneously. This produces a "flow equilibrium", which ensures flexibility in the development of the organization. The more projects and programs a project-oriented organization performs, the greater its management complexity becomes. This is due in part to the complexity and dynamics of individual projects and in part to the diversity of relations between projects.

A project-oriented organization has the following characteristics:

- Management by projects is explicitly defined as an organizational strategy.
- In addition to profit centers and service centers, projects, programs, networks of projects, chains of projects, and project portfolios are objects of management.
- A project portfolio group, a PMO, and expert pools are specific permanent organizational units with integrative tasks.
- Employees have competencies in project, program, project portfolio, and change managing.
- Organizational competencies for project, program, project portfolio, and change managing are ensured by appropriate governance structures.
- A systemic management paradigm characterized, for example, by business value as a success criterion, empowerment, resilience, and sustainable development, etc. is applied.

These characteristics of a project-oriented organization may vary depending on the size of the company or the number of projects performed simultaneously. For example, small businesses like RGC have no project portfolio group or PMO. These tasks can be fulfilled, for example, by the managing director or by the management team.

An organization can be seen from different perspectives, such as the perspective of marketing or technology. "Project orientation" represents a possible construction of reality. By seeing an organization as a project-oriented organization, opportunities for intervention can be created that increase the potential for successful projects and programs.

A project-oriented organization has specific strategies, organizational structures, and cultures for managing projects, programs, and project portfolios. Creating the

Fig. 15.1: "Organizational fit" of the project-oriented organization.

appropriate "organizational fit" between the strategies, structures, and cultures is a challenge in managing a project-oriented organization (see Fig. 15.1).

An organization can also perform projects without providing the strategic, structural, and cultural prerequisites for their success. The characteristics of such a non-project-oriented organization are described in Table 15.1.

Definition: Project-Oriented Organization

An organization which—in addition to performing routine business processes by the permanent organization—performs relatively unique, risky, and comprehensive business processes by projects and programs can be perceived as a "project-oriented organization". A project-oriented organization is characterized by specific strategies, structures, and cultures.

15.1.2 Dynamics of the Project-Oriented Organization

A project-oriented organization is characterized by dynamic boundaries (see Fig. 15.2).

The number, objectives, and size of the projects and programs of a project-oriented organization are constantly changing, as are the required resources and cooperation partners. For example, the annual budgets of IT companies or engineering companies that perform contracts by projects can vary by more than 50 percent.

The contexts of a project-oriented organization are also dynamic. Relations with new project stakeholders have to be managed, and cooperations have to be established as required. The more projects an organization has in its project portfolio and the more diverse these projects are, the greater the dynamics and complexity of the organization.

Table 15.1: Characteristics of a Non-Project-Oriented Organization

Characteristics of a non project-oriented organization
> Projects are defined for many things, including tasks that are not suitable to be defined as projects. This results in an "inflation" of projects. As a consequence, projects do not receive sufficient management attention.
> Project boundaries are defined by department boundaries. This results in too many small projects, with suboptimal results. The required integration cannot be performed by the projects but must be performed by the permanent organization, which is overburdened as a result.
> No one knows what and how many projects are being implemented at the same time. There is no information about the project portfolio. Projects arise informally. Parallel projects with the same objectives are performed. Resources required for projects cannot be managed appropriately.
> No project management methods are used in projects. This leads to a lack of transparency and hinders efficient communication in projects. Creativity is not gained, but lost.
> Individuals decide on the forms of work in projects. This "reinvents the wheel" for each project. Professionalism in project management depends on the qualifications of individuals.
> The objectives and the work packages of projects are agreed upon from one project meeting to the next. Because a "big project picture" is missing, members of the project organization lack orientation.

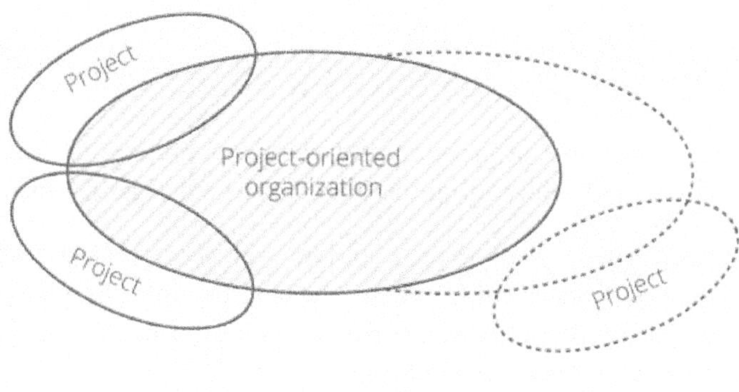

Fig. 15.2: Dynamic boundaries of the project-oriented organization.

To cope with dynamics, a project-oriented organization needs competencies for reflecting, organizational learning, and strategic controlling. Project-oriented organizations have good learning opportunities, but also a high crisis potential. Learning opportunities are mainly provided by the possibility to monitor projects. New approaches are applied in projects, and project stakeholders can be observed during project performance. This information can be used for learning. Crises in a project-oriented organization can result, for example, from inadequate management competencies, too-closely coupled projects, or inadequate perception of the dynamics of the organization itself.

15.1.3 Project-Oriented Organization: Management Objects of Consideration

In addition to the structures of the permanent organization, projects and programs are important objects of consideration for managing the project-oriented organization. Because of the simultaneous performance of multiple projects and programs, the project-oriented organization is strongly differentiated. To fulfill integrative functions, projects and programs can be clustered in project portfolios, project networks, and project chains (see Fig. 15.3). Relations between projects can be considered by means of these clusters. Synergies can be created and possible conflicts avoided in order to enable the realization of the organization's objectives.

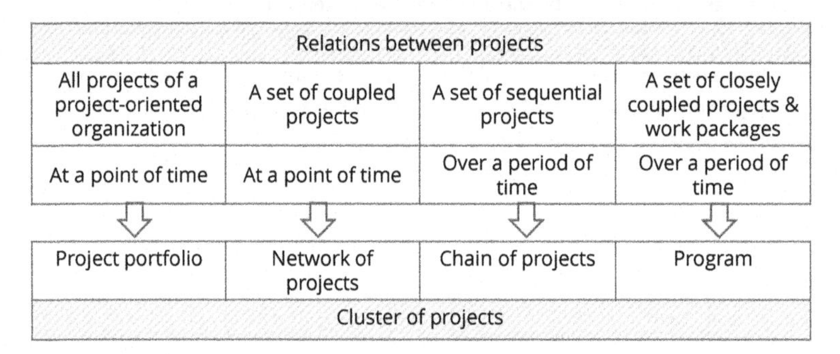

Fig. 15.3: Clusters of projects.

Project portfolios, networks of projects, and chains of projects are thus additional objects of consideration for managing the project-oriented organization.

A project portfolio is the set of all projects (and programs) of a project-oriented organization. The project portfolio includes all current (and possibly also planned, stopped, and closed) projects and programs at a particular point in time. A project portfolio is a construct at a particular point in time.

A network of projects is a set of relatively closely coupled projects. Different criteria, such as the use of a common technology, performance in the same geographic region, or performance for a common customer, can be used to relate projects in a network. The relevant criterion for constructing a network of projects is to be defined in each case. A network of projects is a construct at a particular point in time. An example of a network of projects is the network of the tendering, contract performing, and joint venturing projects of an engineering company in a regional market.

A chain of projects is a set of sequential projects for performing sequential business processes. It is a specific form of a network of projects. A chain is a construct over a period of time. Chains of projects and programs are also possible (e.g., a conceptual project followed by an implementation program). Examples of chains are the chain of projects "Conceptualizing an IT application" and "Implementing an IT application" or the chain "Tendering for a contract" and "Performing the contract".

Unlike projects and programs, project portfolios, networks of projects, and chains of projects are not organizations but clusters of temporary organizations. Individual and organizational competencies are required to manage these clusters.

15.2 Objectives, Strategies, and Business Processes of the Project-Oriented Organization

15.2.1 Objectives and Strategies of the Project-Oriented Organization

A project-oriented organization considers temporary organizations as a strategic option for its organizational design. By defining "managing by projects" as an organizational strategy, the following objectives are to be realized:

- increased organizational flexibility through the use of temporary organizations,
- delegation of management responsibility to projects and programs,
- objectives-oriented work in projects and programs, and
- individual and organizational learning based on monitoring of projects and programs.

"Management by projects" also realizes strategic objectives of human resource management. In projects, the management methods management by objectives, management by delegation, management by motivation, etc., which are often isolated or in competition with one another, can be operationalized and integrated. "Management by projects" uses the motivation and personal development functions of projects as a management strategy.

By defining and performing projects, many organizations implicitly use "managing by projects". The achievement of competitive advantages requires an explicit definition of this strategy and the provision of corresponding structural and cultural requirements.

"Management by projects" is an organizational and personnel strategy of a project-oriented organization. For successful managing, companies use both permanent organizations, such as profit centers or service centers, and temporary organizations—namely, projects and programs.

15.2.2 Business Processes of the Project-Oriented Organization

A project-oriented organization is characterized by the following business processes:

> project initiating and program initiating,
> project managing and program managing,
> project or program consulting,
> change initiating and change managing,
> project portfolio managing,
> networking of projects,
> managing the structures of the project-oriented organization, and
> managing project personnel.

The objectives of "Project initiating" and "Program initiating" are the definition of an adequate organization for the performance of a relatively unique, comprehensive business process and the provision of initial project or program plans. The objectives of the business processes "Project managing" and "Program managing" are contributing to the successful performance of a project or program.

The objective of the business processes "Project consulting" and "Program consulting" is to ensure professional project or program managing. The objective of initiating a change is to define the structures for managing a change. The objective of change managing is to contribute to the successful implementation of a change.

In the business process "Project portfolio managing", the objectives and the resources of the projects of the project portfolio should be aligned with the objectives and strategies of the project-oriented organization so that project portfolio results are optimized. By networking between projects, synergies should be created and conflicts between the projects of a network should be avoided.

The objective of managing the structures of the project-oriented organization is to create organizational structures such as a project portfolio group, a PMO, and expert pools to promote the successful performance of projects and programs. The objective of project personnel managing is to provide the required number of appropriately qualified project personnel.

Project initiating and project managing are described in Chapters 5 and 6, program initiating and program managing in Chapter 13, and change initiating and change managing in Chapter 14. Managing the structures of the project-oriented organization is described in the following section of this chapter. Project and program consulting, project portfolio managing, networking of projects, and managing project personnel are described in Chapter 16.

15.3 Managing the Structures of the Project-Oriented Organization

15.3.1 Flat Structure of the Project-Oriented Organization

The permanent structures of an organization are designed for the efficient performance of routine processes. The organizational structure should provide the employees with orientation by clearly allocating tasks, competencies, and responsibilities, as well as providing concrete guidelines for action. It should also ensure continuity in relations with stakeholders of the organization.

These objectives can be achieved by a stable, permanent organization such as a hierarchical line organization. On the other hand, an organization that regularly implements projects with different objectives and sizes requires flexible, networked organizational structures.

Organizations can be placed on a continuum between an extreme "steep, hierarchical structure" and a "flat, networked structure". The position on this continuum is determined by the relation of routine processes performed by the line organization and processes performed by projects. Though there are no rules for the optimal positioning of an organization, a trend towards flatter, networked structures may be observed.

The use of projects causes organizations to become flatter and more flexible (see Fig. 15.4). The flattening is a result of an increase in breadth of communication and a (partial)

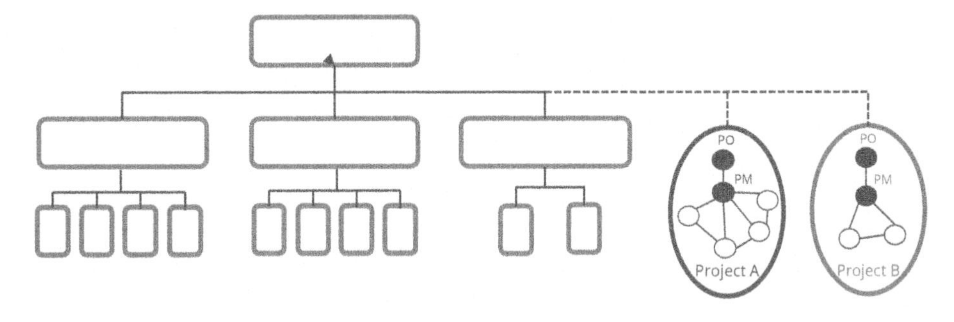

Fig. 15.4: Flattening of organizations by projects.

reduction in the number of hierarchical levels. Flexibility also arises from the temporary character of projects. Projects are established and then dissolved once the project objectives have been fulfilled.

15.3.2 Organization Chart of the Project-Oriented Organization

The organizational structure of a project-oriented organization can be visualized in an organization chart. This makes visible the combination of permanent and temporary organizational units. Permanent organizational units are usually divisions or profit centers as well as service centers and departments. Additional, specific permanent organizational units of project-oriented organizations can be a project portfolio group, a PMO, and expert pools.

Since projects and programs have a high strategic importance for a project-oriented organization, these should also be represented symbolically in the organization chart. Depending on the number of projects performed simultaneously, it might not show individual projects, but rather groups of projects (e.g., by project type) and programs.

An example visualization of a project-oriented organization chart is shown in Figure 15.5.

When designing the organization chart of a project-oriented organization, there are design options regarding the use of symbols for differentiating permanent and temporary organizations, regarding the names of the organizational units, and regarding

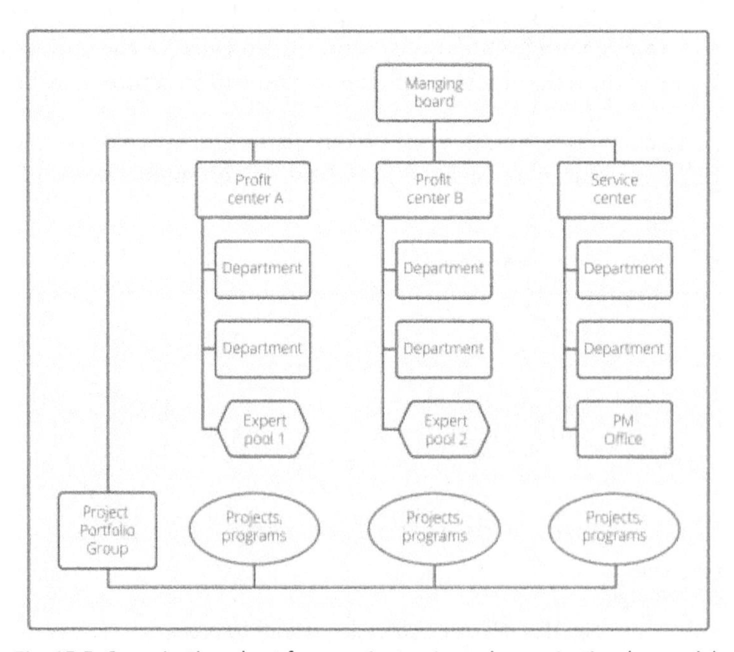

Fig. 15.5: Organization chart for a project-oriented organization (example).

their organizational integration. For example, boxes and ellipses can be used to represent permanent and temporary organizations. The committee for managing the project portfolio can be called a project portfolio group or a project portfolio steering committee, for example. The positioning of a PMO, the project portfolio group, and the expert pools is dependent on the size of the organization and its general structure.

15.3.3 Specific Organizational Units of the Project-Oriented Organization

From the point of view of organizational theory, projects and programs can be understood as instruments for organizational differentiation. Project-oriented organizations also need integrating structures to ensure that the projects and programs follow the objectives and rules of the overall organization.

The specific permanent structures of the project-oriented organization—namely, a project portfolio group, a PMO, and expert pools—are used to integrate projects and programs as well as to manage project portfolios, networks of projects, and chains of projects.

Project Portfolio Group

A project portfolio group is a permanent organizational unit of the project-oriented organization. The primary objectives of a project portfolio group are the strategic design of project portfolio structure, the optimization of project portfolio results, and the optimization of project portfolio risk. A description of the objectives, the organizational position, the tasks, and the formal decision-making authority of the project portfolio group can be found in Table 15.2.

With around 20 to 30 concurrent projects, organizations achieve a complexity that requires independent committees for project portfolio managing. At this point, a project portfolio group should be established to provide appropriate attention to project portfolios. In the case of fewer concurrent projects, these tasks can be performed by existing organizational units, such as the general management team.

A project portfolio group should report directly to the top management of the project-oriented organization. The project portfolio group is authorized to make decisions and is therefore not only an advisory body. Although the integrative structures of organizations, such as leadership circles, are often not represented in organization charts, it is advisable to represent the project portfolio group in the organization chart of the project-oriented organization.

Members of the project portfolio group should be managers of the project-oriented organization. Typical members of the project portfolio group are managers of profit centers and heads of service divisions—for example, the heads of marketing, finance,

Table 15.2: Description of the Role "Project Portfolio Group"

Role: Project Portfolio Group	
Objectives	> Project portfolio structures designed strategically > Project portfolio results optimized > Project portfolio risks optimized > Contributed to the networking of projects > Contributed to managing chains of projects
Organizational position	> Reports to the board of the project-oriented organization > Members: Managers of the project-oriented organization and the manager of the PM Office
Tasks in project initiating and program initiating	> Aligning project objectives with the strategic objectives of the project-oriented organization > Deciding about the organization form to perform a comprehensive, relatively unique business process > Selecting the project owns
Tasks in project portfolio managing	> Coordinating internal and external resources employed in projects > Determining project priorities > Determining overall strategies for designing relations to project stakeholders > Analyzing selected (critical) projects > Deciding about interrupting or stopping projects (for strategic reasons)
Tasks in networking of projects	> Organizing learning of and between projects > Selecting project owners for chains of projects
Formal decision-making authority	> Taking decisions about strategic design of the project portfolio > Accepting or rejecting project proposals > Nominating project owners > Adapting project objectives in project proposals, to align them with the objectives of the project-oriented organization > Taking decisions about strategic design of stakeholder relations > Starting new projects > Interrupting or stopping projects for strategic reasons > Prioritizing projects > Organizing management consulting for projects and programs > Planning the controlling of benefits realization of investments > Deciding upon the networking of projects

IT, and organization. In medium-sized project-oriented organizations, the project portfolio group may also include representatives of top management. A member of the project portfolio group should be nominated as the spokesman of the project portfolio group.

Depending on the number and the complexity of the projects to be coordinated, several project portfolio groups may be required. A project portfolio group should manage project portfolios with a maximum of 20 to 30 projects. Differentiation of project portfolio groups can be performed by organizational units or by project types. An Austrian engineering company, for example, distinguishes between three project portfolio groups: one for tendering and contracting projects, one for product development projects, and one for personnel, organizational, and marketing projects. To ensure that the decisions of these three project portfolio groups are coordinated, individual managers are members of several project portfolio groups.

The project portfolio group is to be supported by the PMO in preparing, performing, and following up on meetings. Above all, project portfolio analyses and reports can be produced by a PMO. PMOs, therefore, provide services not only relating to project and program management, but also relating to project portfolio management.

PMO—Project Management Office

The PMO is also a specific permanent organizational unit of the project-oriented organization. The objective of a PMO is to contribute to professionally managing projects, programs, and project portfolios of the project-oriented organization. Individual, collective, and organizational competencies for this are secured by a PMO.

With the increasing importance of projects and programs in organizations, there is a need to unify approaches and to ensure the quality of project and program managing. There is also the need to perceive and manage project portfolios explicitly.

A PMO is often established when project and program managing are formalized. The PMO represents the institutionalized competencies of an organization in project, program, and project portfolio managing.

The PMO contributes to the achievement of project and program objectives by providing governance structures for project and program managing and by management support. The PMO contributes to the optimization of the project portfolio results by providing governance structures for project portfolio management and by project portfolio management support. The self-understanding of a PMO is that of a service provider and not of a controller. Possible services of the PMO are described in Table 15.3. Non-objectives of the PMO are the fulfillment of the roles of project owner and project manager.

The PMO can use a variety of tools to fulfill its services. An overview of possible tools for project and program managing and project portfolio managing can be found in Table 15.4 (page 419). Because management consultancy and management auditing are becoming increasingly important for the quality assurance of projects and programs, guidelines for this can also be made available.

The permanent organizational unit PMO must be distinguished from the temporary organizational units Project Office and Program Office. The PMO performs services for the project-oriented organization as a whole, while a Project Office or Program Office performs services for a particular project or program.

The basic structure of a PMO is shown in Figure 15.6 (page 420). The role of the PMO manager is described in Table 15.5 (page 420). A key element of the organization chart of the PMO is differentiation of positions which provide services for projects and programs, and positions which provide services for the project portfolio.

Table 15.3: Possible Services of the PM Office

PM Office: Services	
Services for projects and programs	
Providing governance structures for project managing and program managing	> Providing guidelines for project managing and program managing > Providing forms and software for project managing and program managing
Supporting the managing of projects and programs	> Supporting project and program starting, controlling, closing > Supporting project and program transforming and repositioning
Organizing management consulting and management auditing to ensure management quality of projects and programs	> Organizing management consulting for projects and programs > Organizing management auditing for projects and programs
Organizing individual and collective learning for project managing and program managing	> Organizing training on project and program managing for project personnel and for specific projects and programs > Organizing coaching for people with roles in projects and programs > Organizing exchange of experience between project managers and program managers
Organizing the organizational learning for project managing and program managing	> Maintaining the knowledge data base for project managing and program managing > Benchmarking competences for project managing and program managing
Marketing of project managing and program managing	> Performing project and program managing marketing by events > Performing marketing by „Project vernissages“ > Maintaining the PM Office homepage

(continues on next page)

The expert pools “Project Managers”, “Project Management Assistants”, and “Project Management Trainers/Consultants” can be based in the PMO but also in other organizational units.

There are various possibilities for the organizational positioning of a PMO in the project-oriented organization. Figure 15.7 (page 421) shows possible positions in a profit center, a service center, and a staff unit of the management board.

Table 15.3: Possible Services of the PM Office *(cont.)*

Services for the project portfolio	
Providing governance structures for project portfolio managing	> Providing guidelines for project portfolio managing > Providing forms and software for project portfolio managing
Supporting project initiating and program initiating	> Checking quality pf project proposals > Analyzing relations of a new project to the existing project portfolio
Supporting project portfolio managing	> Maintaining the project portfolio data base > Analyzing the project portfolio structure > Developing project portfolio reports > Attending Project Portfolio Group meetings
Organizing and contributing to the networking of projects	> Organizing the networking of projects > Contributing to networking of projects > Ensuring continuity in managing chains of projects

If a number of PMOs exist in different profit centers, they must formally or informally coordinate in order to ensure a joint approach for projects involving the cooperation of several profit centers.

The services of a PMO (see Table 15.3 above) can be differentiated into governance tasks: strategic tasks such as project portfolio managing and operational tasks such as providing management support. Some organizations, therefore, have two PMOs—one

Table 15.4: Possible Tools of the PM Office

Project managing and program managing: Tools
> Guidelines and forms for project and program management > Standard project plans > Defined career path in project and program management > Training program for project and program managing > ICT infrastructure for project and program managing > Facilitation tools > PM Office homepage
Project portfolio managing: Tools
> Guidelines and forms for project portfolio managing > Standards for project proposals > ICT infrastructure for project portfolio managing > Standard project portfolio reports

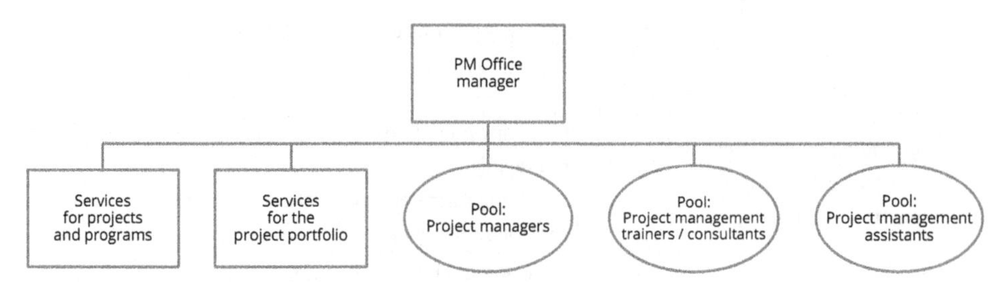

Fig. 15.6: Organization chart of a PMO.

Table 15.5: Description of the Role "PMO Manager"

Role: PM Office Manager	
Objectives	> Professional project managing, program managing and project portfolio managing ensured by providing corporate governance and support services > Process ownership for the business processes project managing, program managing and project portfolio managing performed > Employees of the PM Office directed
Organizational position	> Reports the manager of the business unit, which the PM Office belongs to > Employees of the PM Office report the PM Office manager > Is a member of the Project Portfolio Group
Service-related tasks	> Providing governance structures for project managing, program managing and project portfolio managing > Organizing management support for projects and programs > Organizing management consulting and management auditing of projects and programs > Organizing individual and collective learning for project and program managing > Organizing organizational learning for project and program managing > Performing marketing for project and program managing > Supporting project initiating and program initiating > Supporting project portfolio managing > Organizing and contributing to networking of projects
Management tasks regarding the PM Office	> Organizing the PM Office > Managing the personnel of the PM Office > Managing the budget of the PM Office
Formal decision-making authority	> Taking decisions about the organization, personnel, infrastructure and budget of the PM Office > Giving feedback to initiating teams about project proposals > Organizing the networking of projects

strategic and one operational. A strategic PMO could then be integrated into a staff unit of the management board, and an operational PMO into a profit or service center, as shown in Figure 15.7.[1]

Depending on the size of the project-oriented organization, the scope of the project portfolio, and the services offered, a PMO can consist either of only one person or of several people. An Austrian production and trading company with about 350 employees and a project portfolio of 25 projects on average staffs the PMO with two part-time employees. Larger organizations staff PM offices with multiple full-time employees.

[1] See Ortner G., Stur B., 2015, p. 25ff.

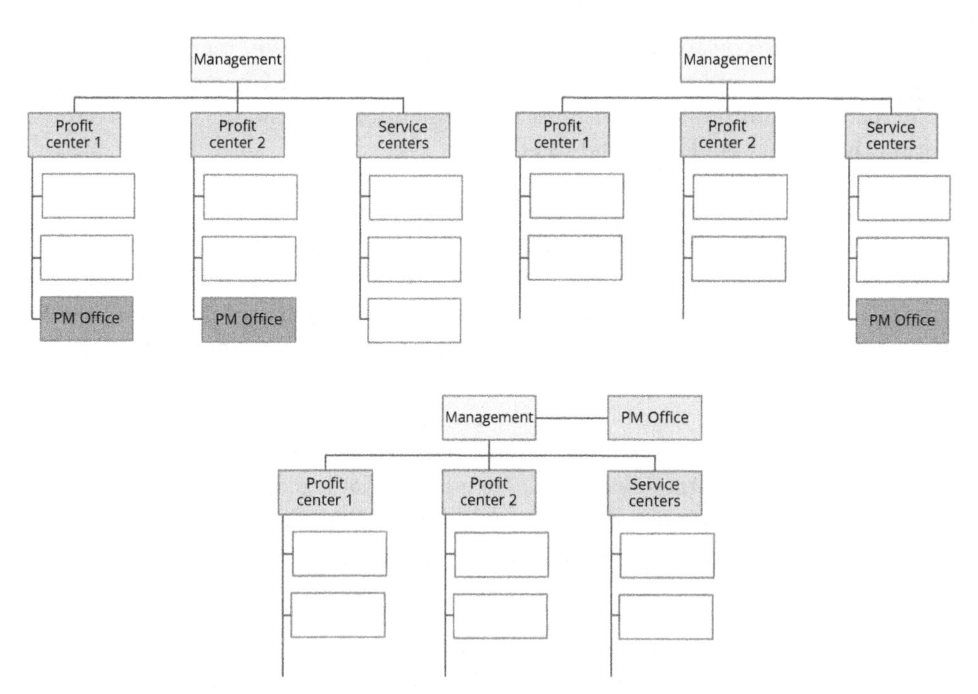

Fig. 15.7: Possible organizational integration of the PM Office.

Expert Pools

The business processes of projects and programs are performed by experts from the various expert pools of the project-oriented organization.[2]

The difference between an expert pool and a traditional department, apart from the empowerment of the experts, is based on the self-understanding of the leaders. The "traditional" head of department is primarily an expert who is responsible for the content-related services of the departmental personnel. The expert pool manager understands him or herself primarily as manager of the pool with personnel and organizational responsibilities, not as the best expert on content. This self-understanding of the expert pool manager, however, does not rule out the possibility that he or she performs expert tasks in projects. He or she is not, however, responsible for the tasks performed by the various experts of the pool.

The objective of an expert pool is to provide qualified experts for projects and programs. It is also an objective to provide appropriate governance structures for the efficient fulfillment of the business processes for which the experts are specialized.

The tasks to be performed in the expert pool are personnel management, process management, and knowledge management. The performance of project-related tasks

[2] Other names are also used in practice instead of "Expert pool"—for example, "Resource pool" and "Center of excellence".

and their quality control do not take place in the expert pool, but in the projects. According to the empowered project organization model (see Chapter 7), the project team members and project contributors are responsible for how services are fulfilled, within the defined guidelines and rules.

Several types of expert pools can be distinguished, depending on the business activities of a project-oriented organization. In an engineering company, for example, there are various technical expert pools (mechanical engineering, electrical engineering, etc.), an expert pool for procuring, an expert pool for assembling, etc. In an IT company, one can differentiate between expert pools for designing, programming, testing, etc. The creation of expert pools is not, however, only useful in organizations which perform contracting projects. People who are qualified to participate in projects can also be assigned to expert pools for internal projects. For example, not all employees of the marketing department of a bank are qualified for project work. However, people who are able to represent departmental interests in projects, who have specialist knowledge and have sufficient project management competencies, can be (virtually) assigned to a marketing expert pool.

There should always be an expert pool of project managers in a project-oriented organization. In larger organizations, the expert pools "Project management support" and "Project management trainers/consultants" can also be defined, as described above in the context of a PMO.

There are different roles in an expert pool. There is a distinction between the expert pool manager and the pool members. The role of the manager of an expert pool of project managers can be fulfilled by the manager of a PMO. The members of an expert pool can have different qualifications. Differentiation into junior experts, experts, and senior experts enables the establishment of an expert career path (see Chapter 16).

A general description of the role of expert pool manager is given in Table 15.6. The expert pool manager should be assessed on the basis of the performance of his or her management services in the expert pool and not on the basis of project and program results.

15.4 Cultures of the Project-Oriented Organization

15.4.1 Organizational Culture and Management Paradigm

An organizational culture can be understood as the values, norms, patterns of behavior, and artifacts that the members of an organization develop and live out. The culture of an organization is not directly graspable but can be observed in the symbols, abilities, and tools used.

In a project-oriented organization, cultural differentiation is enabled by combining the culture of the permanent organization with project and program cultures.

Table 15.6: Description of the Role "Expert Pool Manager"

Role: Expert Pool Manager	
Objectives	> Availability of experts for implementing projects and programs ensured > Governance structures provided for fulfilling the business processes of the expert pools > Knowledge management performed in the expert pool > Expert pool members allocated and coordinated
Organizational position	> Reports to the manager of the profit center or the service center which the expert pool belongs to > Leads junior experts, experts und senior experts in the pool
Service-related tasks	> Providing governance structures for implementing the business processes of the expert pool > Supporting project and program initiating > Supporting project portfolio managing > Allocating personnel of expert pools in projects and programs > Providing the necessary infrastructure for fulfilling the business processes of the expert pool > Providing a knowledge database with information for fulfilling the business processes > Ensuring the compliance with the governance structures of the expert pool > Organizing individual and collective learning of the members of the expert pool
Management tasks regarding the expert pool	> Recruiting personnel for the expert pool > Developing personnel of the expert pool > Managing the budget of the expert pool > Organizing the expert pool
Formal decision-making authority	> Making decisions about the organization, the infrastructure and the budget of the expert pool > Making decisions about the recruitment and development of personnel of the expert pool > Making decisions about the disposition of members of the expert pool in projects and programs > Organizing supervisions or audits about the fulfillment of the business processes of the expert pool

Specific names, logos, rules and norms, artifacts, and behaviors contribute to developing project- or program-specific cultures. Cultural differentiation must be balanced with cultural integration.

The management paradigm of an organization creates a cultural framework. A management paradigm is characterized by the values, objectives, business processes, methods, roles, and stakeholder relations of an organization. A mechanistic management paradigm (see Chapter 2) is characterized by working in functionally defined organizational units, by using the hierarchy as a central integration tool, and by cooperating based on interface definitions. Organizations with a mechanistic management paradigm can make only limited use of the organizational potential of projects and programs.

Successful and efficient application of projects and programs requires a systemic management paradigm. The values of the systemic management paradigm on which the RGC management approaches are based are shown in Figure 15.8 and briefly interpreted. The values of the management concepts agility and resilience are considered

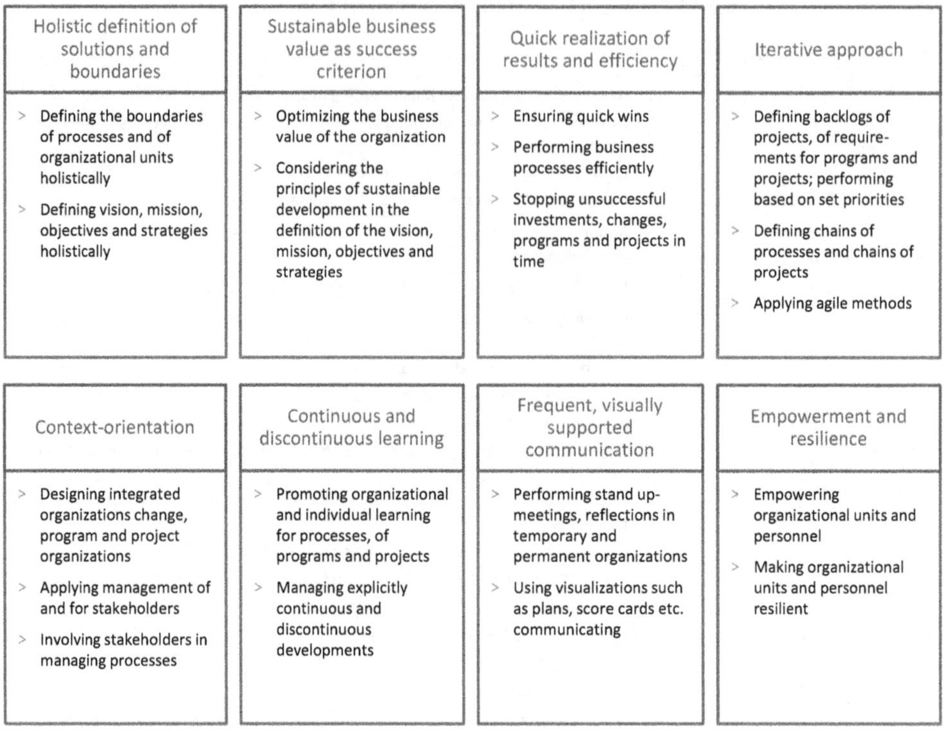

Holistic definition of solutions and boundaries	Sustainable business value as success criterion	Quick realization of results and efficiency	Iterative approach
> Defining the boundaries of processes and of organizational units holistically > Defining vision, mission, objectives and strategies holistically	> Optimizing the business value of the organization > Considering the principles of sustainable development in the definition of the vision, mission, objectives and strategies	> Ensuring quick wins > Performing business processes efficiently > Stopping unsuccessful investments, changes, programs and projects in time	> Defining backlogs of projects, of requirements for programs and projects; performing based on set priorities > Defining chains of processes and chains of projects > Applying agile methods

Context-orientation	Continuous and discontinuous learning	Frequent, visually supported communication	Empowerment and resilience
> Designing integrated organizations change, program and project organizations > Applying management of and for stakeholders > Involving stakeholders in managing processes	> Promoting organizational and individual learning for processes, of programs and projects > Managing explicitly continuous and discontinuous developments	> Performing stand up-meetings, reflections in temporary and permanent organizations > Using visualizations such as plans, score cards etc. communicating	> Empowering organizational units and personnel > Making organizational units and personnel resilient

Fig. 15.8: Values underlying the RGC management approaches.

in this. Detailed interpretations of the values of the RGC management paradigm for the RGC management approaches can be found for project managing in Chapter 6, for program managing in Chapter 13, for change managing in Chapter 14, and for project portfolio managing in Chapter 16.

15.4.2 Agility and Resilience of the Project-Oriented Organization

A project-oriented organization is made agile and resilient by using projects and programs to perform unique and comprehensive short- and medium-term business processes. The combination of permanent and temporary organizations contributes significantly to flexibility and speed as well as to the resilience of a project-oriented organization. Relatively autonomous projects and programs create a high degree of flexibility. Projects are started, closed after objectives have been achieved, or possibly discontinued in the event of a lack of success.

The agility of projects and programs (see Chapters 6 and 13) must be matched by the agility of the permanent organizational units of the project-oriented organization. Agile values and principles (see Chapter 2) can also be used to manage permanent organizational units. Stand-up meetings, visualizations in communications, teamwork, involvement of stakeholders, an iterative approach, etc. can be used in daily managing and contribute to the quality of managing.

"Agile" managing can be achieved either by considering agile values or by considering the agile values plus using agile methods. In both cases, specific responsibilities and working procedures are required. Employees are brought into direct contact with (internal) customers, regardless of their hierarchical position. Managers support teams and employees in accepting responsibility and encourage self-organizing within the framework of guidelines and rules.

Holacracy is a model for implementing agile principles at the organizational level. In the Excursus "Holacracy", an overview of this model is provided.

Excursus: Holacracy

Holacracy is an organizational model that promotes transparency, dynamics, and individual responsibility in organizations. Robertson describes holacracy as a new social methodology for the management and operation of an organization.[3] "From the root 'holarchy', holacracy means governance by the organizational entity itself, not governance by the people within the organization".[4]

An organization is seen as a "holarchy of semi-autonomous circles". Each circle receives a set of objectives from a higher circle and is responsible for their fulfillment. A lower circle is connected to a higher circle by at least two people. These people belong to both circles and have decision-making responsibilities in both circles.

The self-organization of a circle takes place in the context of the system above it. A circle cannot be fully autonomous; the needs of other circles must be taken into account in order to optimize the whole. These structural relations are shown in Figure 15.9.

The holacracy model is based on the strict use of a role model in which one or more roles are assigned to each employee. This results in a highly flexible organizational structure that can quickly adapt to new circumstances.

Transparent rules and clearly defined areas of authority form the framework for a number of autonomously acting teams—so-called "circles". The individual circles are characterized by high problem-solving competency and promote integrative decision making at low hierarchical levels.

As an essential element of the holacracy model, the meeting process depicted in Figure 15.10 describes the work of individual circles.[5]

The clear separation between operational and organizational issues ensures efficient operation. Tactical meetings enable ongoing exchange of information and the handling of friction points. In governance meetings, the focus is on the further development of the organizational structure—for example, roles are adapted or newly created.

[3] See Robertson B., 2016.
[4] Robertson B., 2006, p.4, quoted by holacracy.org.
[5] See holacracy.org (retrieved on January 16, 2017).

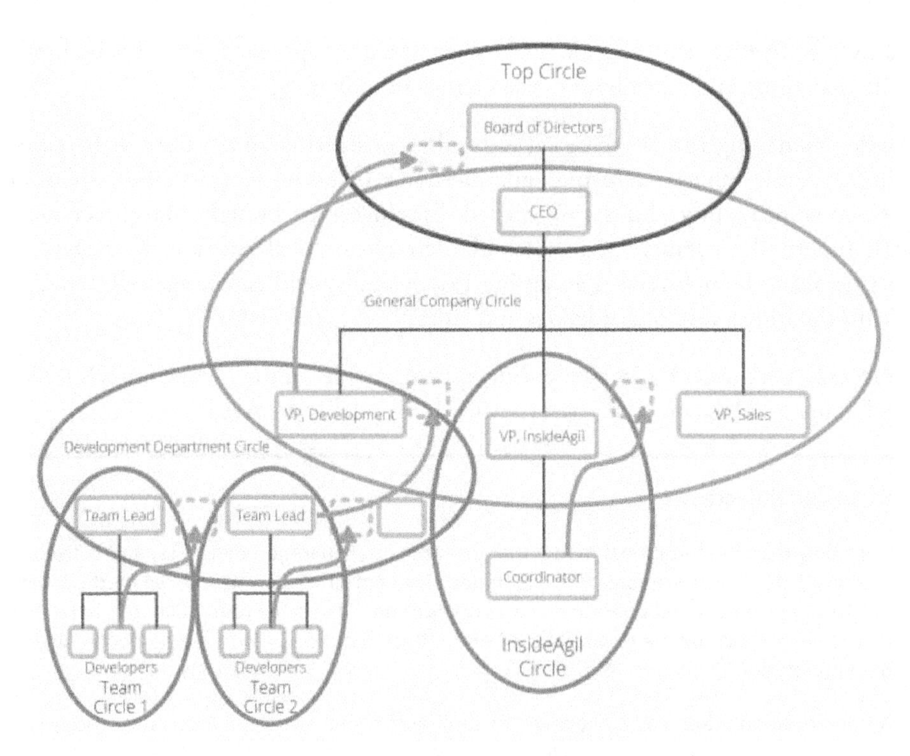

Fig. 15.9: Organization chart with circle structures.[1]

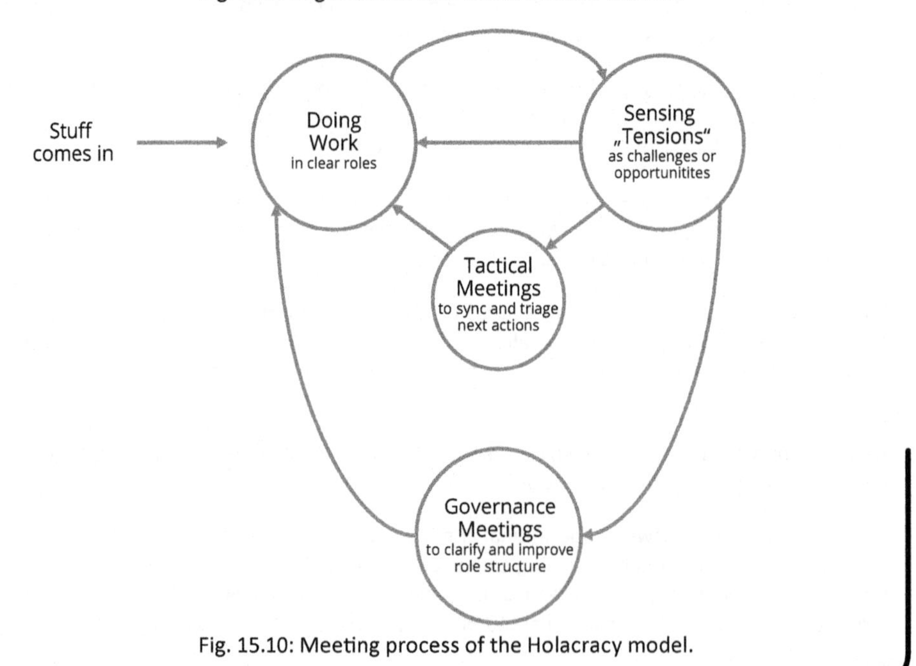

Fig. 15.10: Meeting process of the Holacracy model.

[1] Robertson B., 2006, p.7.

Project orientation provides organizations with flexibility and speed as well as resilience in dealing with complexity and dynamics. An essential management task in agile and resilient organizations is the management of values. Values are to be defined, communicated, interpreted, and exemplified by managers.

15.4.3 Sub-Cultures in the Project-Oriented Organization

The establishment of projects and programs creates sub-systems in the project-oriented organization. As a sub-system, a project can be differentiated from other sub-systems, such as profit centers, service centers, or even other projects, by its specific culture. The cultural differentiation of projects is functional for the project-oriented organization if the autonomies that are created promote the achievement of the objectives.

A prerequisite for the definition of sub-systems is that they are characterized by their own values, rules, and behaviors, which are learned by its members and shared collectively.

Projects are not only tools for cultural differentiation but also tools for integration. Cultural integration takes place, for example, if members from different organizational units cooperate in a project.

The objectives and methods for the development of project- and program-specific cultures are described in Chapter 8.

15.5 Corporate Governance and Developing as a Project-Oriented Organization

15.5.1 Corporate Governance: Definition

Corporate governance can be defined as "a system by which companies are directed and controlled"[6] or "as a system by which companies are strategically directed, integratively managed and holistically controlled in an entrepreneurial and ethical way and in a manner appropriate to each particular context".[7]

The objective of corporate governance is to create guidelines and rules to optimize the risks of an organization and to ensure the quality of its business processes as well as benefits for its stakeholders in the long term. The principles of sustainable development are thereby also fulfilled. Corporate governance should provide the basis for transparent organizational structures and repeatable business processes.

The results of corporate governance are guidelines, plans, ratios, and descriptions that can relate to the objectives, strategies and values, services and products, legal form

[6] Cadbury Committee, 1992, p. 14.
[7] Hilb H., 2012, p.7.

Table 15.7: Examples of Corporate Governance Documents

Type of document	Content
Documents regarding objectives, strategies, and values of the organization	> Guidelines for planning, communicating and monitoring strategic objectives, strategies, and values of the organization > Description of strategic objectives, the mission, strategies, and values of the organization > ...
Documents regarding services and products of the organization	> Regulations about providing services and products for different industries and/or companies > Guidelines and rules for fulfilling services and delivering products > ...
Documents regarding the organizational structures	> Regulations about the development of organization charts, job descriptions, decision rules, description of business processes, etc. > Organization charts, job descriptions, descriptions of business processes, etc. > ...
Documents regarding	> ...

and organizational structures, personnel, infrastructure, finance, and stakeholders of an organization. A list of possible corporate governance documents is presented in Table 15.7. As these examples show, "governing" is not managing. Corporate governance, however, provides rules for daily managing.

Corporate governance is not limited to the provision of documents. Guidelines and rules must be communicated to stakeholders in an appropriate form, and compliance with these requirements must be ensured.

15.5.2 Corporate Governance of the Project-Oriented Organization

A project-oriented organization needs specific governance structures for the governance of projects, programs, and project portfolios. The objective is to ensure management quality in projects, programs, and project portfolios.

The corporate governance structures of a project-oriented organization include descriptions of the business processes for project and program managing, for project portfolio managing, and possibly also for ensuring management quality in projects and programs. Other relevant documents can be values, role descriptions, descriptions of communication structures, and decision-making rules for project, program, and project portfolio managing.

The top management of a project-oriented organization decides on the implementation of corporate governance structures and their controlling. The PMO has operational responsibility for implementation and controlling. The controlling can be performed by means of periodic audits or health checks by internal or external auditors.

Project and program owners contribute to the compliance of projects and programs with corporate governance structures. However, they are not responsible for implementing and controlling corporate governance structures. Responsibility for the corporate governance of the project-oriented organization is therefore to be distinguished from management responsibility for projects and programs. The task of corporate governance is to define how projects or programs are to be managed in general, but not to manage them.

Responsibility for the implementation and controlling of the specific corporate governance structures of a project-oriented organization must be distinguished from management responsibility for projects and programs. It is the task of corporate governance to define how projects or programs are to be managed. The managing of a project or program is the task of the corresponding project or program organization. There are therefore no corporate governance tasks for individual projects or programs.

15.5.3 Guidelines and Standards of the Project-Oriented Organization

Corporate governance documents of the project-oriented organization are guidelines and standard project plans. Typical guidelines for a project-oriented organization are guidelines for project and program managing, for project portfolio managing, and possibly also for project and program consulting.

Guidelines for Project and Program Managing

The objectives of guidelines for project and program managing are to unify the approach and to ensure the quality of project and program managing. The guidelines should provide employees with guidance on the managing of projects and programs, and the efficiency of performing projects and programs should be ensured. Guidelines for project and program managing make an essential contribution to the organizational competencies of a project-oriented organization.

Guidelines for project and program managing should define when business processes should be organized as projects or programs, which methods are to be used to manage them, and which roles are to be fulfilled in projects and programs. Sample project documentation and standard project plans can also be made available.

As an example, the table of contents of a guideline for project and program managing is shown in Table 15.8.

Table 15.8: Table of Contents of a Guideline for Project and Program Managing (Example)

Table of contents
1 Introduction 1.1.Objectives of the guidelines for project and program managing 1.2.Updating the guidelines
2 Definitions 2.1 Definitions: Small project, project, program 2.2 Definition: Project managing and program managing 2.3 Project types and program types
3 Project initiating 3.1 Business process: Project initiating 3.2 Roles for project initiating 3.3 Methods for project initiating
4 Program initiating 4.1 Business process: Program initiating 4.2 Roles for program initiating 4.3 Method for program initiating
5 Project managing 5.1 Business process: Project managing 5.2 Subprocess project starting 5.3 Sub-process project controlling 5.4 Sub-process project coordinating 5.5 Sub-process project transforming/project repositioning 5.6 Sub-process project closing
6 Designing of project organizations 6.1 Project organization chart 6.2 Project roles 6.3 Project communication formats 6.4 Projects and values 6.5 Methods for project managing
7 Program managing 7.1 Business process: Program managing 7.2 Organization of programs 7.3 Sub-processes of program managing 7.4 Methods for program managing 7.5 Tools for program managing
8 Annex 8.1 Forms for project and program managing 8.2 Glossary 8.3 Links to standard project plans

Guidelines for project and program managing should be kept as short as possible (about 20–30 pages without appendices). The use of the business processes and methods described in the guidelines should be compulsory. Guidelines for project and program managing usually also include a set of templates that support the application of methods for project and program managing.

Guidelines for Project Portfolio Managing

The objectives of guidelines for project portfolio managing are to unify approaches and assure the quality of project portfolio managing. This creates organizational competencies. In guidelines for project portfolio managing, the business processes of project portfolio managing, the roles, and the methods for project portfolio managing are described. In guidelines for project portfolio managing, the business process "Networking of projects" (see Chapter 16) can also be described.

Table 15.9: Table of Contents of Guidelines for Project Portfolio Managing and Networking of Projects

Table of contents
1 Introduction 1.1 Objectives of the guideline 1.2 Updating the guideline
2 Definitions 2.1 Definition: Project portfolio, network of projects, chain of projects 2.2 Definition: Project portfolio managing, networking of projects
3 Business processes 3.1 Business process: Project portfolio managing 3.2 Business process: Networking of projects
4 Roles for project portfolio managing and networking of projects 4.1 Expert pool manager 4.2 Project Portfolio Group 4.3 PM Office 4.4 Project manager
5 Methods for project portfolio managing and networking of projects 5.1 Methods for project portfolio managing 5.2 Methods for networking of projects
6 Annex 6.1 Forms for project portfolio managing and networking of projects 6.2 Glossary

The roles for project portfolio managing are the project portfolio group, the PMO, and expert pools (see above). An example of the table of contents of guidelines for project portfolio managing and networking of projects is shown in Table 15.9.

Standard Project Plans

Standard project plans can be used to manage repetitive projects such as tendering projects, contracting projects, product development projects, event organization projects, etc. Standard project plans can, for example, be standard work breakdown structure, standard milestone plans, standard work package specifications, standard responsibility charts, and standard organization charts.

The use of standard project plans reduces the planning effort for initiating and starting projects and makes it possible to make use of prior experience.

A risk when using standard project plans is their "linear" application, which does not appropriately consider the specifics of a new project. Standard project plans should therefore be adapted as appropriate to take project specifics into account.

15.5.4 Developing as a Project-Oriented Organization

Objectives and Objects of Consideration of Developing as a Project-Oriented Organization

Frequent use of projects and programs requires appropriate organizational competencies as a project-oriented organization, in addition to specific individual competencies. Organizational competencies can be formalized by corporate governance structures (see above).

According to a systemic management approach, the development of an organization can relate to its structural dimensions—namely, services and products, organizational structures and cultures, personnel structures, infrastructures, budget and financing—as well as to its contexts. Relevant contextual dimensions of organizations are their history and expectations for the future, their stakeholders, and the larger social system to which the organization contributes (see also Chapter 3).

These structural and contextual dimensions can be the objects of consideration of developing as a project-oriented organization (see Table 15.10). The extent of the change is to be planned. Objectives can be either establishing or further developing a project-oriented organization.

Establishment as a Project-Oriented Organization

When establishing a project-oriented organization, it is assumed that the organization has as yet no or only a few individual and organizational competencies for project, program, and project portfolio managing. The development of basic competencies is pursued through personnel development measures and through organizational structure and organizational process, infrastructure-related, stakeholder-related, and budgetary measures.

The establishment can be achieved by developing basic structures as a project-oriented organization and implementing them in project, program, and project portfolio managing. In order to create the basic structures, the following measures must be implemented:

Table 15.10: Objects of Consideration of Developing as a Project-Oriented Organization

Dimension	Object of consideration for the development
Services	> Project managing and program managing as differentiation criterion and as a service
Organizational structure	> Project Portfolio Group, PM Office, expert pools > Project and program managing as tasks in job descriptions of managers
Business processes	> Project initiating, program initiating > Project managing, program managing > Project or program consulting > Project portfolio managing > Networking of projects
Personnel	> Project management career path > Developing the project personnel > Project-related incentive systems > Project objectives as part of MbOs of managers
Infrastructure	> Software for project and program managing > Software for project portfolio managing > Facilitation material > Project rooms
Stakeholder	> Involving customers, partners, suppliers in projects and programs
Budget	> Project budget and program budget > PM Office budget > Budget for consulting projects and programs

> developing guidelines for project, program, and project portfolio managing,
> developing standard project plans,
> creating personnel development plans and a career path for project managers,
> providing software for project, program, and project portfolio managing, and
> possibly establishing a PMO and a project portfolio group.

The implementation of these structures in project, program, and project portfolio managing can be performed by means of the following measures:

> training and possibly coaching owners, managers, and team members of projects and programs,
> developing a project portfolio database,
> preparing project portfolio reports,
> training and possibly coaching managers for project portfolio managing,

> holding exchange of experience workshops for project and program managing,
> management consulting for projects and programs,
> performing project management and program management marketing.

Further Developing as a Project-Oriented Organization

The objectives of further developing as a project-oriented organization can be the optimization of already established business processes and the establishment of additional structures. The analysis of competencies as a project-oriented organization is of particular importance in defining the objectives of further developing.

When analyzing a project-oriented organization, both individual and organizational competencies are assessed. Methods for analyzing the individual competencies of project personnel are described in Chapter 16. Analysis of organizational competencies as a project-oriented organization can be performed with maturity models for project-oriented organizations. Maturity models offered by project management associations are, for example, OPM3®[8] and Delta.[9] In the Excursus "RGC Model for Analyzing Maturity as a Project-Oriented Organization", an RGC model is briefly described.

Excursus: RGC Model for Analyzing Maturity as a Project-Oriented Organization

The questionnaire for the analysis of maturity as a project-oriented organization is structured according to business processes of a project-oriented organization. The business processes considered can be seen in the "spider net" of the maturity model (see Fig. 15.11).

The "maturity" of each business process can be analyzed by means of questions. As examples, questions about the use of methods (never, rarely, etc.) are shown in Table 15.11 (page 436) for project starting and for project portfolio managing.

The maturity model of the project-oriented organization is based on an algorithm which enables the calculation of a "maturity area" in the spider net and a "maturity ratio". When interpreting the results of the analysis, it is necessary to consider not only the individual business processes, but also their relations. For example, maturity in project portfolio managing can be only "high" if the basis for it—namely, the maturity of project managing—is "very high".

The results of the analysis can be compared by benchmarking against the maturities of other project-oriented organizations.

Measures for further developing as a project-oriented organization can be differentiated into optimizing measures and establishing measures. Optimizing measures relate to already existing business processes of project, program, and project portfolio

[8] See Project Management Institute, 2013. OPM3 is a registered trademark of the Project Management Institute.
[9] See Gesellschaft für Projektmanagement, 2017.

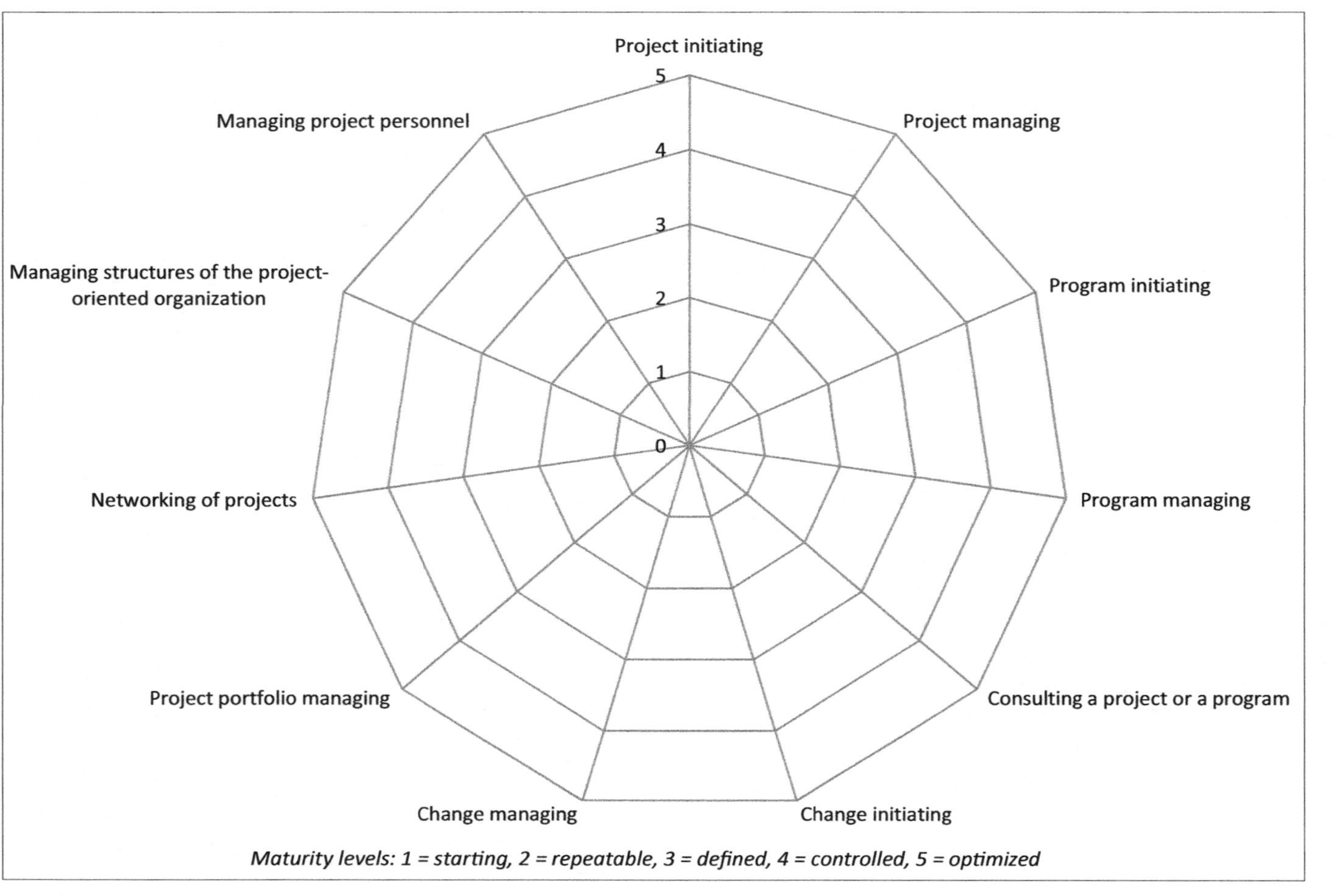

Fig. 15.11: Spider net visualizing maturity as a project-oriented organization.

Table 15.11: Questions for Analyzing Maturity as a Project-Oriented Organization

Project starting: Methods	Value (0-5)
Planning project objectives	
Planning objects of consideration	
Developing a project work breakdown structure	
Specifying a work package	
Developing a project bar chart	
Developing a CPM project network (if needed)	
Planning project resources (if needed)	
Planning project cash flows (if needed)	
Planning project costs	
Performing a project risk analysis	
Performing a project scenario analysis (if needed)	

Project portfolio coordinating: Methods	Value (0-5)
Developing / maintaining a project portfolio data base	
Developing a project portfolio score card	
Developing a project portfolio list	
Developing a project portfolio bar chart	
Developing a project portfolio matrix	
Developing a projects interdependency chart	
Analyzing project proposals	
Analyzing selected project progress reports	

0 = no answer, 1 = never, 2 = rare, 3 = sometimes, 4 = often, 5 = always

managing. Possible measures for establishing additional competencies and structures are, for example:

> promoting the networking of projects,
> developing guidelines for project and program consulting, and
> developing potential consultants and auditors of projects or programs.

15.6 Further Developing as a Project-Oriented Organization: Case Study RGC

RGC Situation in 2015

Strengths of RGC are customer orientation, innovation, and professional organization, which makes it possible to work efficiently and effectively. Even with relatively few employees, good content-related and financial results can be achieved. The management solutions which are offered to customers and implemented by them are also used internally. This makes it possible to offer customers an authentic approach: "We practice what we preach".

A vision, values, a mission statement, objectives, and strategies and an organization chart for RGC existed in October 2015 as a result of strategic managing and of corporate governance. The vision, the mission statement, and the values were also available on the website for the information of stakeholders. Since 2010, the sustainable objectives plan has been sent annually to stakeholders for discussion.

A process map and descriptions of most business processes existed as a result of process managing. A project portfolio database and a variety of project portfolio reports existed as a result of project portfolio managing. The book *Happy Projects!* by Roland Gareis was seen as a corporate governance document for project, program, and project portfolio managing.

The values, the process map, and the organization chart for RGC are shown in Figures 15.12 to 15.14 as examples of the governance documents, which were available at that time.

The business processes were differentiated into primary, secondary, and tertiary processes, depending on the proximity to the customer. The shaded business processes may have needed to be performed by projects. So, for example, small events were developed by working groups, and the HAPPYPROJECTS conference was organized as a small project.

In the organization chart, the permanent organizational structures profit centers and service centers were represented by squares, expert pools by diamonds, and temporary organizations by ellipses.

Quality
- Meeting expectations of stakeholders by delivering high process & result quality
- Applying best management practices

Transparency
- Making decisions transparent
- Informing stakeholders comprehensively

Sustainable development
- Considering the principles of sustainable development in cooperations
- Developing structures and context-relations sustainably

Innovation
- Optimizing structures
- Exploring new potentials systematically

Empowerment
- Increasing the degree of autonomy
- Delegating responsibility and ensuring its acceptance

Fig. 15.12: RGC values as of October 2015.

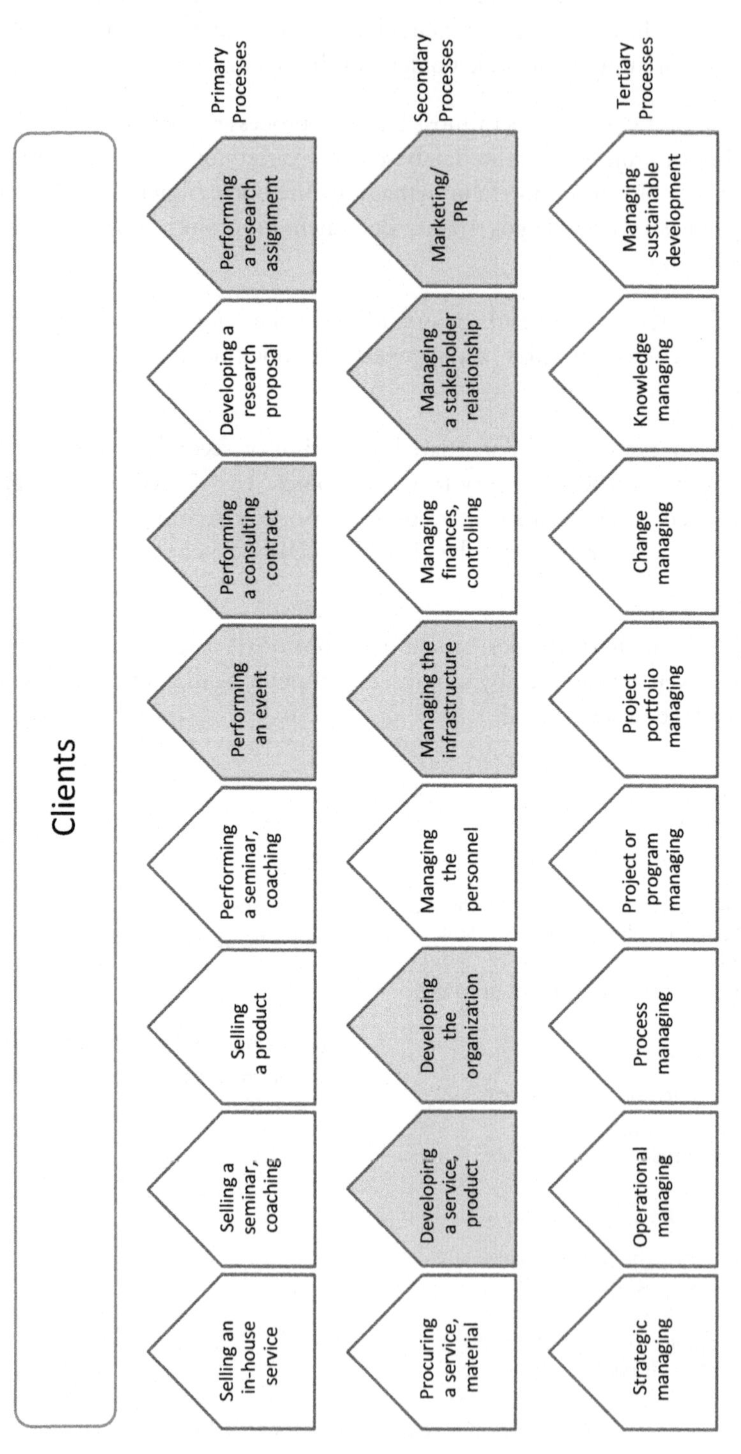

Fig. 15.13: RGC process map as of October 2015.

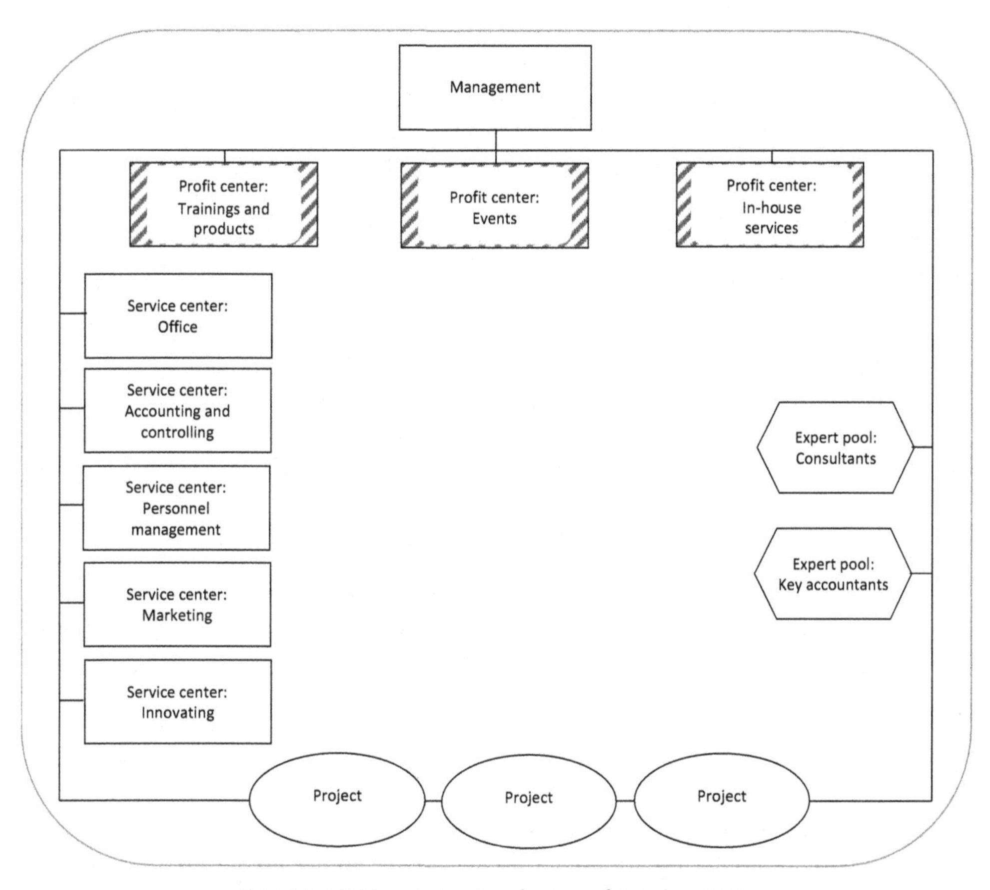

Fig. 15.14: RGC organization chart as of October 2015.

Intervention in 2015

In October 2015, it was decided to develop RGC management approaches, services, and products as well as the management practices of RGC on the basis of the change vision "Values4Business Value". In the project "Developing Value4Business Value", a phase "Prototyping and further planning" was defined, which was to enable the creation of "minimum viable products" in the form of services for customers as well as solutions for managing RGC. In addition, work packages for developing, implementing, and reflecting new management solutions were defined in parallel with the preparation of chapters of the book *PROJECT.PROGRAM.CHANGE* (see Fig. 9.3: Work breakdown structure for the project "Developing Values4Business Value", starting on page 195).

An accompanying opportunity for learning through reflections was created by defining the further development of the RGC as an investment and change "Values4-Business Value".

Further Developing RGC

The change had two main objectives: First, the RGC management approaches were to be further developed and documented in a book publication. The concepts of agility, resilience, sustainable development, benefit realization management, holacracy, etc. were to be considered. Second, the underlying values were to be defined and interpreted for process, project, program, project portfolio, and change managing. In Figure 15.8 (page 424), the further developed values are presented as the basis for RGC management approaches.

Management practices were to be further developed. Existing practices and documents had to be optimized, and new practices and behavior had to be established. The structures of the RGC objectives plan, its organization chart, and role descriptions, as well as the structure of the project portfolio list, were to be optimized. *Happy Projects!* was to be superseded by the publication *PROJECT.PROGRAM.CHANGE* as a governance document for project, program, project portfolio, and change managing.

An excerpt from the sustainable objectives plan 2017 of RGC, which is structured according to customer services, open training, events and products, organization, personnel, infrastructure, financing, and stakeholder relations, can be found in Table 15.12.

The economic and ecological objectives of the program "inter-company training and products" for 2017 are listed and partly quantified in the excerpt from the objectives plan. The corresponding social objectives are summarized in the section "stakeholder relations", which is not presented here. For those objectives for which projects are necessary, the corresponding projects are listed in the "project" column.

The new organization chart, which implements the concept "holacracy", is shown in Figure 15.15 (on page 442).

This organization chart provides a new, additional view at the RGC organization. Different representations of organizations in organization charts can provide different information. A new understanding of the "RGC management circle" is a result of implementing the holacracy model. The "circle" was understood in the past as a communication structure and is now perceived as a formal management structure, which empowers managers of profit and service centers.

Regular "stand-up meetings" were newly established as an additional communication format. To support communication, a white board is often used for visualizing decisions (see Fig. 15.16 on page 443).

The newly structured project portfolio reports are presented in Chapter 16. These illustrate a stronger stakeholder orientation and business value orientation as well as the consideration of the principles of sustainable development in project portfolio managing.

Table 15.12: Excerpt from the RGC Objectives Plan 2017

Objectives 17 per 12/2016	Project 17
Seminars, coachings: Economic objectives	
X seminars with Y participants in 2017 performed, revenue of EUR xxx,- and contribution margin of EUR xxx,- reached	
X certification coachings with Y participants in total in 2017 performed, revenue of EUR xxx,- and contribution margin of EUR xxx,- reached	
Sales initiatives for key accounts performed	
2 sales campaigns performed	
App used for seminars/trainings	RGC digitalisation
Events, products: Economic objectives	
HAPPYPROJECTS 17 with X participants performed; turnover of xxx,-, contribution margin of xxx,- reached	HAPPYPROJECTS 17
HappyProjects 18 planned and marketing started	
Symposium project audit with X participants performed	Symposioum project audit 17
Symposium project audit 2018 planned	
2 analysis workshops performed	
Exhibition at pma focus performed	
Presentations at external events performed	
sPROJECT internally used, externally sold	
App used for events	RGC digitalisation
Seminars, events, products: Ecological	
Event location optimized ecologically (optimizing the mobiliy of the participants)	

Outlook

A key objective for the next few years, as for many other organizations, is digitalization. In this context, the RGC management approaches, services, and management practices are again to be further questioned and developed. We are looking forward to it . . .

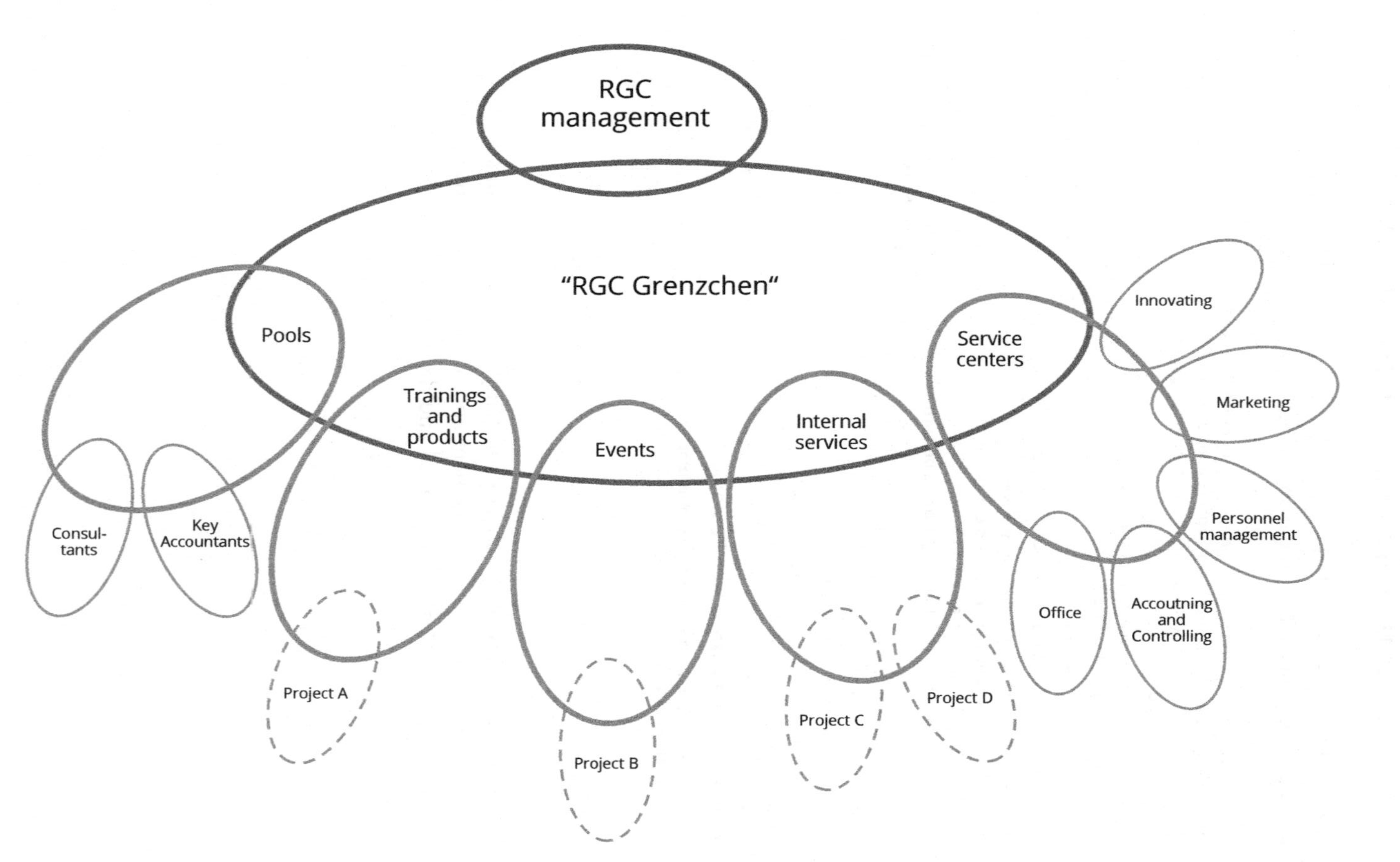

Fig. 15.15: RGC organization chart as of 15 February 2017.

Fig. 15.16: RGC stand-up meeting.

Literature

Cadbury Committee: *Report of the Committee on the Financial Aspects of Corporate Governance,* Gee and Co., London, 1992

Gesellschaft für Projektmanagement (GPM): Assessments für Organisation (IPMA Delta), downloaded from https://www.gpm-ipma.de/lightbox_seiten/ipma_delta.html (18.01.2017).

Hilb, M.: *New Corporate Governance: Successful Board Management Tools,* Springer, Berlin Heidelberg, 2012.

Ortner, G., Stur, B.: *Das Projektmanagement-Office,* 2nd edition, Springer, Berlin, Heidelberg, 2015.

Project Management Institute (PMI): *Organizational Project Management Maturity Model (OPM3®),* 3rd edition, PMI, Newton Square, PA, 2013.

Robertson, B. J.: *Holacracy: Ein revolutionäres Management-System für eine volatile Welt,* Franz Vahlen, Munich, 2016.

Robertson, B. J.: Holocracy: A Complete System for Agile Organizational Governance and Steering, *Agile Project Management Executive Report,* 7(7), Cutter Consortium, 2006.

16 Business Processes of the Project-Oriented Organization

A project-oriented organization is characterized by specific business processes. The processes of project initiating, project managing, program initiating, program managing, change initiating, change managing, and managing the structures of the project-oriented organization have already been covered in previous chapters.

The business processes "Project portfolio managing", "Networking of projects", "Project consulting", and "Managing project personnel" are described below.

The objectives of project portfolio managing are to ensure an appropriate project portfolio structure and to achieve good project portfolio results. The objectives of networking of projects are to create synergies, to avoid conflicts, and to organize learning of the projects of the network.

Management consulting of a project or a program contributes to ensuring its management quality. Corresponding consulting processes include consulting a project in performing a project management sub-process, management auditing of a project, and performing a short project intervention.

Business processes for managing project personnel at the level of the project-oriented organization (and not at the project level) are recruiting, assigning, evaluating, developing, and releasing project personnel.

The business processes of a project-oriented organization discussed in this chapter are highlighted in the following maturity model of the project-oriented company (see next page).

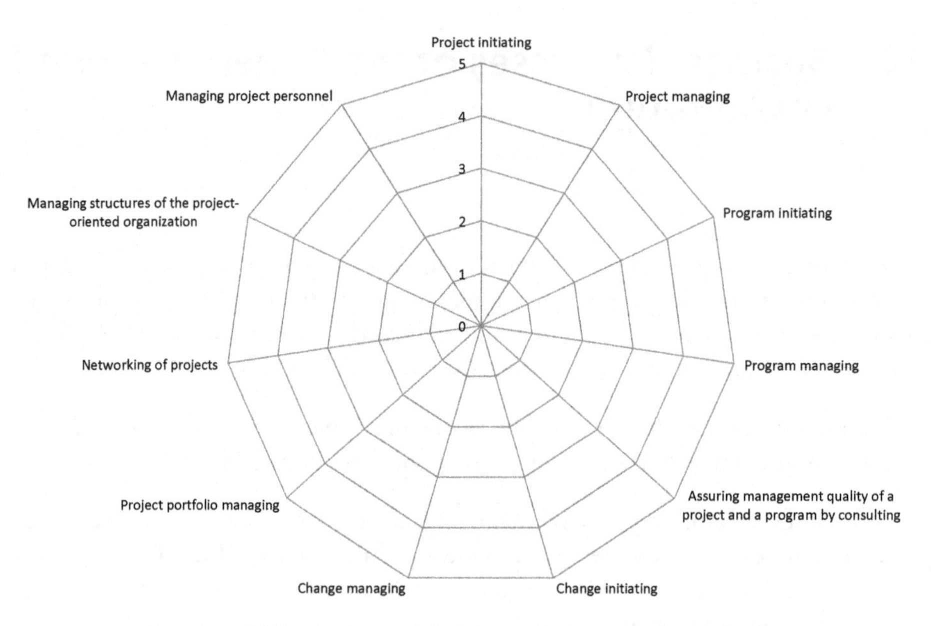

Overview: RGC maturity model of the project-oriented organization.

16 Business Processes of the Project-Oriented Organization

16.1 Structures of Project Portfolios

Project portfolios are integration instruments of the project-oriented organization. The organizational differentiation achieved by defining projects and programs is balanced by the integrative view provided by project portfolios.

The project portfolio of a project-oriented organization is the set of all projects and programs performed simultaneously. A project portfolio can include either all the projects of an organization or a subset of them. If, for example, a project-oriented organization performs more than 30 projects at the same time, it is beneficial to form project sub-portfolios to reduce management complexity. Sub-portfolios can be structured according to organizational units, project types, top projects and other projects, and external and internal projects.

To ensure the agility of project portfolios, the duration of the projects considered must be kept as short as possible. The realization of this objective is supported by the formation of chains of projects.

The projects included in a project portfolio, their objectives, and their status are subject to change. Starting new projects and closing completed projects make portfolios dynamic. On the other hand, the basic structures of project portfolios are stable. The project types included, the number of projects, the stakeholders of the projects, and, for example, their risk types are relatively stable. Discontinuities of project portfolios may be caused by discontinuities of projects of the portfolio or by changes in the environment of the organization holding the portfolio of projects.

Because project-oriented organizations are continually performing projects, project portfolios are not limited in time but exist throughout the lifetime of a project-oriented organization.

16.2 Business Process: Project Portfolio Managing

16.2.1 Project Portfolio Managing: Objectives

The objects of consideration of project portfolio managing are the projects and programs ongoing and planned at a particular point in time, the projects and programs stopped and closed during a considered period, as well as the relations between these projects and programs.

The objective of managing a project portfolio is to optimize the project portfolio results. The objective is not the optimization of the results of individual projects or programs but an optimization of the portfolio from the point of view of the project-oriented organization. This objective can conflict with the optimization of the objectives of individual projects.

To achieve good project portfolio results, it must be ensured that the structure of the project portfolio is appropriate. On the other hand, the objectives of the projects must be aligned with the objectives and strategies of the project-oriented organization; the internal and external resources used in the projects must be coordinated; learning from and between projects must be organized; and priorities must be established.

In managing the project portfolio, the objectives and the resources required for the organization's daily business are to be considered as contexts. The projects and programs of the project portfolio must be coordinated with ongoing business activities. Strategic managing provides an integrated view of the objectives and resources required for daily business and those required for projects and programs. A company's strategic objectives plan and its implementation plan, therefore, consider both its daily business activities and its projects and programs.

16.2.2 Project Portfolio Managing: Process

Project portfolio managing includes the sub-processes "Collecting & analyzing information", "Preparing project portfolio reports", "Preparing for a project portfolio group meeting", "Conducting a project portfolio group meeting", and "Following up a project portfolio meeting". The business process "Project portfolio managing" is shown in Figure 16.1 as a flow chart and in Table 16.1 as a responsibility chart.

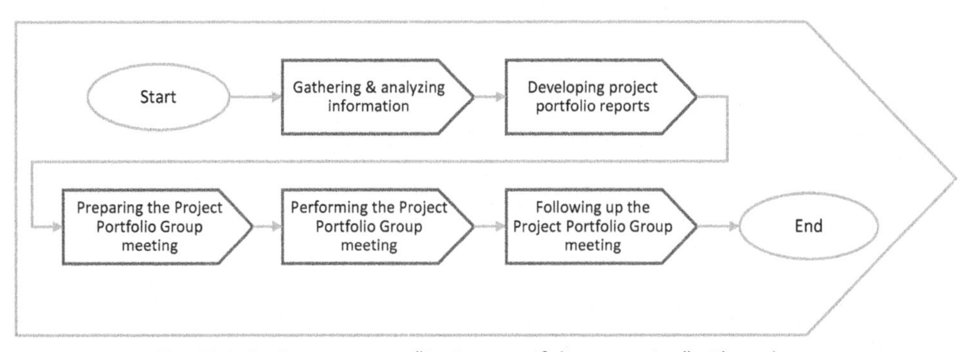

Fig. 16.1: Business process "Project portfolio managing"—Flow chart.

A cycle of project portfolio managing should not take more than seven to ten days. It starts with the (permanent) assignment of periodic project portfolio managing and ends with informing the employees of the project-oriented organization about the results of project portfolio managing.

The frequency of project portfolio managing depends on the size and dynamics of the project portfolio to be managed. It may be necessary to hold a two- to four-hour project portfolio group meeting either once a month or every two weeks.

Optimizing the structures of a project portfolio and of the project portfolio results can be assured by project portfolio analyses, the calculation of project portfolio ratios,

Table 16.1: Business Process "Project Portfolio Managing"—Responsibility Chart

Sub-process: Project portfolio managing

Tool/Document:
1 ... Selected project progress reports
2 ... New project proposals
3 ... Current project portfolio database
4 ... Project portfolio reports
5 ... Invitation for Project Portfolio Group meeting
6 ... Minutes of the meeting of the Project Portfolio Group meeting
7 ... Adapted project portfolio database
8 ... Adapted project portfolio reports

Legend:
P ... Performing
C ... Contributing
I ... Being informed
Co ... Coordinating

Process tasks		Roles: Project Portfolio Group	PM Office	Selected project owners	Employees, project managers	Initiation team	Expert pool managers	Tool/Document
1	Gathering & analyzing information							
1.1	Collecting project progress reports		P		C			1
1.2	Collecting project proposals		P			C		2
1.3	Updating the project portfolio database		P					3
1.4	Analyzing the existing project portfolio		P	C	C		C	
1.5	Analyzing project progress reports		P		C			
1.6	Analyzing project proposals		P			C	C	
2	Developing project portfolio reports							
2.1	Identifying project portfolio ratios		P					
2.2	Developing project portfolio reports		P					4
2.3	Gathering additional information		P		C	C	C	

(continues on next page)

and the preparation of project portfolio reports. Project portfolio analyses are used to produce an overall view of the projects of a project portfolio and their relations.

Analyses of project portfolios are periodic and performed in accordance with the dynamics of a project portfolio, as determined by information in the project portfolio database. Analyses are to be performed mainly at the point of a planned starting or closing or stopping of a project in order to determine the consequences for the project portfolio. Analysis of the consequences of the stopping or closing of a project enables the transfer of know-how, reallocation of personnel, allocation of freed-up resources to other projects, etc.

Table 16.1: Business Process "Project Portfolio Managing"—Responsibility Chart *(cont.)*

3	Preparing the Project Portfolio Group meeting							
3.1	Inviting participants to the Project Portfolio Group meeting	I	P	I				5
3.2	Planning objectives, agenda of the Project Portfolio Group meeting	C	P					
3.3	Providing documents for the project portfolio group meeting	I	P	I	I		I	
4	Performing the Project Portfolio Group meeting							
4.1	Alligning project objectives with the strategic objectives of the organization	P	C	C	C			
4.2	Optimizing the project portfolio structure	P	C	C	C		C	
4.3	Optimizing project portfolio risks	P	C	C	C			
4.4	Optimizing stakeholder relations	P	C	C	C		C	
4.5	Organizing the learning of and between projects	P	C	C	C		C	
4.6	Prioritizing projects of the project portfolio	P	C					
4.7	Developing minutes of the meeting		P					6
4.8	Organizing networking of projects	C	P	C	C		C	
5	Following up the Project Portfolio Group meeting							
5.1	Adapting the project portfolio database		P					7
5.2	Adapting project portfolio reports		P					8
5.3	Providing project portfolio information	I	P	I	I	I	I	

To ensure an appropriate project portfolio structure, different ratios can be considered and different optimization measures can be implemented. The various possibilities are shown in Table 16.2.

To achieve good project portfolio results, project portfolios can be optimized by means of the following measures:

> managing the internal and external resources used in the portfolio,
> managing relations with project and program stakeholders from a portfolio point of view,
> coordinating the progress of projects and programs,
> organizing learning from and between projects,
> ensuring management quality of projects and programs, and
> controlling the benefits realization of the investments implemented by projects or programs.

Table 16.2: Possible Measures for Ensuring an Appropriate Project Portfolio Structure

Ratios for ensuring an appropriate project portfolio structure	Possible measures for optimizing the project portfolio structure
Optimal number of projects of each project type in the project portfolio	> Postponing/prioritizing newly started projects > Merging projects > Stopping projects > Interrupting projects
Minimum and maximum project portfolio budget	> Postponing/prioritizing newly started projects > Merging projects > Stopping projects > Interrupting projects
Optimal allocation of bottleneck resources in the project portfolio	> Increasing bottleneck resources > Postponing/prioritizing newly started projects > Merging projects > Stopping projects > Interrupting projects
Maximum number of cooperation projects with one supplier	> Defining risk optimizing measures > Using different suppliers instead of only one regular supplier > Postponing/prioritizing newly started projects > Merging projects > Stopping projects > Interrupting projects
Maximum number of projects with the same person as project manager	> Reallocating project managers > Postponing/prioritizing newly started projects > Merging projects > Stopping projects > Interrupting projects

To manage the resources used, priorities can be established for the access of projects to scarce internal resources. Relations with project stakeholders can be managed, for instance, by agreeing on the customer strategies to be implemented in the corresponding projects. The assignment of multiple roles can be used to organize learning

Table 16.3: Methods for Project Portfolio Managing

Project portfolio managing: Methods	
Maintaining a project portfolio database	Must
Developing a project portfolio score card	Must
Developing a project portfolio list	Must
Developing project portfolio matrices	Must
Developing a project interdependencies matrix	Must
Analyzing a project proposal	Must
Analyzing selected project progress reports	Must

between projects; the observance of guidelines for project and program managing can be controlled; and decisions can be made regarding the use of management consultants for ensuring management quality.

16.2.3 Project Portfolio Managing: Methods

Essential methods for project portfolio managing are shown in Table 16.3 and are described in the following.

Creating and Maintaining a Project Portfolio Database

The project portfolio database, which provides aggregated information about projects and programs, provides a basis for project portfolio managing. The project portfolio database is assembled using data from individual projects. To compare and collect data from projects, minimum requirements for the documentation of projects must be established. Assuming a systemic project management approach, a project portfolio database should include the information shown in Table 16.4.

By analyzing the project portfolio database, project portfolio ratios can be determined, project portfolio reports can be produced, and networks of projects can be

Table 16.4: Information to be Included in a Project Portfolio Database

Information type	Information in the project portfolio database
Information about the project organization	> e.g. Project owner, project manager, selected project team members
Information about stakeholders	> e.g. Customers, suppliers and partners
Information about products and markets	> e.g. Product type, technology, region
Information about project types	> e.g. Ongoing, planned, cancelled, interrupted and closed projects, external and internal project
Information about relations of a project to other projects	> e.g. Affiliation to a program
Information about the investment which is implemented by the project	> e.g. Costs and benefits of the investment; economic, ecological and social consequences of the investment
Information about project ratios	> e.g. Project start date, project end date, project costs, project profit/loss, project risk, project progress and overall status of the project

constructed. Thus, the project portfolio database is not only a tool for the Project Portfolio Group and the PM Office. It also benefits the project and program managers of project-oriented organizations. It provides them with the "big picture" of the project portfolio, "empowers" them by providing relevant information, and enables them to optimize the results of their projects using this information.

Determining Project Portfolio Ratios

Project portfolio ratios can be defined in the first instance and determined as part of project portfolio managing. Examples of relevant ratios are shown above in Table 16.2.

Target values for project portfolio ratios can be defined for corporate governance. Actual ratios can be compared with the organization's target values. This can contribute to quality assurance. The ratio "number of projects in the project portfolio" can be used to regulate the organization's project load. The ratio "number of projects led by a person" can, for example, be used to review decisions about the allocation of personnel.

When analyzing the costs and income or benefits of the projects of a portfolio, a distinction must be made between contracting projects and internal projects. The costs and income of contracting projects can be calculated by means of economic ratios such as costs and contribution margins. For internal projects, social costs and benefits must be considered from the perspective of different stakeholders (see Chapter 3).

Project Portfolio List, Project Portfolio Bar Chart, Project Portfolio Interdependences Matrix, and Project Portfolio Scorecard

A project portfolio list contains selected information from the project portfolio database in a list format. Different project portfolio lists can include information for different target groups. For example, it is interesting to provide an overview differentiating planned, ongoing, stopped or interrupted, and closed projects. The set of planned projects can be defined as a "project backlog" from which projects to be started can be selected.

A project portfolio bar chart provides a visualization of the duration and timing of the projects of a project portfolio. If necessary, a project portfolio bar chart can also indicate scheduling dependencies among projects.

A project portfolio interdependences matrix shows dependencies between projects. It can be seen from the graphical representation whether dependencies exist. If they do, they must be specified. Dependencies between projects can relate to the project objectives, the content of the project, the project progress, the project deadlines, the project resources, the project risks, and the project stakeholders.

A project portfolio scorecard can be used to assess and document the status of a project portfolio at a controlling date.

The following criteria can be considered when assessing a project portfolio:

> the project portfolio structure,
> ratios of project types,
> the contributions of the projects to the realization of the organizational objectives,
> the quality of the relations of projects to selected stakeholder groups, and
> assurance of the management quality of projects.

In the following case study of project portfolio managing, possible targets for the assessment criteria of a project portfolio are shown in Figure 16.5 (page 460).

By using traffic light colors for scoring the criteria on a scorecard, the Project Portfolio Group's attention can be drawn to critical developments. The scoring of the individual criteria of the scorecard requires interpretation. The status of a project portfolio is to be seen in a temporal context.

Changes in the project portfolio over time can be analyzed by comparing the status at different reference dates. By displaying the status of the project portfolio at various points in time, the success of project portfolio controlling measures can also be assessed.

Case Study: Managing the RGC Project Portfolio

Building on the basic information in Chapter 15 about RGC as a project-oriented organization, this case study describes its project portfolio managing. On the one hand, project portfolio managing takes place in controlling workshops, the so-called "RGC management circles", which take place every six weeks. Operational controlling of the current business activities and strategic controlling are performed in these workshops. In strategic controlling, the current project portfolio is analyzed and measures for optimizing the project portfolio are agreed upon. If necessary, decisions to start new projects or discontinue ongoing projects are also made.

More basic and formal strategic controlling, usually involving extensive changes to the project portfolio, is performed twice a year. In addition to the project portfolio, the vision, the organizational objectives, the budget, and the implementation plan are further developed in strategic workshops, the so-called "RGC strategy workshops".

The methods used for project portfolio managing—namely, a project portfolio list, a project portfolio bar chart, a project portfolio interdependences matrix, and a project portfolio scorecard—are shown as examples in Figures 16.2 to 16.6.

The project portfolio list shown is differentiated according to ongoing, planned, and cancelled projects.

RGC Project Portfolio List (15.02.2017)

Ongoing projects				
Project	Project type	Project size	Project owner	Project manager
Customer contract A	Contracting	Project	Customer	L.Gareis
Customer contract B	Contracting	Small project	Customer	M. Stummer
Sales 17	Marketing/Sales	Project	L. Gareis	V. Riedling
Values4Business Value	Innovating	Project	R. Gareis + L. Gareis	P. Ganster
sPROJECT release 2.0	Innovating	Small project	L.Gareis	L. Gareis
HAPPYPROJECTS 17	Event	Small project	R. Gareis	V.Riedling
2nd Symposium Project Auditing	Event	Small project	R. Gareis	V.Riedling

Planned projects				
Project	Project type	Project size	Project owner	Project manager
Project mining	Innovation	Small project	R. Gareis	P. Ganster
Values4Business Value English	Innovation	Small project	R. Gareis + L. Gareis	P. Ganster
ICB 4.0	Innovation	Small project	M. Stummer	P. Ganster
RGC digitalisation	Organisation	Project	R. Gareis	V.Riedling

Cancelled projects				
Project	Project type	Project size	Project owner	Project manager
pm test	Innovating	Small project	R. Gareis	L. Gareis

(continues on next page)

Fig. 16.2: RGC project portfolio list as of 15 February 2017 (example).

Projects that have been closed since the last controlling date are defined as "closed". The project portfolio includes seven current projects, two of which are external contracting projects, and five internal projects.[1]

The relatively large number of internal projects expresses RGC's orientation towards innovation.

Although the information contained in the project portfolio list seems trivial, a relatively high level of "maturity" of project and program managing is required to provide the data. For example, the project portfolio list enables clear differentiation of project types and includes key information and project ratios for each project.

[1] RGC defines only client contracts in which RGC also formally conducts the project management as contracting projects. Consultants also work alongside clients on project content. However, these contracts are not defined as RGC projects. If this distinction would be made consistently in consulting, there would be relatively few contracting projects in the consulting sector.

Project start	Project end (planned)	Project end (adapted)	Person days	Project risk	Benefits-costs-difference
Oct. 15	Oct. 16	March 17	630	medium	high
Feb/ 16	Feb/ 17		340	low	medium
Feb/ 17	Jun/ 17		150	low	high
Oct. 15	Dez/ 16	Mai/ 17	2130	high	high
Apr/ 16	Jun/ 16	Apr/ 17	120	medium	medium
Sep/ 16	Mai/ 17		140	medium	medium
Sep/ 16	March 17		90	medium	low

Project start	Project end (planned)	Project end (adapted)	Person days	Project risk	Benefits-costs-difference
Sep/ 17	Feb/ 18		220	low	medium
Jul/ 17	Dez/ 17		130	low	medium
Jul/ 17	Dez/ 17		80	medium	medium
Jul/ 17	March 18		350	high	high

Project start	Project end (planned)	Project end (adapted)	Person days	Project risk	Benefits-costs-difference
Nov/ 15	Jun/ 16	Dez/ 16	170	medium	medium

(continues on next page)

Fig. 16.2: RGC project portfolio list as of 15 February 2017 (example) *(cont.)*.

The project portfolio bar chart from the controlling date indicates that the ongoing projects would end in May and June 2017, and the starting of planned projects would only take place thereafter.

From the project portfolio interdependences matrix, it can be seen that the project "Values-4Business Value" project had the greatest number of relations to other projects. The contexts are briefly interpreted in the following.

Interpretation:

1) and 2) use of the same senior consultant
3) and 4) sales at customers A and B as part of the sales initiative
5) HAPPYPROJECTS 17 to support Sales 17 (invitations, etc.)
6) Invitation of representatives of customers to the Peer Review Group
7) New developments in Values4Business Value considered in sPROJECT

Ecological cost-benefits difference	Social cost-benefits difference	Essential project stakeholders	Project progress	Project status
low	hoch	Customer	80%	2
low	medium	Cusomter	90%	1
low	medium	Diverse customers	5%	1
medium	hoch	Reader, publisher	70%	2
low	low	Programmer	80%	4
low	high	Participants, speaker	40%	1
low	high	Participants, speaker	60%	2

Ecological cost-benefits difference	Social cost-benefits difference	Essential project stakeholders	Project progress	Project status
low	medium	WU Vienna		
low	medium	Reader, publisher		
low	medium	PMA, consultants		
high	high	RGC staff		

Ecological cost-benefits difference	Social cost-benefits difference	Essential project stakeholders	Project progress	Project status
medium	medium	Programmer	40%	

Fig. 16.2: RGC project portfolio list as of 15 February 2017 (example). *(cont.)*

8) and 9) New developments in Values4Business Value presented at HAPPYPROJECTS 17 and 2nd symposium
10) and 11) Developed new sPROJECT release marketed at HAPPYPROJECTS 17 and 2nd symposium
12) to 15) Customers A and B offered the opportunity to attend lectures and to get reduced participation fees for HAPPYPROJECTS 17 and 2nd symposium

In assessing the status of the project portfolio, RGC orients itself using the target values listed in Figure 16.5 (page 460), which form part of the corporate governance structures. These targets also provide a basis for the interpretation of the scores on the project portfolio scorecard.

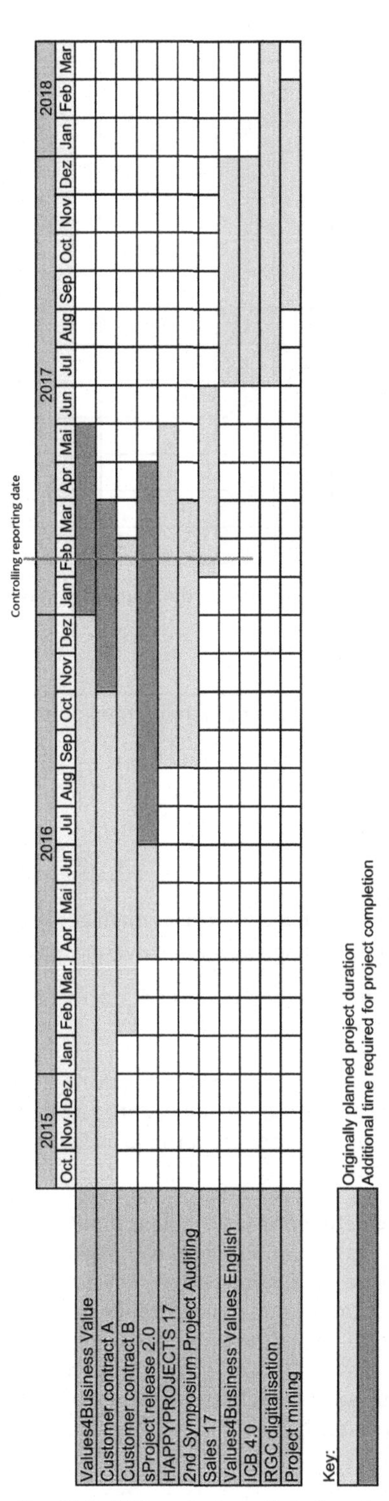

Fig. 16.3: RGC project portfolio bar chart (example).

Project	Customer contract A	Customer contract B	Sales 17	Values4Business Value	sPROJECT release 2.0	HAPPYPROJECTS 17	2nd Symposium project auditing
Customer contract A	x	1)					
Customer contract B	2)	x					
Sales 17	3)	4)	x			5)	
Values4Business Value	6)			x	7)	8)	9)
sPROJECT release 2.0					x	10)	11)
HAPPYPROJECTS 17	12)	13)				x	
2nd Symposium Project auditing	14)	15)					x

Fig. 16.4: RGC project portfolio interdependences matrix.

Number of ongoing projects in the RGC project portfolio: maximum 10
Number of ongoing projects per project type
Contracting projects: maximum 4
Marketing and sales projects: maximum 2
Innovating projects: maximum 1
Event projects: maximum 2
Organization projects: maximum 1
Number of projects one person shall manage as project manager: maximum 2
Number of projects one person shall manage as project owner: maximum 4

Fig. 16.5: Target values for the structures of the RGC project portfolio in 2017.

The project portfolio scorecard shows that, based on the relatively large number of ongoing projects in February 2017, the project portfolio structures were assessed as average. Because of the large number of ongoing projects, the project portfolio budget and resource utilization were also higher than in December 2016.

The ratios of the project types also deteriorated in comparison with December as a result of a large requirement for bottleneck resources. Relations with stakeholders were stable. The important relations with clients were good, since several "key accounts" could be involved in the innovation project and the event projects as cooperation partners. The organizational objectives had been appropriately realized by the current project portfolio. An exception was the realization of the financial objectives, because essential capacities for consultancy were tied up in internal projects and thus could not fulfill contracts to the usual extent. Employee-related objectives were very well realized by involving employees in the internal projects with resulting personnel development.

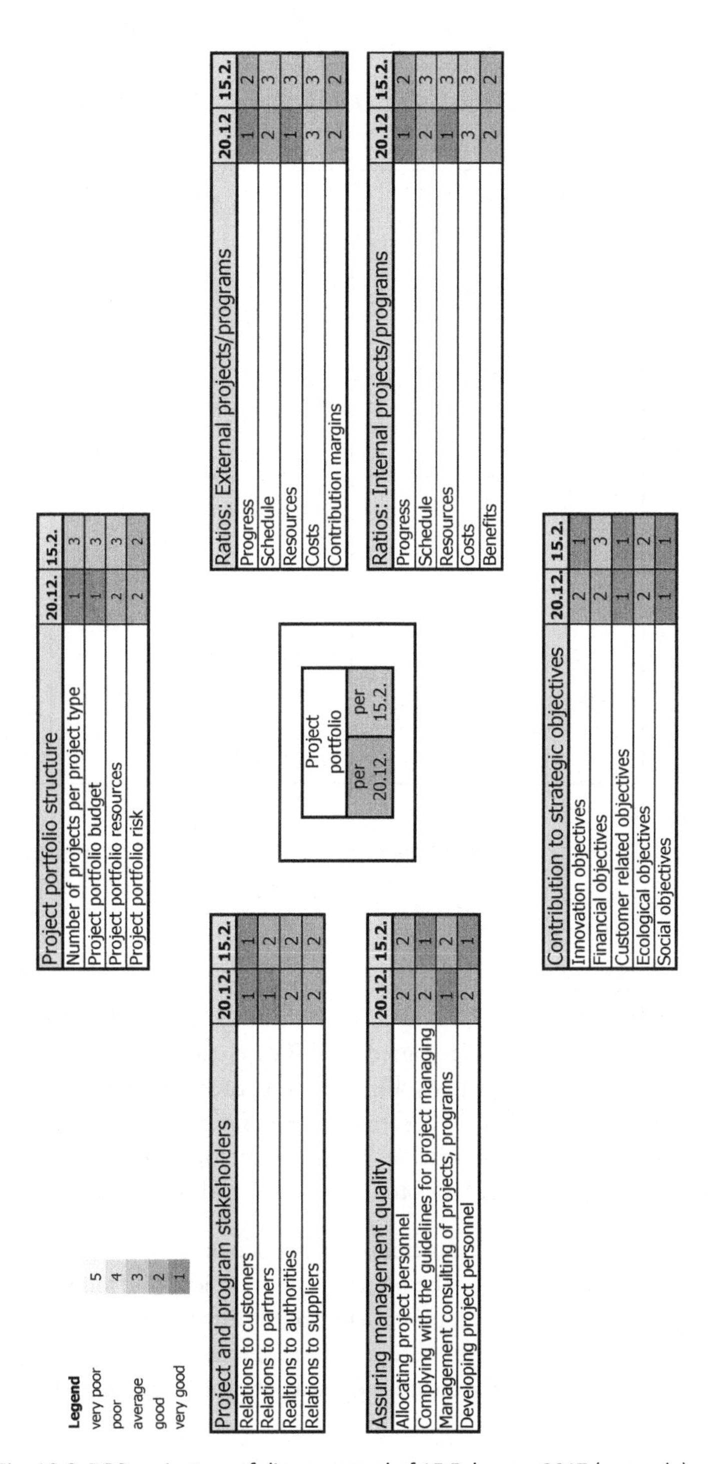

Legend	
very poor	5
poor	4
average	3
good	2
very good	1

Project portfolio structure	20.12.	15.2.
Number of projects per project type	1	3
Project portfolio budget	1	3
Project portfolio resources	2	3
Project portfolio risk	2	2

Project and program stakeholders	20.12.	15.2.
Relations to customers	1	1
Relations to partners	1	2
Realtions to authorities	2	2
Relations to suppliers	2	2

Assuring management quality	20.12.	15.2.
Allocating project personnel	2	2
Complying with the guidelines for project managing	2	1
Management consulting of projects, programs	1	2
Developing project personnel	2	1

Project portfolio	
per 20.12.	per 15.2.

Ratios: External projects/programs	20.12	15.2.
Progress	1	2
Schedule	2	3
Resources	1	3
Costs	3	3
Contribution margins	2	2

Ratios: Internal projects/programs	20.12	15.2.
Progress	1	2
Schedule	2	3
Resources	1	3
Costs	3	3
Benefits	2	2

Contribution to strategic objectives	20.12.	15.2.
Innovation objectives	2	1
Financial objectives	2	3
Customer related objectives	1	1
Ecological objectives	2	2
Social objectives	1	1

Fig. 16.6: RGC project portfolio scorecard of 15 February 2017 (example).

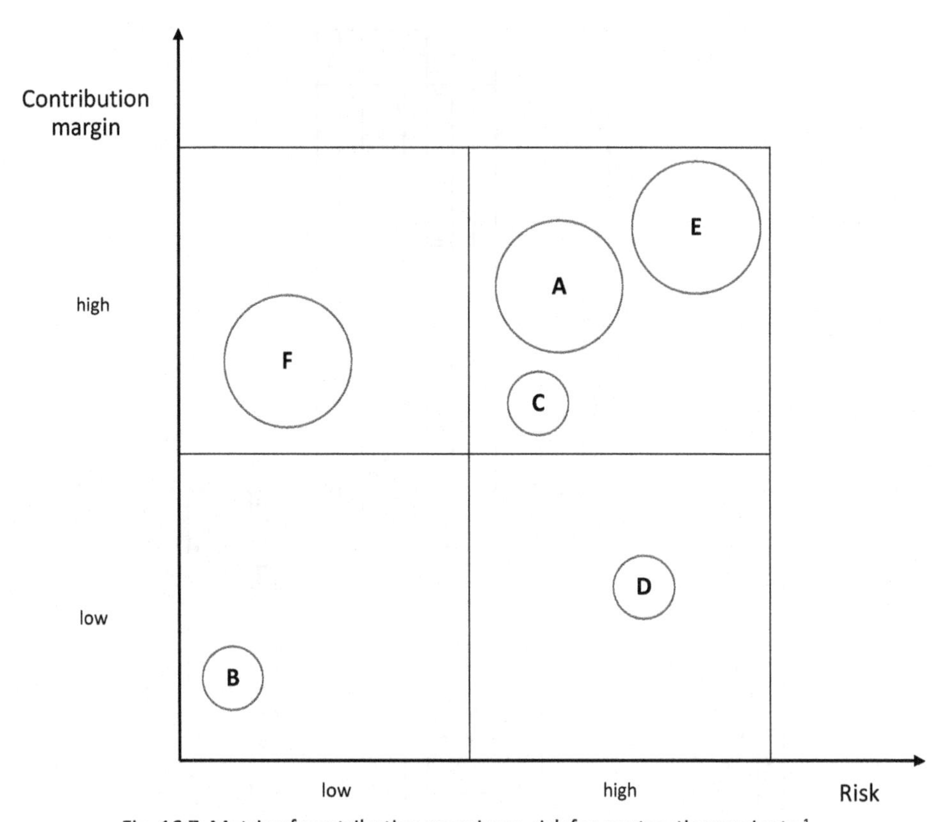

Fig. 16.7: Matrix of contribution margin vs. risk for contracting projects.[1]

Other Project Portfolio Reports

Other project portfolio reports are, for example, analyses of project portfolio resources and project portfolio matrices. A resource analysis of the project portfolio is necessary if several projects utilize a (scarce) resource. In this case, independent resource planning for individual projects is no longer possible. Planning, coordinating, and prioritizing are required at the project portfolio level. Various project ratios can be visualized by project portfolio matrices. Possible combinations of ratios are, for example, contribution margin and project risk (see Fig. 16.7) or project duration and project risk.

The informative value of project portfolio reports depends on the quality of the data in the project portfolio database. The preparation of project portfolio reports requires specific competencies. Only appropriate selection of information relevant to decision making, the visualization of information, its interpretation, and the use of symbols such as "traffic light colors" to indicate the status of projects, ensures the informative value of the reports.

[1] The contribution margin indicator is defined as "revenue minus variable costs". Risk is defined here as "negative deviation from financial target".

16.2.4 Project Portfolio Managing: Organization

The Project Portfolio Group and a PM Office (see Chapter 15) play key roles in project portfolio managing. The Project Portfolio Group or a similar organization is responsible for optimizing project portfolio results. On the basis of project and program proposals, it decides on appropriate organizational forms for implementing business processes and nominates project and program owners. The Project Portfolio Group may also recommend controlling of benefits realization.

The PM Office supports the Project Portfolio Group in fulfilling these tasks and promotes networking between projects. Project portfolio ratios and project portfolio reports can be communicated differently to different target groups. The analysis and processing of information are performed either by a PM Office or by the user of the information. To communicate the relatively complex relations in project portfolios, it is advisable to use multiple forms of visualization (lists, matrices, bar charts, tables, etc.)

16.3 Business Process: Networking of Projects

16.3.1 Network of Projects and Chain of Projects: Definition

A network of projects is a set of coupled projects. Projects in networks can be coupled by having the same cooperation partners, suppliers, or customers; by being implemented in the same region; by using the same technology; etc. Networks of projects are social networks (see Excursus: Social Networks).

A chain of projects is a set of sequential projects for performing sequential business processes. A chain is a specific form of a network of projects. A chain of projects is considered over a period of time. Chains made up of both projects and programs are also possible.

Excursus: Social Networks

The word "network" means a set of nodes and connections between these nodes. Every node is connected to other nodes either directly or indirectly. All parts of the network are flexible. Parts of a network do not stand alone but can influence other parts of the network and can also be influenced by them.

The word "social" refers to the network concept of social systems. "A social network is a set of social units together with the social relations between these units. [. . .] The social units can be people, positions or roles as well as groups, organizations or even whole societies. The social relations may vary accordingly, so at the individual level they are about sympathy, communication or role obligations and at the group level about overlapping memberships, financial dependencies or trade relations between states".[2]

[2] Endruweit, G., 1989, p. 465.

The boundaries of social networks are open and unclear. Where interactions end, where and how they arise, change, and lapse, is too complex to be understood even by network participants. Individual interactions have situational endpoints, but these are only the end of a communication chain. They simply come to an end when further interaction no longer appears to be meaningful or possible, but they may at any time extend or retract as conditions change.

"The boundaries of networks are often blurred and their activity often seems to turn on and off with no discernible regularity. [. . .] Instead of being held together within a boundary, a network coheres to form shared values, goals, and objectives. A network is recognized by its clusters of interaction and channels of communication, rather than by a fixed boundary that includes and excludes".[3]

This indeterminacy also applies to the inner order (functions, levels, roles) of a social network. Unclear boundaries and subdivisions make it difficult for the human mind to recognize social networks as coherent, functioning social structures. That is why objects with no fixed boundaries, such as social networks, create orientation problems.[4]

"Unlike a hierarchy, whose internal parts and external boundaries can be crisply mapped on a flow chart, a network has few inner divisions and has indistinct borderlines. A network makes a virtue out of its characteristic fuzziness, frustrating outside observers determined to figure out where a network begins and ends".[5]

The cohesion of a social network is produced by the common intentions and values of its members. A social network must represent certain basic intentions. These intentions form the basis of a social network. They are the cement that holds the network together.

"If a network could be drawn on a paper, its lines of coherence would consist of the ideas that the participants agree upon, manifested in commitments to similar ideals".[6]

These collective basic intentions characterize the network. If these intentions—the sense of the network—are represented by a name or a symbol, this facilitates the identification of the network as well as the identification of the participants with the network.

"In contrast to bureaucracies, whose existence hinges on members who perform highly specialized tasks and who are totally dependent on one another, networks are composed of self-reliant and autonomous participants".[7]

It is a particular advantage of social networks that they not only authorize this self-reliance and the individual interests of the members, but also promote them.[8] For the different intentions and the autonomy of the network participants do not, as one might assume, endanger the existence of a network. On the contrary, they make it more robust.

"For governance, 'network' implies a non-hierarchical system of equal, self-sustaining members. Unlike a bureaucracy a network is dependent on no one of its parts. No organ performs a specialized task necessary for the function of the whole. A net has no center. It is made up of links between parts. [. . .] It is precisely this attribute of self-sustaining parts that gives the network form its remarkable resiliency and its adaptability to stress. Segmentation explains

[3] Lipnack, J., Stamps, J., 1982, p. 229 f.
[4] Lipnack, J., Stamps, J., 1982, Chapter 1.1.3.
[5] Lipnack, J., Stamps, J., 1982, p. 8.
[6] Lipnack, J., Stamps, J., 1982, p. 9.
[7] Lipnack, J., Stamps, J., 1982, p. 7.
[8] Lipnack, J., Stamps, J., 1982, p. 8.

why, for example, underground political movements are so difficult to suppress. Squashing one node does little to impair the effectiveness of the net as a whole".[9]

Relations in a social network are based on three characteristics that distinguish social networks from social systems: autonomous participants, flexible and decentralized organization, and voluntary participation in the network.

Elements of networks of projects are projects and programs that can communicate with each other. The common intention of the network is the creation of synergies in the network and the organization of learning. Networking takes place, for example, in workshops, meetings, and talks. Communication can be supported by common databases, chatrooms, and links. Networks of projects have no clear boundaries, which means that the projects of clients, partners, or suppliers of a project-oriented organization can be considered, provided that their consideration can contribute to the realization of common objectives.

The need for networking of projects can be recognized by project managers, a Project Portfolio Group, or a PM Office. Because of the need to concentrate on project objectives in daily project work, the potential of networking with other projects is not always apparent to the members of project organizations. Promoting networking of projects is therefore also a task for the Project Portfolio Group and the PM Office.

Networking of projects can occur as a situation requires, or it can be established as a periodic form of communication for a network over a period of time. Situations requiring ad hoc networking are, for example, project discontinuities. The consequences for other projects of a project crisis or chance must be analyzed, and necessary measures must be agreed upon. Situational criteria for coupling projects must be defined. Communication can then be carried out in networking workshops and meetings.

Networking of projects requires a cooperative organizational culture. An active information policy, the possibility of horizontal communication, and mutual trust are central values for networking.

16.3.2 Networking of Projects: Objectives and Process

The objectives of networking of projects are to create synergies and avoid conflicts, as well as to promote learning in a network of projects.

The need for networking can be recognized by project managers, the Project Portfolio Group, or the PM Office. They can initiate the networking of projects. Project networking starts when the project managers involved decide to network. The process ends when the agreed-upon measures for the use of synergies and the avoidance of conflicts between projects have been performed.

[9] Lipnack, J., Stamps, J., 1982, p. 223 and 225.

To prepare for networking, the integrating criterion of the network must be identified, members of the individual project organizations must be invited, and relevant information about the projects must be provided.

It is advisable to use various communication formats for networking. As a communication format, a workshop enables direct interaction between the members of the different project organizations. The relations between the projects must be analyzed. A graph of the network of projects can be used to visualize these relations. This supports the collective construction of the network at a particular point in time.

On the basis of an analysis of synergetic and conflictual relations between projects, strategies and measures can be defined for using synergies and avoiding conflicts. Implementation of the agreed-upon measures is a task for the project managers of the networked projects.

The business process "Networking of projects" is represented as a responsibility chart in Table 16.5.

The results of networking are an understanding of the relations between the projects and measures for using synergies or avoiding conflicts in the network. The following measures are possible in principle:

> redefining project objectives in case of conflicting objectives of projects,
> reallocating personnel in case of resource conflicts between projects, or possibly changes in project priorities,
> reshaping of project stakeholder relations on the basis of a holistic view resulting from the consideration of multiple projects,
> risk balancing between projects by altering contractual relations with clients, partners, and suppliers,
> transferring know-how between projects, and
> establishing communication structures for periodic coordination between projects.

In an extreme case, the findings of an analysis of the relations in a network of projects can also lead to projects being stopped or interrupted.

16.3.3 Networking of Projects: Methods

Workshops and Meetings for Networking of Projects

Networking of projects is facilitated by establishing specific communication formats. In workshops, representatives of the networking project organizations and of project stakeholders can interact directly.

A workshop for networking can take half a day to one day. Usually only one workshop is necessary. If necessary, workshops can be supplemented by periodic meetings for networking.

Table 16.5: Business Process "Networking of Projects"—Responsibility Chart

Business process: Networking of projects									
Tool/Document: 1 ... Invitation to the networking workshop 2 ... Network of projects graph 3 ... List with measures Legend: P ... Performing C ... Contributing I ... Being informed Co ... Coordinating Process tasks		Roles							
		Project Portfolio Group	PM Office	div. representatives of project owners	div. Project managers	Project team members	Project stakeholders	Expert pool managers	Tool/Document
1	Planning networking of projects								
1.1	Collecting project progress reports		C		P				
1.2	Concreting the demand for networking of projects			C	P	C			
2	Preparing the networking of projects								
2.1	Inviting participants for the networking		P	I	I	I	I	I	1
2.2	Collecting information about the considered projects		P		C	C	C		
3	Performing a workshop for networking of projects								
3.1	Exchanging information between projects		C	C	P	C	C	C	2
3.2	Developing a network of projects graph		C	C	P	C	C	C	
3.3	Planning measures based on the provided information		C	C	P	C	C	C	
3.4	Agreeing on the planned measures	I	C	C	P		I	C	3
4	Following up the networking workshop								
4.1	Organizing for the performance of the measures	I	P	I	P	I	I		
4.2	Controlling the measures		P		P				

Graph of a Network of Projects

Relations between the projects of a network can be visualized in a graph. This graph represents the networked projects and the relations between them. Circles of different sizes and colors can be used to represent the projects. The symbols used need to be explained in a legend.

Projects with similarities can be gathered together in various groupings. Lines between the projects (but no directional arrows) can be used to represent the relations. Relations between projects are the primary focus of networking.

Relations between individual projects can be interpreted by qualitative statements. As an example, Figure 16.8 shows the RGC network of projects as of 15 February 2017.

Case Study: RGC Networking of Projects

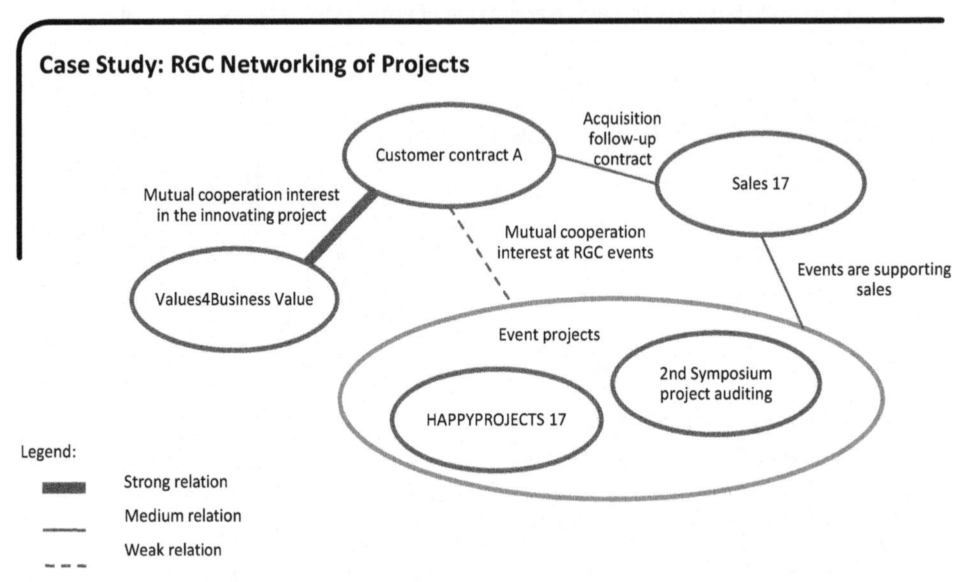

Fig. 16.8: RGC network of projects for managing relations with customer A.

Interpretation:

> The network of projects for managing relations with customer A was constructed in order to plan a strategy and measures for managing the customer relationship.
> Projects that seemed relevant to managing the relationship with customer A have been included in the network graph. Relations between the projects were analyzed and briefly described, as shown in the graph.
> The following customer binding strategy was formulated: "Wide-ranging cooperation, with a focus on securing business value for the customer".
> The following measures for customer binding were planned on the basis of the analysis:
- Inviting representatives of customer A to the peer review group,
- Inviting representatives of customer A to give lectures and to participate in the RGC events HAPPYPROJECTS 17 and the 2nd symposium "Project Auditing",
- Introducing customer representatives to RGC consultants at events, and
- Use experiences from the project "Values4Business Value" by the customer representatives in their own practice.

16.3.4 Managing a Chain of Projects

The projects of a chain of projects are closely coupled by belonging to the same investment process. Typical chains are chains formed by a conceptual and a realization project, from a tender offer and a contracting project, or from a pilot and a follow-up project.

Managing chains of projects has the objective of ensuring continuity in the management of two or more successive projects or programs undertaken to implement an investment. Managing a chain of projects is not a separate business process of a project-oriented organization, but a special case of networking between (successive) projects.

To manage chains of projects, personnel management measures and organizational measures must be implemented. An essential personnel management measure is the involvement of members of the project organization of an ongoing project in the project organization of a follow-up project. There should be overlaps between the members of the project owner teams and the project teams. The Project Portfolio Group is responsible for the selection of the members of the project owner teams as part of project portfolio managing.

The objective of integrating two successive projects into a chain can be realized in the project closing process of the ongoing project and in the project starting process of the follow-up project. The planning of the structures of the follow-up project and its documentation in an initial project handbook are work packages of the ongoing project. Potential members of the project organization of the follow-up project must be involved in the project closing process in order to help plan the structures of the follow-up project and to meet representatives of project stakeholders. To ensure the transfer of know-how into the follow-up project, members of the project organization and representatives of project stakeholders of the previous project should be invited to participate in the project starting process.

No specific methods are required to manage chains of projects. The (organizational) measures for the integration of the two projects are presented in the project plans, especially in the project organization documents.

16.4 Project Portfolio Managing and Networking of Projects: Values

The values of the RGC management paradigm are shown in Figure 16.9 and are interpreted as follows for project portfolio managing and networking of projects.

Holistic Definition of Solutions and Boundaries

To be able to optimize the structures and results of a project portfolio, it must be considered holistically. To this end, the boundaries of the organization whose project portfolio is to be managed must be adequately defined, and all the projects of this organization should be considered.

The boundaries of the organization must be defined in such a way that as many relations as possible between projects can be considered. When structuring project portfolios, organizational units constitute the first criterion and project types the second criterion. Structuring by project type alone does not enable the consideration of

Holistic definition of solutions and boundaries	Sustainable business value as success criterion	Quick realization of results and efficiency	Iterative approach
> Defining holistic boundaries of the organization whose project portfolio is managed > Considering all project types in the project portfolio	> Using the business value as criterion for prioritizing projects in the project portfolio > Considering the principles of sustainable development for project portfolio managing	> Ensuring quick wins by appropriately prioritizing projects in the project portfolio > Using appropriate methods for project portfolio managing > Organizing the project portfolio managing efficiently	> Defining a project backlog and managing it iteratively > Deciding about the application of agile methods in projects by the Project Portfolio Group

Context-orientation	Continuous and discontinuous learning	Frequent, visually supported communication	Empowerment and resilience
> Managing of and for stakeholders > Considering projects of other organizations with relations to own projects > Involving stakeholders in project portfolio managing	> Coping with continuous and discontinuous changes of the project portfolio > Learning by networking of projects > Reflecting on project portfolio managing and on networking of projects	> Using project portfolio reports for visualized communication > Performing stand up meetings in the PM Office for networking	> Empowering the Project Portfolio Group > Ensuring resilience of the project portfolio

Fig. 16.9: Project portfolio managing, networking of projects, and values.

essential relations between projects of different types. This results in an inappropriate reduction of management complexity, which permits only suboptimal solutions.

This means, for example, that a larger organization with several profit centers needs both a sub-portfolio for the strategic projects of the organization and sub-portfolios for the individual profit centers. A structuring of sub-portfolios by project type is therefore not recommended.

However, the data in the project portfolio database can be presented in various ways, using matrix logic to provide information for the management of organizational units, as well as additional information for managing the project portfolios of the different sproject types.

To provide an adequate basis for strategically "managing by projects", organizations must consider projects of all types in their project portfolios. Some organizations that perform large contracting projects, such as construction, engineering, or IT companies, reduce their project portfolio to contracting projects. Or IT departments consider only IT projects and therefore do not analyze their relations to projects of other types. As a result, these companies lose significant opportunities for optimization in management.

Holistic project portfolio managing means, above all, to see a project portfolio not only as a set of projects and programs, but to also consider their networking. The analysis and management of relations in project portfolios can be promoted by adequately considering information in the project portfolio database about stakeholders, products, markets, or technologies. In risk managing for project portfolios, the project portfolio risk is therefore not to be seen as the sum of the risks of the individual projects or programs. As in the financial sector for securities portfolios, relations between projects must also be considered—for example, for risk compensation. There is an analogy here to risk managing in programs (see Chapter 13).

Sustainable Business Value as a Success Criterion

Business value orientation is a key strategic element of project portfolio managing. Project portfolio managing should help secure sustainable business value for the organization. Although the making of investment decisions is not a project portfolio managing task, contributions to securing business value can be made by:

> selecting appropriate organizational forms (small project, project, program) for performing comprehensive business processes,
> selecting owners who give their projects and programs appropriate management attention,
> managing stakeholder relations from the point of view of the project portfolio,
> dissolving close links between projects in the project portfolio to reduce risk for the organization,
> prioritizing projects and programs in the project portfolio, and
> arranging for networking between projects.

The contribution of a project to business value is a key criterion for prioritizing projects of the project portfolio. Taking into account the principles of sustainable development, business value can be understood as not only economic but also ecological and social, not only in the short term but also in the medium term, and not just locally but also regionally.

The principles of sustainable development also influence the business process "Project portfolio managing" and its methods. So, for example, sustainability principles can be taken into account in the project portfolio database and project portfolio reports, and stakeholders can be involved in project portfolio managing.

Quick Realization of Results and Efficiency

Quick project portfolio results can be achieved by prioritizing projects appropriately. Quick results can also be achieved by keeping project portfolios "lean". Lean project portfolios can be ensured by minimizing the number of concurrent projects, planning projects with short project durations, and differentiating sequential projects in chains of projects.

The planning of quick wins of projects can be influenced by the objectives of project portfolio managing. Rapid and easy-to-achieve results can be planned and controlled across project boundaries.

The quality of the information in the project portfolio about individual projects and programs is an essential basis for good project portfolio results and efficient project portfolio managing. Professional project and program managing is a prerequisite for efficient and effective project portfolio managing.

Efficiency in project portfolio managing does not require an integrated project and project portfolio database. For project portfolio managing, only aggregated project data is necessary—for example, project type, project owner, project manager, project budget, project stakeholders, project start and end dates, etc. No detailed information about work packages, costs, schedules, risks, and responsibilities for work packages and stakeholder relations is required.

If needed, this information can be obtained from the documentation and tools for managing the individual projects. Detailed project planning and project controlling is decentralized and used for empowerment at the project level.

In project portfolio managing, the focus is on the availability of relevant data for decision making, on communication and networking, to be able to deal with the complexity of project portfolios. A relatively simple stand-alone software solution is not optimal but might be sufficient for project portfolio managing.

Iterative Approach

Project portfolio managing is a strategic management task that is performed periodically. Depending on the dynamics of the respective project-oriented organization, projects and programs are iteratively proposed, started, prioritized, discontinued, etc. The planned projects of an organization can be perceived as their "project backlog". From the backlog, projects to be started are selected in periodic project portfolio meetings.

Iterative planning of the project portfolio is performed alongside the iterative planning of projects to be started. The strategic objectives of the project-oriented organization might also be adapted iteratively on the basis of feedback from the projects in the project portfolio.

It can be observed that the cycles of project portfolio managing are becoming shorter as a result of dynamism in the economic environment. Some industries not only prepare project proposals once a year as part of budgeting, but also decide on project proposals on a quarterly or even monthly basis and perform corresponding rolling budgeting. These short-term planning cycles require projects with short project durations, which result from defining chains of projects.

Context Orientation

The content-related context of project portfolio managing is the ongoing business activity of a project-oriented organization. Thus, in strategic managing, the objectives, measures, and resource needs of the ongoing business activities and the project portfolio must be coordinated in an integrative way.

In project portfolio managing, the temporal context is taken into account by considering changes in the project portfolio over time. The dynamism of the project portfolio is supported by appropriate project portfolio management structures—that is, the frequency of meetings and workshops, the involvement of stakeholders, the types of analysis, the ratios to be considered, and the project portfolio reports to be prepared.

Context orientation in the social sense means that when constructing networks of projects, projects and programs are considered which are performed outside the organization under consideration—for example, the projects of customers, suppliers, or partners—provided that this results in additional benefits.

Managing stakeholder relations from the point of view of project portfolios differs from stakeholder managing in individual projects by being more strategically oriented. Strategies are needed to manage relations with stakeholders who are simultaneously stakeholders of several projects. All relations are analyzed and strategically managed. Conflicts of interest of stakeholders regarding different projects and programs may come to light in this process. This aspect of the complexity of project portfolios must be dealt with. Management "for" stakeholders can also be implemented in project portfolio managing (see Chapter 6).

The involvement of selected external stakeholders in project initiating, project portfolio managing, and, above all, in networking between projects is challenging but can reduce risks for the organization and create opportunities for optimization.

Continuous and Discontinuous Learning

Project portfolio managing can involve continuous and discontinuous learning. Learning in project portfolio managing is achieved by means of observation, reflection, and feedback. For example, the following can be observed and reflected upon:

> whether the structures of the project portfolio are appropriate,
> whether the organizational forms defined for performing business processes are adequate,

> whether the projects and programs of the project portfolio contribute to the implementation of the organizational objectives,
> whether stakeholder relations are appropriately managed, and
> whether controlling measures were successful at the project portfolio level.

If external stakeholders are involved in portfolio managing, it is possible to learn from their differing experiences.

Performing project portfolio managing periodically makes it possible to optimize the business process by reflection and feedback.

Project portfolio managing can also contribute to learning in projects and programs and to strategic managing. Management quality in projects and programs can be ensured by checking whether the corporate governance structures for project and program managing are implemented. The project portfolio manager contributes to the further development of the strategic plans on the basis of the findings of projects and programs.

Frequent, Visually Supported Communication

The dynamism of project portfolios necessitates frequent communications in project portfolio managing. Members of the initiation team, the Project Portfolio Group and project and program organizations, employees of the PM Office, and representatives of stakeholders are to be involved, according to pressing needs.

The objective of using project portfolio managing to perform an integrative function in a project-oriented organization can be supported by various communication formats, such as workshops of project initiation teams, project coordination meetings, and workshops and meetings for networking of projects.

Frequent, short-term communications of a Project Portfolio Group are usually not possible. Stand-up meetings of initiation teams, of the members of the PM Office, and also of groups for networking can be held and are useful.

An important function of the communication formats of project portfolio managing is to contribute to the development of the culture of a project-oriented organization. Openness, trust in the exchange of information, and learning from each other (in networking) are to be assured.

Project portfolio ratios and project portfolio reports can be used as visual aids for communication. Project portfolio lists, project portfolio bar charts, and project portfolio scorecards are particularly suitable. Appropriate room infrastructure and ICT infrastructure must be provided to support project portfolio communication.

Empowerment and Resilience

The Project Portfolio Group and the PM Office are to be empowered in project portfolio managing. In larger organizations, there might be the specific role "Project Portfolio Manager" in the PM Office. Project portfolio managing is institutionalized by this role.

Empowering the Project Portfolio Group and the PM Office in project initiating means that they can make the following decisions:

> accepting or rejecting project proposals,
> nominating project owners,
> adapting project objectives in project proposals to make them consistent with the organization's objectives, and
> initiating management consulting for projects and programs.

In implementing an investment, these decisions are made in the context of investment decisions to be made by an investment decision committee or the Project Portfolio Group.

Empowering the Project Portfolio Group and the PM Office in project portfolio managing means that they can make the following decisions:

> designing the project portfolio strategically,
> starting new projects,
> stopping or interrupting projects for reasons of organizational strategy,
> prioritizing projects,
> strategically managing stakeholder relations,
> initiating management consulting for projects and programs,
> initiating controlling of benefits realization for investments, and
> initiating networking of projects.

In empowering the Project Portfolio Group, relations to the decision-making powers of project and program owners must be respected.

A key objective of project portfolio managing is to ensure appropriate project portfolio structures. These contribute to the resilience of a project-oriented organization. The stability of a project-oriented organization should not be jeopardized if individual projects or programs of the project portfolio have problems and unforeseen developments occur.

The resilience of a project-oriented organization can be supported by agile and redundant structures in the project portfolio. In project portfolio managing, a project-oriented organization can prove that it is capable of reacting to dynamics and implementing change.

16.5 Project or Program Consulting[10]

Project Quality: Definition

Project quality can be defined as "meeting all defined and stipulated solution requirements".[11] Quality can also be understood as meeting the expectations of external or internal customers. It is the requirements at the time of providing the solution to the customer that are to be met, not the initially defined requirements.

In projects, a distinction can be made between the quality of the project results to be achieved and the process quality of project managing. Appropriate quality management is to be performed to ensure quality (see Excursus: Quality Management).

Excursus: Quality Management

Quality management originated at the beginning of the 20th century. The division of labor and the mass production introduced by Frederick W. Taylor led to a need to control the quality of products.

Since then, quality management has evolved from product-related quality control to an organization-oriented total quality management approach with the objective of continuously improving business processes.[12]

Deming is one of the founders of quality management.[13] He pointed out that quality and productivity do not contradict one another but positively correlate with each other, provided that production is perceived as a process that takes account of customer relations and customer feedback. Deming describes the following chain reaction: improving quality results in lower production costs, as fewer error corrections are necessary. Fewer errors lead to better utilization of machines and materials. Lower production costs lead to better productivity and enable better quality at lower prices. This leads to a higher market share, which in turn ensures the existence of the company and thus of its jobs. Deming introduced the Deming cycle, which takes into account both error correction and error prevention. The Deming cycle consists of the steps "plan, do, check, act—PDCA" (see Fig. 16.10).

Deming's approach has been further developed by Ishikawa and Taguchi. Ishikawa postulated that all divisions of the company and all employees are responsible for quality. The quality is defined by the customer—whereby customer is understood not only as the end customer who pays for the product, but also as the next person in a business process. Each employee is therefore both a customer and a supplier.[14]

Deming's approach and its further development by Ishikawa and Taguchi corresponds to a total quality management approach that can be characterized by seven principles[15]:

[10] In the following, primarily the management quality of projects will be considered. The objectives, processes, methods, and organizational structures for programs are similar.

[11] Society for Project Management, 2005, p. 27 ff.

[12] See Seaver, M., 2003.

[13] See Deming, W.E., 1992.

[14] See Ishikawa, K., Lu, D., 1985.

[15] See Bounds, G. M., Dobbins, G. H., Fowler, O. S., 1995; Krczal, A., 1999, p.399 ff.

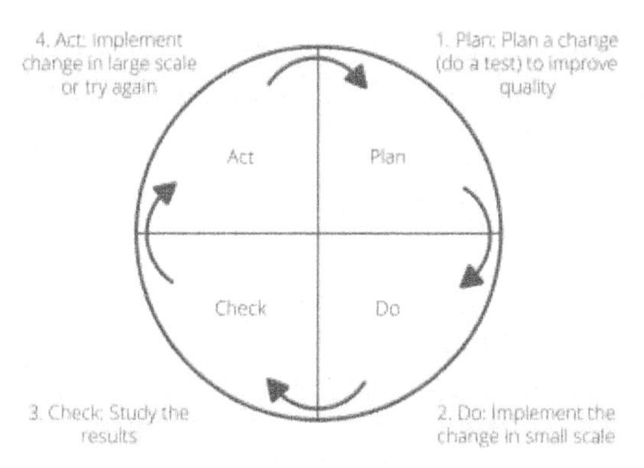

Fig. 16.10: Deming's PDCA cycle.

- customer orientation,
- continuous improvement of systems and processes,
- process management,
- search for the true cause of errors,
- data collection and statistical methods,
- employee orientation, and
- team orientation.

Projects and Programs as Clients of Consulting

Traditionally, the objects of consulting are permanent organizations such as companies, profit centers, or service centers. The complexity and dynamics of projects and programs might necessitate support in the fulfillment of content-related processes and management processes. The perception of projects and programs as temporary organizations (see Chapter 1) allows them to be defined as social systems to be consulted. Projects and programs can be "client systems" of consulting.

Ensuring the Quality of the Project Results

Professional project managing contributes to ensuring the quality of the results of a project. If necessary, content-related consulting can be performed to ensure this quality. The objective of content-related consulting of a project is to contribute to solving content-related problems. In content-related consulting, a consultant cooperates with the members of the project organization as a project contributor. In the case of more extensive activity, the consultant can also be a project team member.

Ensuring Management Quality of a Project

Management quality of a project is to be assured by professional project managing. If necessary, management quality of a project can also be assured by management consulting.

Management consulting of a project can be differentiated according to the following services and business processes:

- consulting a project when implementing a project management sub-process,
- management auditing of a project, and
- performing a short project intervention.

Furthermore, the project-oriented organization—but not the project, as it does not yet exist—can be consulted for project initiation. The objectives of consulting in project initiating are to ensure that an appropriate organizational form for performing a business process is selected and that initial plans for starting a project are prepared. The objective of consulting a project when performing a project management sub-process is (further) developing the project management competence of the project. The objectives of management auditing of a project are to analyze the management competency of a project and develop recommendations for the further development of this competency. The objective of performing a short project intervention is to contribute to the resolution of a specific management problem of a project.

In management consulting, the values of a systemic management approach (see Chapter 2), such as stakeholder orientation, securing quick wins, optimizing the business value, using an iterative approach, ensuring sustainable development, etc., are considered. In performing a short project intervention, quick wins are to be achieved with a small consulting effort. Efficiency in management consulting is achieved by differentiating, as described later, between the client system, the consulting system, and the consultant system.

Management consulting in a project is to be distinguished from project management training, coaching of the members of project organizations, and temporary project management. Coaching enables the transfer of the contents of training into the daily practice of training participants and the solving of current "on the job" challenges of managers. In temporary project management, the project manager assumes management responsibility. Management consulting in projects is often combined with project management training and coaching of the members of project organizations.

16.5.1 Consulting of an Organization in Project Initiating

When a project is initiated, strategic project decisions are made which define the basic structures for the project to be performed (see Chapter 5). A professional project initiation is therefore important.

The objective of consulting an organization in project initiating is to optimize the results of project initiating. It must be ensured that

- an adequate organizational form for performing a business process is selected,

> appropriate initial project plans are available for starting the project,
> an appropriate project owner is nominated, and
> a project manager and a project team are assigned.

Consulting in project initiating can include the following services:

> supporting the definition of project boundaries and the development of initial project plans,
> supporting the definition of initial project strategies,
> analyzing the local, regional, and global ecological and social impact of the project,
> analyzing the relations of the project with other projects and programs,
> supporting the preparation of the project proposal, and
> supporting the involving of project stakeholders in the initiating process.

The project initiator assigns a consultant with the performance of consulting services in project initiating. The social systems involved in the consulting process are the client system "Organizational unit initiating a project", the consulting system "Project initiating", and the consultant system "Home base of consultants" (see Fig. 16.11). Aside from the project initiator, the initiation team, and the Project Portfolio Group, the client system can also include representatives of the PM Office and stakeholders. The consultant and representatives of the client system cooperate in the consulting system. The consultant's "home base" may be internal to the organization (e.g., the expert pool of "project managers") or external (e.g., consulting companies).

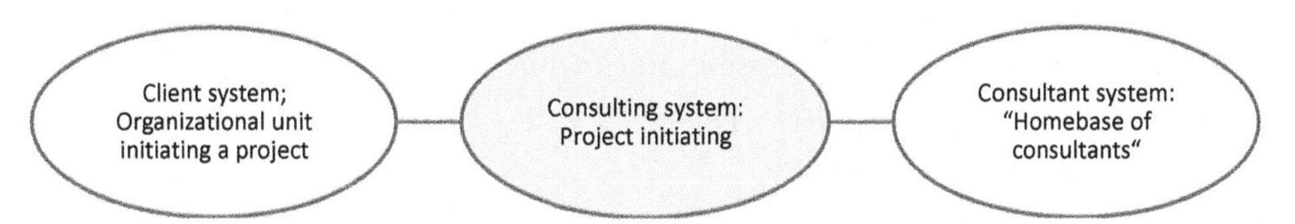

Fig. 16.11: Systems involved in consulting an organization in project initiating.

16.5.2 Consulting of a Project in Performing a Project Management Sub-Process

Consulting a Project in Performing a Project Management Sub-Process: Objectives and Objects of Consideration

The objective of consulting in a project management sub-process is the (further) development of the project management competence of the project. It develops not only the competencies of the project manager and/or the project team, but also of the project as a temporary organization. The project is the client system in the consulting process. Consulting can be used to implement the requirements of guidelines for project and program managing.

Consulting on a project can occur during the implementation of the sub-processes "Project starting", "Project controlling", "Project coordinating", "Project transforming" or "Project repositioning", and "Project closing". Consulting can relate to one or to several of these sub-processes. For example, consulting on a project is frequently performed in project starting and subsequently during several controlling cycles in order to ensure the sustainability of the project structures developed in project starting.

In consulting, the following objects must be considered for each project management sub-process:

> the design of the sub-process in the context of the design of the project management process,
> the quality of the individual project plans and the consistency of the different plans,
> the design of the project organization, and
> the design of project context relations.

Project consulting in project starting can include the following services:

> analyzing existing project management documents,
> interviewing project stakeholders,
> supporting the selection of project team members,
> preparing and reflecting the performance of a project start workshop,
> supporting the development of detailed project plans and a project manual,
> supporting the use of an appropriate project infrastructure (rooms, ICT, etc.),
> supporting the managing of project context relations,
> supporting the planning of possible quick wins,
> preparing and reflecting on the performance of a project owner meeting, and
> observing project meetings.

The results of management consulting can be assessed on the basis of the further development of management competency of a project achieved. Any improvements in quality can be observed in the competencies of members of the project organization, in the efficiency of meetings, in the content and form of the project management documentation, in the business value orientation of members of the project organization, in the image of the project, and in relations with project stakeholders.

Consulting a Project in a Project Management Sub-Process: Organization

Management consulting for a project should be assigned to a consultant by the project owner and not, for example, the general management of an organization. Managers of the permanent organization can propose consulting, but the project owner should be convinced of the usefulness of the consulting, want it, and agree to it with the members of the project organization.

Management consulting is performed by a consulting system. This represents an intermediate system between the project as a client system and the organization from which the consultants come. The system "Project management consulting" has specific objectives, processes, roles, and methods, which differ, for example, from the client system "Project".

The organizations from which consultants come (their "home base") can be either consulting companies or organizational units within the project-oriented organization. Figure 16.12 shows these relations for management consulting in a project.

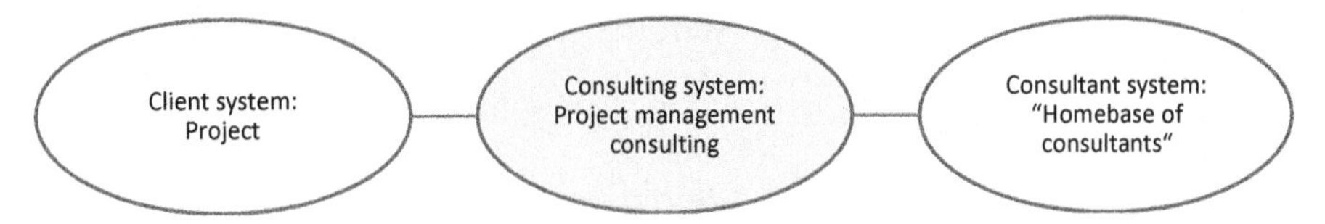

Fig. 16.12: Consulting on a project for performing a project management sub-process—social systems involved.

The representatives of the project in the consulting system are experts in the project and its contents and problems, and they support the transfer of the results of the consulting system into the project. The consultant cooperates as project management expert with the representatives of the project in the consulting system.

The "home base of consultants" can influence the success of consulting. The acceptance of an internal consultant—for example, from an expert pool of "project managers"—is likely to be higher than that of a consultant whose "home base" is an expert pool of "moderators" in a HR department.

16.5.3 Management Consulting of a Program

Management consulting of programs differs from management consulting of projects because of the longer duration and higher complexity of programs. Specific services in management consulting of programs are, for example, the establishment of a program office, support for program marketing, and the development of integrated program standards (see Chapter 13).

To increase the efficiency of consulting, management consulting of a program can be performed in combination with management consulting of individual projects of the program.

16.5.4 Management Auditing of a Project or a Program

"An audit is a systematic and independent investigation to determine whether the quality-related activities and their related results conform with the planned instructions and whether these instructions are effectively implemented and suitable for achieving the objectives".[16] Auditing is a quality assurance method in which the observance of

[16] DIN EN ISO 8402:1995_08 bzw. DIN EN ISO 9000:2015-11.

predefined procedures and standards as well as their effectiveness and appropriateness are examined.

Terms which are also used in practice instead of the term "audit" are "revision", "review", or "health check". "Audit" and "revision" are often associated with a higher formalization of the process and its consequences than "review" and "health check".

In a project auditing, the management quality as well as the quality of the results of a project are the objects of auditing. Auditing should be a learning opportunity for the audited project.

Management Auditing of a Project: Objectives

The objective of a management audit of a project or program is to assess management competencies and provide recommendations for the further development of these competencies. In this audit the project management quality is considered—that is, the organizational competency of the project, the collective competencies of the teams of a project, and the individual management competencies of members of the project organization.

Management auditing of a project can be planned for or can be performed based on a specific demand. It can only be performed during the performance of a project, as a project does not exist before starting it and does not exist anymore after project closing.

A management audit of a project may include the following:

- analysis of project management documents,
- interviews with project stakeholders,
- observation of project meetings,
- assessment of the management competency of the project,
- benchmarking of the management competency of the project with other projects, with "best practices" and with a defined project management standard,
- development of recommendations for further developing the management competencies of the project,
- preparation of an audit report,
- performance of an audit workshop, and
- presentation of the results of the management audit.

The project management standard which is the basis for a management audit of a project is to be agreed upon when the audit is assigned. This basis can be an organization-specific project management standard, existing in the form of internal guidelines and project-related corporate governance rules, or a generic project management standard, such as the ISO standard or DIN for project management, *the PMBOK® Guide* from PMI, or the RGC project management approach.

Management Auditing of a Project: Organization

Management auditing of a project begins with an assignment to the auditor or auditors. The demand for a management audit of a project can be expressed by either a member of the permanent organization or the project owner. The client of a project audit is the project.

Management auditing of a project is carried out by an intermediate auditing system, in which the auditors, representatives of the project either and representatives of stakeholders of the audited project cooperate. The quality of the audit depends on the willingness of the representatives of the project to be audited, the scope and quality of the information provided, the time available, and the availability of resources.

Auditors can be recruited from the project-oriented organization or externally. In some project-oriented organizations, there is an expert pool of "project auditors". Auditing tasks can be performed by senior project managers as a "job enlargement". To perform the auditing tasks well, it is important that the respective auditor have an appropriate distance to the client system "project".

Not Project Audits

Quality assurance measures at different points in time allow different objectives to be pursued (see Fig. 16.13). In the case of a quality check before a project begins, it can be checked whether the necessary basis for starting a project exists. A quality check after the end of a project does not benefit the project anymore, but it enables learning in other projects and in the project-oriented organization.

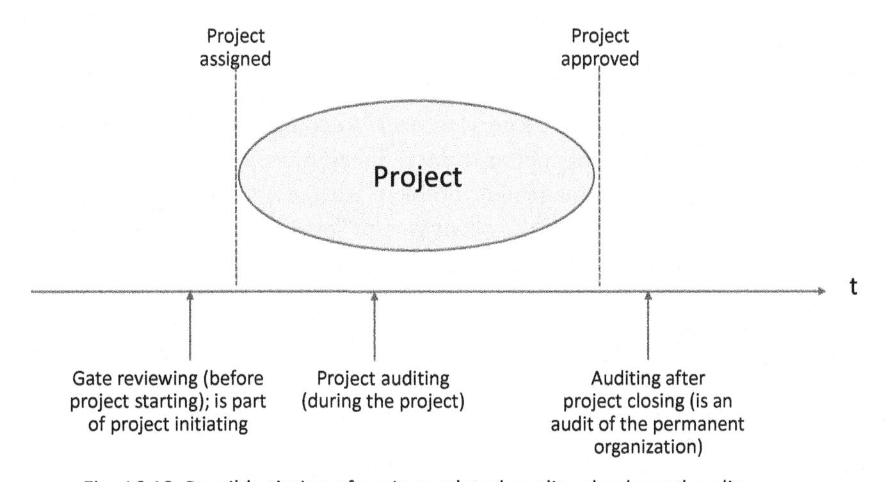

Fig. 16.13: Possible timing of project-related quality checks and audits.

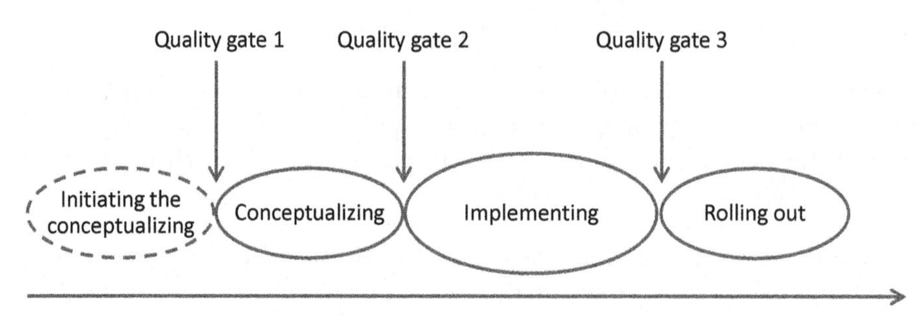

Fig. 16.14: Quality gates in chains of projects.

In chains of projects, it is possible to perform several quality checks before starting a new project in the chain (see Fig. 16.14). The check points are called "quality gates" or "toll gates".[17]

A quality gate can be passed after a successful quality check. For passing a gate, existing results of the preceding phase are analyzed and an assessment is made as to whether an appropriate basis exists for a project to be performed. If the required quality is not guaranteed, further preparatory work is necessary. By detecting shortcomings at an early stage, costs of future errors should be reduced. Quality gates are defined by events and not by fixed dates. This distinguishes them from milestones.

For repetitive project chains, such as the project chains of international oil companies for exploring oil, or of developers for developing buildings, quality gates can be standardized and quality criteria can be defined for each gate. Quality gates synchronize process steps and ensure that all requirements defined for a gate are fulfilled. For quality gate reviews in repetitive project chains, organizations often establish permanent gate review teams.

16.5.5 Short Project (or Program) Intervention

The objective of a short project intervention is to contribute to the further development of its project management competency. Short interventions can quickly contribute to solving an urgent management problem with low use of resources. Rapid and flexible solutions are created for the client system "project" and the resulting solutions are reflected upon.

Members of the project organization are thus granted an opportunity to become familiar with new types of reflection and new forms of work. Possible forms of work for short interventions are expert feedback, reflecting positions, reflecting team, project simulation, and systemic constellation.

Short project interventions may include the following:

[17] See Cooper, R. G., 2010.

- analysis of the situation,
- definition of the management problem or challenge,
- brief analysis of the problem or challenge,
- suggestion of alternative problem solutions, and
- feedback on the analysis and suggestions by the representatives of the project organization.

A short project intervention is performed by a consultant or a team of consultants in cooperation with representatives of the project organization.

16.5.6 Institutionalizing the Management Consulting of Projects or Programs

Management consulting of projects or programs in project-oriented organizations to ensure management quality is gaining in importance. It is developing into a new external but also internal service.

In some organizations, the development of individual and organizational competencies makes it possible to institutionalize management consulting of projects and programs. Organizations develop internal consultants and establish expert pools, which include internal and possibly also external consultants as members. The role "Internal project management consultant" can be integrated into a project management career path.

The development of internal project management consultants may include the following steps:

- attaining project management competencies (training and certification in project management, getting experience as a project manager),
- attaining social competence through group dynamics training, teamwork, etc.,
- attaining understanding of one's role as a consultant through training and networking with project management consultants,
- gathering experience as a project management coach,
- cooperating with senior consultants in project management consulting,
- project management consulting under the supervision of a senior consultant, and
- independent responsibility for project management consulting.

Organizational competencies for management consulting of projects and programs are developed by specifying consultancy services, describing the consulting processes and methods, and describing the social systems and roles involved in consultancy. All this can be summarized as a corporate governance document in guidelines for the consulting of a project or program. The table of contents of the guidelines of an Austrian oil and gas company for consulting a project or program is shown as an example in Table 16.6.

Table 16.6: Table of Contents of "Guidelines for Management Consulting of a Project or Program" (Example)

Table of contents: Guidelines for consulting of a project or a program
1 Introduction
2 Definitions
2.1 Content-related consulting of a project or a program
2.2 Management consulting of a project or a program
2.3 Management auditing of a project or a program
3 Business processes of management consulting
3.1 Consulting of project initiating
3.2 Management consulting of a project or a program
3.3 Management auditing of a project or a program
3.4 Short-term interventions in a project or a program
4 Roles in management consulting
4.1 Owner of the management consulting
4.2 Management consultant
4.3 Representatives of the project or the program in management consulting
5 Methods for management consulting
5.1 Methods used in management consulting
5.2 Description of methods
5.3 Forms and checklists for management consulting
6 Annex

16.6 Managing Project Personnel of the Project-Oriented Organization

16.6.1 Roles and Careers in the Project-Oriented Organization

Roles in the Project-Oriented Organization

In the project-oriented organization, a distinction can be made between permanent and temporary roles. Permanent management roles are, for example, general manager, manager of a profit center or service center, manager of an expert pool, member of the Project Portfolio Group, and manager of the PM Office. Permanent expert roles are, for example, member of an expert pool or PM Office member. Temporary roles are project and program owner, project and program manager, project and program team member, project and program contributor, and project and program consultant (see role descriptions in Chapter 7, page 123).

Not only people who perform roles as project and program managers, but all people who are involved in projects, programs, changes, and project portfolio managing, can be defined as project personnel.

Definition: Project personnel

All people in the project-oriented organization who perform roles in projects, programs, and changes, as well as in project portfolio managing, can be defined as project personnel. These people must have competencies in project, program, change, and project portfolio managing.

Career Paths in the Project-Oriented Organization

Traditionally, "career" is understood as the hierarchical ascent of a person within an organization, but also across several organizations. Schein distinguishes vertical, horizontal, and centripetal career movements.[18] In vertical career movements, promotion involves hierarchical ascent. In horizontal career moves, there is no hierarchical ascent. Centripetal career movements are movement towards the inner core of the organization. An example of a centripetal career movement in a project-oriented organization is becoming a member of the Project Portfolio Group.

Because of the flat organizational structure, career movements in a project-oriented organization do not necessarily involve reaching a higher position in the hierarchy. A career in a project-oriented organization can therefore be understood as the development process of a person.

In project-oriented organizations, a management career, a project management career, and an expert career can be distinguished. A management career is understood as the assuming of roles with increasing responsibility for leading personnel and management. A management career in a project-oriented organization also includes assuming roles as a project or program owner. A project management career progression is not a requirement for this. Selection as an owner primarily requires competencies connected with the management role in the permanent organization.

The stages of the project management career path—namely "junior project manager", "project manager", and "senior project manager"—are shown in Figure 16.15. The possibility of performing different project-related roles is linked to the different career stages.

The basic options for the performing of project roles are dependent on the "maturity" of the project-oriented organization. The implementation of projects and programs, the offering of management consulting for projects and programs, and the existence of a PM Office and a Project Portfolio Group influence these options. This also involves a need for the profession "project manager", which is created when employees are primarily performing project and program managing roles.

As the specialist knowledge of experts is an important competitive factor for project-oriented organizations, there should also be opportunities to develop a career as an

[18] Schein, E., 1978, p. 37 ff.

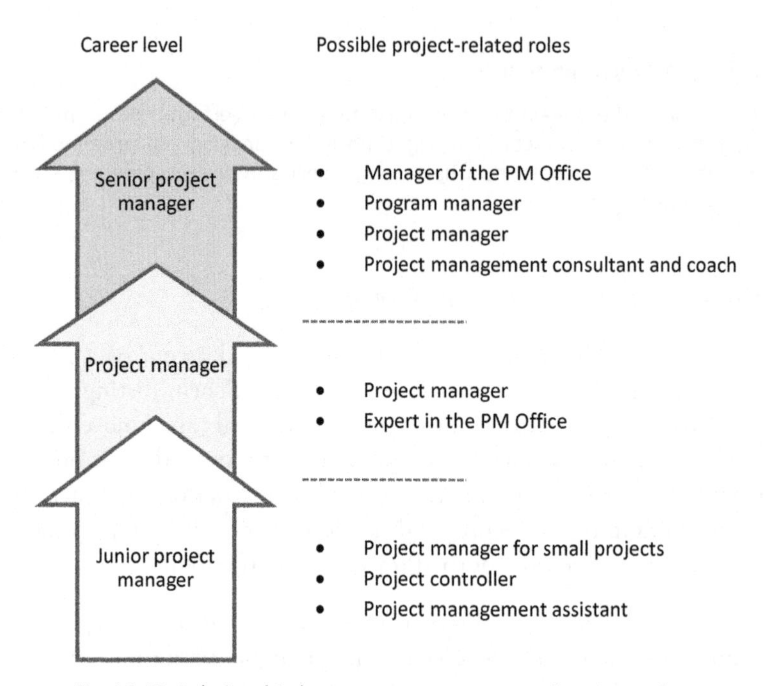

Fig. 16.15: Relationship between career stages and project roles.

expert. The stages of an expert career path can be junior expert, expert, and senior expert. Appropriately positioning the expert career in the organization should avoid devaluation of experts by unilaterally promoting project management careers. The equivalence of expert and project management careers should be ensured.

Management, expert, and project management career paths exist side by side in the project-oriented organization and complement each other. To guarantee flexibility in the personnel management of a project-oriented organization, permeability between the individual career paths must be ensured.

16.6.2 Managing Project Personnel: Processes

Project personnel is to be managed both within projects and in the project-oriented organization. Recruiting, allocating, leading, assessing, developing, and releasing project personnel in a project are management tasks and thus an integral part of project managing. Objectives and methods for the fulfillment of these tasks are described in Chapters 8 and 9.

Recruitment, allocation, leadership, assessment, development, and release of project personnel also needs to be performed on a non-project basis. Future project managers can be recruited without anticipating their assignment to a particular project; the assessment and training of project personnel can serve overall human resource development objectives; and project managers can be released when the organization no longer needs their services. These business processes of a project-oriented organization for managing project personnel (see Fig. 16.16) are described below. Providing corporate governance structures for managing project personnel, such as the definition of a career path, the definition of personnel development structures, and the establishment of general incentive systems, is a general management task.

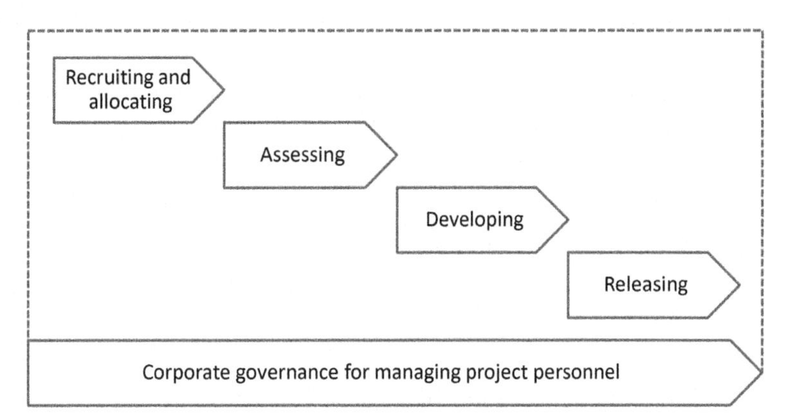

Fig. 16.16: Business processes of the project-oriented organization for managing project personnel.

Business Processes: Recruiting and Allocating a Project Manager

Recruiting is the procurement of personnel.[19] Recruiting a project manager is expanding the expert pool "project managers" of the project-oriented organization. It can be combined with allocating the recruited project manager to a specific project.

The needs of an organization for project managers are defined by demand planning for the expert pool. The qualification profile and general incentive models for project managers must be established as the basis for recruiting, and it must be decided whether personnel should be recruited from the internal or external personnel market. In the case of internal recruitment, experts who have also gained project management competency through their work as a project team member and are interested in a project management career can apply for the position. In external recruitment, the external personnel market is accessed. Possible methods for selecting personnel are application letters, job interviews, tests, and assessment centers for project managers.

Business Process: Assessing the Competencies of a Project Manager

The objectives of assessing the competencies of a project manager are the analysis of his or her performance and the planning of his or her further activity in the organization.

[19] Lueger, G., 1996, p. 338.

In the assessment, feedback can be given on the performance and relations with stakeholders can be reflected on.

Needs for further development can be identified, and further development measures can be planned. Future assignments to projects should also be planned, and salary and incentives should be reviewed.

Project managers are to be evaluated periodically by their superiors in the permanent organization. A superior can be, for example, the manager of a profit center, an expert pool, or a PM Office. When evaluating a project manager, feedback from project owners, project team members, and project stakeholders of projects managed by the project manager can be taken into account.

In a project-oriented organization, not only project managers, but also project owners and project team members should be assessed.

Business Process: Developing a Project Manager

The objective of developing a project manager is primarily the further development of his or her competencies for managing projects and programs. Development measures can include training, coaching, and mentoring. In addition to competencies for project and program management, competencies for change management and social competencies can be developed.

Development measures can be offered internally or externally, "on the project" or independently of projects, individually or in groups. Freelance project managers working for an organization can also be considered in development measures.

In a project-oriented organization, further development should be provided not only to project managers but also to project team members and project owners to ensure a common management understanding of all roles cooperating in projects.

Business Process: Releasing a Project Manager

The releasing of personnel covers measures designed to reduce personnel surpluses. The need to release a project manager is ascertained from the planning of an organization's demand for project managers.

Before a project manager is released, other possible assignments can be considered. For example, there may be the option of assigning him or her as a (temporary) employee to the PM Office or to a non-project management role in the permanent organization.

When freelance project managers are released from the organization, interest in future cooperation shall be discussed.

16.6.3 Methods for Managing Project Personnel

Specific methods for managing project personnel are methods for assessing project management competencies, for (further) developing competencies, and for designing project-related incentive models.

Methods for Assessing Project Management Competencies

Possible methods for assessing the project management competencies include self-assessment, external assessment, and assessment centers.

The objective of assessing individual project management competencies is to evaluate a person's project management knowledge and project management experience. The assessment can be performed by the person as self-assessment and/or by a third party.

Table 16.7 shows an excerpt from a questionnaire for self-assessing project management competencies. An external assessment is based on the results of a self-assessment. The external assessor investigates the self-assessment in a personal interview and analyzes project plans prepared by the assessed project manager. This enables project management knowledge and experience to be evaluated based on concrete documents. Interviews with colleagues of the candidate can also contribute to the assessment.

Project Management Assessment Center

In an assessment center, candidates are placed in simulated working situations in order to assess how competent they are to fulfill a project management role. An assessment center usually takes between two and five days.[20]

In project-oriented organizations, assessment centers can be used for selecting and further developing project and program managers. Methods used in general management assessment centers have to be adapted to the specific work situations in projects. Examples are shown in Table 16.8.

Basic and Advanced Project Management Training

The target groups for training in a project-oriented organization are not only project managers, but also project team members, project contributors, and project owners. The content and duration of basic and advanced trainings have to be designed specifically for these different target groups.

Training measures can be organized "on the job" or "off the job", internally or externally. "On-the-job" training of project managers can be performed by means of internships, job rotation, and individual coaching.

[20] Kompa, A., 1999, p.31ff.

Table 16.7: Questionnaire for Self-Assessment of Project Management Competencies (Excerpt)

Self-Assessment: Project management competence		
Please indicate below how high you would estimate your project management know-how and your project management experiences. Please use the scale of 1 to 5 whereby 1 stands for the lowest and 5 for the highest know-how and experience.		
Methods for project planning		
1=non, 2=little, 3=average, 4=much, 5=very much	Know-how	Experience
Planning project objectives		
Planning project strategies		
Planning objects of consideration		
Developing the project work breakdown structure		
Specifying a work package		
Planning project milestones		
Developing a project bar chart		
Developing a CPM project network		
Planning project resources		
Planning the project budget		
Performing a cost-benefit analysis od business case analysis		
Developing a project assignment		
Developing a project organization chart and listing project roles		
Developing a project responsibility chart		
Planning the project communication formats		
Defining project rules		
Defining a project name		
Developing a project logo		
Methods for project risk managing		
1=non, 2=little, 3=average, 4=much, 5=very much	Know-how	Experience
Analyzing project risks		
Analyzing project scenarios		
Planning project alternatives		
Methods for designing the project context relations		
1=non, 2=little, 3=average, 4=much, 5=very much	Know-how	Experience
Analyzing project stakeholders		
Describing the pre-project phase and the post-project phase		
Analyzing relations of a project to other projects		
Analyzing relations of a project to the strategic objectives of the organization		
Performing project marketing		

Table 16.8: Methods for a Project Management Assessment Center

Assessment center methods	Project-related application
Presentation	Presentation of project objectives in a project starting workshop
Group discussion	Discussion: Relations between the role project manager and the role project owner
Role play	Role play project owner meeting: Defining a project crisis
Analysis	Analysis of a project progress reports or a project score card

The advantage of holding lectures and seminars internally is that the designs can be specifically adapted to the needs and structures of an organization. The advantage of external "open" trainings is familiarization with the practices of other organizations and the possibility of networking with other project managers.

Coaching and Mentoring a Project Manager

A coach contributes to the self-help of a project manager by giving advice on how to deal with concrete work situations. In contrast to consulting, the client of coaching is a person in a specific role and not an organization. It is performed when a role player needs a solution to a problem. Coaching a project manager can, for example, take place after project management training, to support the project manager in implementing learned content in a current project.

Mentoring provides long-term guidance and advice to young project managers by means of regular discussions with a mentor. Experienced managers who are not the direct superiors of the mentored project manager can act as mentors. Support can be provided by sharing basic information for a better understanding of the structures and cultures of the organization. Young project managers can have experienced project managers as mentors.

Designing Incentive Models in a Project-Oriented Organization

The objectives of applying incentive models are to support the recruitment of competent employees, to create a framework for self-motivation, and to secure the loyalty of employees to the organization. From a systemic point of view, social systems can only control themselves. Therefore, in order to be effective, "incentives" must be perceived as such by the person or the team. One cannot motivate anyone—motivation is always self-motivation.

In a project-oriented organization, a distinction can be made between incentive models for the employees and teams of the permanent organization and of projects and programs. It is important to ensure that these different incentive models complement each other.

Possible project-related incentives are the project work itself, project bonuses, and gifts, but also special forms of appreciation as rewards for special achievements. Project work itself is one of the intrinsic incentives that appeal to people's motivation to achieve, to acquire competence, and to establish professional relationships.

It is the responsibility of managers of project-oriented organizations to provide challenging and exciting projects. These are incentives for employees to work in the organization.

Often project-specific incentive models are reduced to project bonuses. Project bonuses can be agreed upon either with individual members of the project organization—for example, the project manager and an expert—or with the project team. The performance to be achieved for the payment of project bonuses, their amount, and the distribution of project bonuses in the project team are to be agreed upon in the project starting process. The payment of project bonuses should occur at the time of project closing, dependent on the project success.

Non-monetary incentives are often of a symbolic nature. Expressing personal appreciation—for example, by expressing praise in a project team meeting—can have a great impact on the motivation of employees.

16.6.4 Organization for Managing Project Personnel

In a project-oriented organization, the tasks of recruiting and assigning, evaluating and developing, as well as releasing project personnel are performed by the managers of profit centers, expert pools, a PM Office, a Project Portfolio Group, and the personnel department. As several organizational units are usually involved in managing project personnel, networking and communication between them is important.

The general tasks for managing project personnel, such as the establishment of the profession "project manager", the development of a project management career path, the organization of training, and possibly supporting project management certifications, are to be performed jointly by the PM Office and the manager of the expert pool "Project managers". If these organizational units do not exist in small organizations, these tasks must be performed by the general management or a profit center manager in collaboration with the personnel department.

Literature

Bounds, G. M., Dobbins, G. H., Fowler, O. S.: *Management—A Total Quality Perspective,* South-Western Publication, Cincinnati, OH, 1995.

Cooper, R. G.: *Top oder Flop in der Produktentwicklung: Erfolgsstrategien: Von der Idee zum Launch,* Wiley-VCH, Weinheim, 2010.

Deming, W. E.: *Out of the Crisis,* Cambridge University Press, Cambridge, 1992.

DIN EN ISO 8402: 1995-08: *Qualitätsmanagement—Begriffe,* Beuth, Berlin, Vienna, Zürich, 1995.

DIN EN ISO 9000: 2015-11: *Qualitätsmanagement—Grundlagen und Begriffe,* Beuth, Berlin, Vienna, Zürich, 2015.

Endruweit, G. (ed.): *Wörterbuch der Soziologie,* Dt. Taschenbuch-Verlag, München, 1989.

Gesellschaft für Projektmanagement (GPM): Projektqualität: Begriffliche und konzeptionelle Grundlagen des Qualitätsmanagements in Projekten, *projektManagement aktuell,* 3, 2005.

Ishikawa, K., Lu, D.: *What is Total Quality Control? The Japanese Way,* Prentice Hall, Englewood Cliffs, NJ, 1985.

Kompa, A.: *Assessment Center—Bestandsaufnahme und Kritik,* Rainer Hampp, Munich, 1999

Krczal, A.: Von der Qualitätskontrolle zur kontinuierlichen Qualitätsverbesserung, in: Eckardstein, D. v., Kasper, H., Mayrhofer, W. (Eds.), *Management. Theorien—Führung—Veränderung,* Schäffer-Poeschel, Stuttgart, 1999.

Lipnack, J., Stamps, J.: *Networking. The First Report and Directory,* Doubleday, New York, 1982.

Lueger, G .: Beschaffung und Auswahl von Mitarbeitern, in: Kasper, H., Mayrhofer, W. (Eds.): *Personalmanagement—Führung—Organisation,* 338–387, Linde, Vienna, 1996.

Schein, E.: *Career Dynamics: Matching Individual and Organizational Needs,* Addison-Wesley, Reading, MA, 1978.

Seaver, M. (ed.): *Gower Handbook of Quality Management,* Gower, Aldershot, 2003.

17 Message and Vision

Applying a traditional, mechanistic approach for project and program managing often leads to the failure of projects and programs. Key messages characterizing systemic project, program, and change managing are presented. By using a systemic management approach "business value" can be created.

A society that frequently performs projects and programs for realizing changes in the profit and non-profit sectors and whose institutions offer services for project, program and change managing can be perceived as a project-oriented society. Maturity as a project-oriented society can be analyzed and further developed in order to create competitive advantages for the society.

17.1 Message from *PROJECT.PROGRAM.CHANGE*

Traditional, mechanistic project and program managing is still frequently represented in the management community. In projects, thinking is in terms of the "magic triangle" of project scope, project costs, and project schedule. However, this leads to too many projects not being successful.

RGC has been advocating a systemic project management approach since 1990. This approach is sometimes assessed in the community as theoretical, as not easily applicable in practice. However, so that failures, flops, difficulties, shipwrecks, crises, discontinuities, etc. in projects, programs, and changes can be reduced, *PROJECT.PROGRAM.CHANGE* presents a new start for the "business value" of holistic, context-oriented, and constructivistic managing.

We are convinced of this and know that applying a systemic management approach will make a significant contribution to ensuring project, program, and change success. We do not, however, only want to advocate a systemic management approach, but we also . . .

> integrate into the management approaches new concepts such as agility, benefits realization management, requirements management, and sustainable development,
> consider the relations between project, program, and change managing,
> define the values underlying these management approaches, and
> describe the epistemological positioning of the management approaches.

The message of *PROJECT.PROGRAM.CHANGE* is summarized by the following statements:

Changes by Projects!

Projects and programs are not performed for their own sake. They realize changes. The objectives, processes, methods, roles, and communication formats of project, program, and change managing must therefore be coordinated. Projects, programs, and changes need to contribute to the business value of the project-oriented organization.

Professional Project Initiating Is Strategically Important!

Professionally initiating projects creates the basis for the successful performance of these projects. Strategic decisions that create framework conditions for projects are made in project initiating. In projects one is limited in decision making by the framework established in the pre-project phase.

Programs Are Already Here!

Many organizations that implement *de facto* programs try to reduce the management complexity by not perceiving them as programs. Therefore, programs are not managed as programs, but as "large projects". However, the project structures used are not suitable for managing the actual complexity. This leads to major problems. Programs need to be perceived as such and managed accordingly.

Values for Business Value!

Value-oriented managing enables sustainable "business values" to be secured for organizations. Values are the basis for managing—that is, for the objectives, processes, methods, and roles for managing. Values as the basis of a systemic management approach are, for example, holistic boundaries, context orientation, an iterative approach, empowerment, and resilience.

The Project Manager as Intrapreneur!

Project managers, program managers, and change managers should no longer be perceived as performers or administrators of projects, programs, or changes. They need to develop a self-understanding as intrapreneurs in the project-oriented organization and communicate that appropriately.

Epistemological Positioning of Project, Program, and Change Managing!

RGC's management approaches are based on social system theory and radical constructivism. Permanent and temporary organizations are perceived as social systems in their contexts. Boundaries, stakeholders, crises, etc. are understood as social constructs created by communicating and consensus finding. Thus the models and methods of social systems theory, constructivism, and organizational theory can contribute to the approaches of project, program, and change managing.

Appropriate Dealing with Complexity and Dynamism with a Systemic Management Approach!

Complexity, dynamism, and self-reference are characteristics of social systems. Projects, programs, and changes are by definition complex, dynamic, and self-referential and must be managed accordingly. Social systems theory provides models and methods for dealing appropriately with complexity and dynamism.

17.2 Vision of the Project-Oriented Society

Construct: Project-Oriented Society

In many national societies, projects and programs are performed not only in companies, but also in other organizations such as municipal councils, associations, schools, and even families, to realize changes. "Management by projects" becomes an organizational strategy for societies to deal with increasing complexity and dynamism. The globalization of the economy, new technologies with ever-shorter product life cycles, and the application of a new management paradigm encourage the use of projects, programs, and changes. Projects and programs are seen as suitable organizational forms for realizing changes—not only in industry, but also in nonprofit organizations.

The structure of a society, its history, and its expectations of the future influence the development of the competencies of a society. The perception of a society as a project-oriented society is a construct. It requires seeing society through a "specific lens"—namely, the lens of project orientation. The focus is on those structures of a society which are related to projects, programs, and changes.

A society which frequently uses projects and programs to implement changes and whose institutions offer services for project, program, and change managing can be perceived as a project-oriented society.

Maturity as a Project-Oriented Society

For the successful implementation of projects, programs, and changes, a society requires competencies and sufficient "maturity".

In the RGC model "maturity of a project-oriented society", the practices of project-oriented organizations in project managing, program managing, project portfolio managing, project personnel managing, managing organization structures, and change managing are considered. Also considered are institutional services which support the application of project, program, and change managing in industry. Related services are provided by training, consulting, research, and marketing institutions.

The maturity of a project-oriented society can be visualized using a spider network (see Fig. 17.1). The axes of the spider network represent the dimensions of the practices of project-oriented organizations and the services of institutions for project, program, and change managing. The "maturity" can be assessed on the basis of the scores of the individual dimensions.[1]

The dimensions of practice in the maturity model correspond to the dimensions of the maturity model of a project-oriented organization presented in Chapter 15, some of the processes being aggregated. Further, service-related dimensions are considered in the maturity model.

Formal training in project, program, and change managing can be offered by private and public training institutions. They can lead to an academic degree. At the Romanian University SNSPA (*Şcoala Naţională de Studii Politice şi Administrative*),

[1] The area in Figure 17.1 is an example of the results of an analysis of the maturity of a project-oriented society.

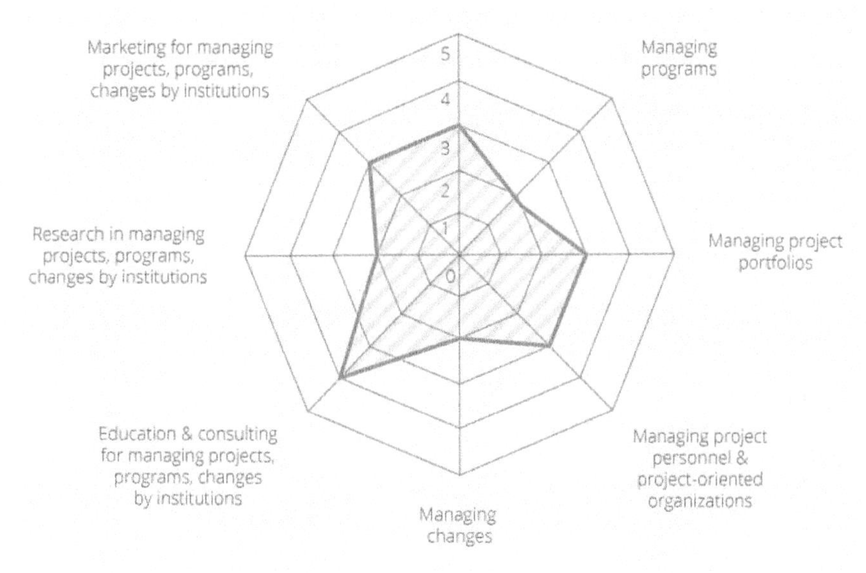

Fig. 17.1: Maturity model of a project-oriented society.

for example, in addition to a "Master's Program in Project Management", a "Master's Program in Program Management" was established in 2016. This shows growing awareness about the relevance of program and change management.

Consulting in project, program, and change managing is offered by national and international consulting companies. Projects, programs, and changes are supported, and structures for successfully managing projects, programs, and changes are created.

Relevant services offered by research institutes are research projects and research programs, publications, and research events for project, program, and change managing. Also, institutions may provide specific financing for research in project, program, and change managing.

Marketing tasks in a project-oriented society are performed by universities, colleges, training and consulting companies, and professional project management and change management associations. Services provided by associations are, for example, membership services, certifications of individuals and of organizations, hosting of events, etc.

A maturity analysis for the individual dimensions of the model can be performed with the help of a questionnaire. The maturity of six project-oriented societies was analyzed and compared in a research program at the WU—the Vienna University of Business Administration.[2] In Table 17.1, two questions from the questionnaire for analyzing a project-oriented society are presented as an example.

[2] See Gareis, R., Huemann, M., 1999.

Table 17.1: Questions for Analyzing the Maturity of a Project-Oriented Society (Examples)

How many of the following institutions are offering formal project management education programmes?	
Secondary schools (such as high schools, trade schools, ...)	
Colleges	
Universities	
Continuing education institutions	
Consulting companies	
Other educational institutions (please state)	

How many of the following institutions perform project management-related research?	
Colleges	
Universities	
Continuing education institutions	
Consulting companies	
Other educational institutions (please state)	

1 ... none of them, 2 ... few of them, 3 ... some of them, 4 ... many of them, 5 ... all of them

An analysis of the maturity of a project-oriented society provides a basis for the further development of its competencies. A high maturity in project, program, and change managing creates competitive advantages for a society. The use of project, program, and change managing in the nonprofit sector, in urban development, and in regional development offers great opportunities for efficiency and effectiveness. The perception of social developments as changes has to be encouraged, and change management also has to be practiced at the societal level.

Literature

Gareis, R., Huemann, M., PM—Competence of the Project-Oriented Society, *Project Management,* 5(1), 28–29, 1999.

References

Adams, W. M.: The Future of Sustainability: Re-Thinking Environment and Development in the Twenty-first Century, *Report of the IUCN Renowned Thinkers Meeting,* Volume 29, 2006.

Albach, H.: *Investition und Liquidität: die Planung des optimalen Investitionsbudgets,* Betriebswirtschaftlicher Verlag Dr. Th. Gabler, Wiesbaden, 1962.

Asendorpf, J. B.: *Persönlichkeitspsychologie,* 3rd edition, Springer, Heidelberg, 2009.

Ashby, W. R.: *An Introduction to Cybernetics,* 5. Auflage, University Paperbacks, London, 1970.

Association for Project Management (APM): *APM Body of Knowledge,* 6th edition, APM, Buckinghamshire, 2012.

Barnat, R.: *The Nature of Strategy Implementation,* abgerufen von http://www.introduction-to-management.24xls.com/en201 (30.9.2016), 2005.

Bateson, G.: *Geist und Natur—Eine notwendige Einheit,* Suhrkamp, Frankfurt am Main, 1990.

Beck, K., Beedle, M., van Bennekum, A. et al.: *The Agile Manifesto,* 2001.

Botta, C.: Das Role Model Canvas—Rollen schnell und gemeinsam definieren, *Projekt Magazin,* 07, 2016.

Boulding, K.E.: Time and Investment, *Economica,* Volume 3, London, 1936.

Bounds, G. M., Dobbins, G. H., Fowler, O. S.: *Management—A Total Quality Perspective,* South-Western Publication, Cincinnati, OH, 1995.

Bradley, G.: *Benefit Realisation Management: A Practical Guide to Achieving Benefits Through Change,* 2nd edition, Gower, Surrey, Burlington, VT, 2010.

Cadbury Committee: *Report of the Committee on the Financial Aspects of Corporate Governance,* Gee and Co., London, 1992

Čamra, J. J. (Ed.): *REFA-Lexikon: Betriebsorganisation. Arbeitsstudium, Planung und Steuerung,* 2nd edition, Beuth, Berlin, 1976.

Cleland, D. I., King, W. R.: *Systems Analysis and Project Management,* McGraw Hill, New York, NY, 1968.

Cooper, R. G.: *Top oder Flop in der Produktentwicklung: Erfolgsstrategien: Von der Idee zum Launch,* Wiley-VCH, Weinheim, 2010.

Dandridge, T. C.: Symbols' Function and Use, in: Pondy, L. R., Frost, P., Morgan, G. (Eds.), *Organizational Symbolism,* pp. 69–79, Jai Press, Greenwich, CT, 1983.

Davidson, J.: Sustainable Development: Business as Usual or a New Way of Living? *Environmental Ethics,* 22(1), 45–71, 2000.

Daxner, F., Gruber, T., Riesinger, D.: Werteorientierte Unternehmensführung: Das Konzept, in: Auinger, F., Böhnisch, W. R., Stummer, H. (Eds.), *Unternehmensführung durch Werte. Konzepte—Methoden—Anwendungen,* p. 3–34, Deutscher Universitäts-Verlag, Wiesbaden, 2005.

De Janasz, S., Dowd, K. O., Schneider, B. Z.: *Interpersonal Skills in Organizations,* 5th edition, McGraw-Hill Education, New York, NY, 2015.

Deming, W. E.: *Out of the Crisis,* Cambridge University Press, Cambridge, 1992.

DIN EN ISO 8402: 1995-08: *Qualitätsmanagement—Begriffe,* Beuth, Berlin, Vienna, Zürich, 1995.

DIN EN ISO 9000: 2015-11: *Qualitätsmanagement—Grundlagen und Begriffe,* Beuth, Berlin, Vienna, Zürich, 2015.

DIN 69901-5:2009-01: *Projektmanagement—Projektmanagementsysteme*—Teil 5: Begriffe, 9. Auflage, Beuth, Berlin, Wien, Zürich, 2009.

Duncan, W.R.: *A Guide to the Project Management Body of Knowledge (PMBOK® Guide),* Project Management Institute (PMI), Newton Square, PA, 2000.

Endruweit, G. (ed.): *Wörterbuch der Soziologie,* Dt. Taschenbuch-Verlag, München, 1989.

Erpenbeck, J., Von Rosenstiel, L. (Eds.): *Handbuch Kompetenzmessung: Erkennen, verstehen und bewerten von Kompetenzen in der betrieblichen, pädagogischen und psychologischen Praxis,* 2. Auflage, Schäffer-Poeschel, Stuttgart, 2007.

Freeman, R. E.: *Strategic Management: A Stakeholder Approach,* Pitman, Boston, MA, 1984.

Freeman, R. E., Harrison, J. S., Wicks, A. C.: *Managing for Stakeholders: Survival, Reputation, and Success,* Yale University Press, London, New Haven, CT, 2007.

Gaitanides, M.: *Prozessmanagement—Konzepte, Umsetzungen und Erfahrungen des Reengineering,* Hanser, München, 1994.

Gareis, R., Huemann, M., Martinuzzi, A.: *Project Management & Sustainable Development Principles,* Project Management Institute (PMI), Newton Square, PA, 2013.

Gareis, R., Huemann, M., Martinuzzi, A., Weninger, C., Sedlacko, M.: *Project Management & Sustainable Development Principles,* Project Management Institute (PMI), Newtown Square, PA, 2013.

Gareis, R., Huemann, M., PM—Competence of the Project-Oriented Society, *Project Management,* 5(1), 28–29, 1999.

Gareis, R.: Changes of Organizations by Projects, *International Journal of Project Management,* 28(4), 314–327, 2010.

Gareis, R., Stummer, M.: *Prozesse und Projekte,* Manz, Wien, 2007.

Gesellschaft für Projektmanagement (GPM): Projektqualität: Begriffliche und konzeptionelle Grundlagen des Qualitätsmanagements in Projekten, *projektManagement aktuell,* 3, 2005.

Gesellschaft für Projektmanagement (GPM): Assessments für Organisation (IPMA Delta), downloaded from https://www.gpm-ipma.de/lightbox_seiten/ipma_delta.html (18.01.2017).

Gester, P.: Warum der Rattenfänger von Hameln kein Systemiker war? Systemische Gesprächs- und Interviewgestaltung, in: Schmitz C., Gester P., Heitger B. (Eds.), *Managerie—Systemisches Denken und Handeln im Management,* 136–164, 1. Jahrbuch, Carl Auer, Heidelberg, 1992.

Gilson, L. L., Maynard, M. T., Young, N. C. J., Vartiainen, M., Hakonen, M.: Virtual Teams Research: 10 Years, 10 Themes, and 10 Opportunities, *Journal of Management,* 41(5), 1313–1337, 2015.

Global Reporting Initiative: Sustainability Reporting Guidelines, Version 3.1, Amsterdam, 2011.

Gloger, B.: *Scrum Produkte zuverlässig und schnell entwickeln,* 5th edition, Carl Hanser, München, 2016.

Greenwood, M.: Stakeholder Engagement: Beyond the Myth of Corporate Responsibility, *Journal of Business Ethics,* 74(4), 315–327, 2007.

Habermann, F.: Der Project Canvas—rojekte interdisziplinär definieren, *Projekt Management aktuell,* 1, 36–42, 2016.

Habermann, F., Schmidt, K.: *The Project Canvas. A Visual Tool to Jointly Understand, Design, and Initiate Projects, and Have More Fun at Work,* Gumroad E-Book, Berlin, 2014.

Hanisch, R.: *Das Ende des Projektmanagements: Wie die Digital Natives die Führung übernehmen und Unternehmen verändern,* Linde, Vienna, 2013.

Hasso Plattner Institute (HPI): *Was ist Design Thinking?,* accessed at https://hpi-academy.de/design-thinking/was-ist-design-thinking.html (16.01.2017).

Henderson, B.: *The Experience Curve—Reviewed IV. The Growth Share Matrix or the Product Portfolio,* The Boston Consulting Group, Boston, MA, 1973.

Henderson, L. S.: The Impact of Project Managers' Communication Competencies: Validation and Extension of a Research Model for Virtuality, Satisfaction, and Productivity on Project Teams, *Project Management Journal,* 39(2), 48–59, 2008.

Highsmith, J.: *Agile Project Management,* 2nd edition, Addison-Wesley, Boston, MA, 2010.

Highsmith, J.: *Agile Software Development Ecosystems,* Addison-Wesley, Boston, MA, 2002.

Hilb, M.: *New Corporate Governance: Successful Board Management Tools,* Springer, Berlin Heidelberg, 2012.

Hill, W., Fehlbaum, R., Ulrich, P.: *Organisationslehre 1: Ziele, Instrumente und Bedingungen der Organisation sozialer Systeme,* 5. Auflage, UTB, Stuttgart, 1994.

Hillier, F. S., Lieberman, G. J.: *Introduction to Operations Research,* McGraw-Hill, Boston, MA, 2001.

Hommel, U., Scholich, M., Vollrath, R. (Hrsg.): *Realoptionen in der Unternehmenspraxis: Wert schaffen durch Flexibilität,* Berlin, Heidelberg, Springer, 2013.

Hüsselmann, C.: Agilität im Auftraggeber-Auftragnehmer-Spannungsfeld: Mit hybridem Projektansatz zur Win-Win-Situation, *Projekt Management aktuell,* 25(1), 38–42, 2014.

International Institute of Business Analysis (IIBA): *A Guide to the Business Analysis Body of Knowledge® (BABOK® Guide)* 3.0, IIBA, Toronto, 2015.

International Project Management Association (IPMA): ICB. *IPMA-Kompetenzrichtlinie,* Version 3.0, Nijerk, 2006.

Ishikawa, K., Lu, D.: *What is Total Quality Control? The Japanese Way,* Prentice Hall, Englewood Cliffs, NJ, 1985.

Jann, B., *Einführung in die Statistik,* Oldenbourg, München, Wien, 2002.

Jenner, S., Kilford, C.: *Management of Portfolios,* TSO (The Stationary Office), Norwich, 2011.

Julian, S. D., Ofori-Dankwa, J. C., Justis, R. T.: Understanding Strategic Responses to Interest Group Pressures, *Strategic Management Journal,* 29(9), 963–984, 2008.

Juran, J. M.: *Handbuch der Qualitätsplanung,* Moderne Industrie, Landsberg/Lech, 1991.

Kaplan, R., Norton, P.: *Balanced Scorecard, Strategien erfolgreich umsetzen,* Schäffer-Poeschel, Stuttgart, 1997.

Kaplan R., Norton D.: The Balanced Scorecard—Measures that Drive Performance, *Harvard Business Review* (1–2), 1992.

Kasper, H.: Vom Management der Organisationskulturen zur Handhabung lebender sozialer Systeme, in: Helmut Kasper (Ed.), Post Graduate Management Wissen: *Schwerpunkte des Führungskräfteseminars der Wirtschaftsuniversität Wien,* 189–224, Wirtschaftsverlag Carl Ueberreuter, Wien, 1995.

Kasper, H.: *Die Handhabung des Neuen in organisierten Sozialsystemen,* Springer, Wien 1990.

Kendall, N.: *What is Strategic Management?* abgerufen von http://www.applied-corporate-governance.com/what-is-strategic-management.html (25.10.2016).

Kompa, A.: *Assessment Center—Bestandsaufnahme und Kritik,* Rainer Hampp, Munich, 1999

Komus, A.: *Studie: Status Quo Agile. Verbreitung und Nutzen agiler Methoden—Ergebnisbericht (Langfassung),* Hochschule Koblenz, Koblenz, 2012.

Kotter, J. P.: Leading Change, *Harvard Business Review Press,* Boston, MA, 2012.

Kotter, J. P.: Leading Change, *Harvard Business Review Press,* Boston, MA, 1996.

Krczal, A.: Von der Qualitätskontrolle zur kontinuierlichen Qualitätsverbesserung, in: Eckardstein, D. v., Kasper, H., Mayrhofer, W. (Eds.), *Management. Theorien—Führung—Veränderung,* Schäffer-Poeschel, Stuttgart, 1999.

Levin, G., Green, A. R.: *Implementing Program Management: Templates and Forms Aligned with the Standard for Program Management,* 3rd edition, CRC Press, 2013.

Lewin, K.: Frontiers in Group Dynamics. Concept, Method and Reality in Social Science: Social Equilibria and Social Change, *Human Relations,* 1(1), 5–40, 1947.

Lipnack, J., Stamps, J.: *Networking. The First Report and Directory,* Doubleday, New York, 1982.

Lockwood, T.: Forward, in: Lockwood, T. (ed.), *Design Thinking: Integrating Innovation, Customer Experience and Brand Value,* pp. vii–xvii, Allworth Press, New York, 2010.

Lueger, G .: Beschaffung und Auswahl von Mitarbeitern, in: Kasper, H., Mayrhofer, W. (Eds.): *Personalmanagement—Führung—Organisation,* 338–387, Linde, Vienna, 1996.

Luhmann, N.: *Soziale Systeme: Grundriss einer allgemeinen Theorie,* Suhrkamp, Frankfurt am Main, 1984.

Luhmann, N.: Komplexität, in: Grochla, E. (Hrsg.), *Handwörterbuch der Organisation,* 2. Auflage, Poeschel Verlag, Stuttgart, 1980.

Luhmann, N.: *Funktionen und Folgen formaler Organisation,* Duncker und Humblot, Berlin, 1964.

Malik, F.: *Systemisches Management, Evolution, Selbstorganisation: Grundprobleme, Funktionsmechanismen und Lösungsansätze für komplexe Systeme,* 4th edition, Paul Haupt, Bern, 2004.

Martens, P.: Sustainability: Science or fiction? *Sustainability: Science Practice and Policy,* 2(1), 36–41, 2006.

Martinuzzi, A., Krumay, B.: The Good, the Bad and the Successful—How Corporate Social Responsibility Leads to Competitive Advantage and Organizational Transformation, *Journal of Change Management,* 13(4), 424–443, 2012.

Meadowcroft, J.: Who Is in Charge Here? Governance for Sustainable Development in a Complex World, *Journal of Environmental Policy and Planning,* 9(3), 299–314, 2007.

Mitchell, R. K., Agle, B. R., Wood, D. J.: Toward a Theory of Stakeholder Identification and Salience: Defining the Principle of Who and What Really Counts, *Academy of Management Review,* 22(4), 853–886, 1997.

Němeček, P., Kocmanová, A.: *Management Paradigm, 5th International Scientific Conference "Business and Management",* 559–564, Vilnius, 2008.

Polzin, B., Weigl, H.: *Führung, Kommunikation und Teamentwicklung im Bauwesen,* 2nd edition, Springer, Wiesbaden, 2014.

Ortner, G., Stur, B.: *Das Projektmanagement-Office,* 2nd edition, Springer, Berlin, Heidelberg, 2015.

Pondy, L. R., Frost, P., Morgan, G. (Eds.): *Organizational Symbolism,* JAI Press, Greenwich, CT, 1983.

Porter, M. E.: *Competitive Advantage: Creating and Sustaining Superior Performance,* Free Press, New York, NY, 1985.

Porter, M. E., Kramer, M. R.: The Big Idea: Creating Shared Value. *Harvard Business Review,* 89, 1–2, 2011.

Prahalad, C. K., Hamel, G.: The Core Competencies of the Corporation, *Harvard Business Review,* 68(3), 79–91, 1990.

Project Management Austria (PMA): *pm baseline,* Version 3.0, Wien, 2008.

Project Management Institute (PMI): *Code of Ethics and Professional Conduct,* downloaded at http://www.pmi.org/-/media/pmi/documents/public/pdf/ethics/pmi-code-of-ethics.pdf?sc_lang_temp=en (21 Feb. 2017).

Project Management Institute (PMI): *PMI Professional in Business Analysis (PMI-PBA)® Handbook,* PMI, Newton Square, PA, 2016.

Project Management Institute (PMI): *A Guide to the Project Management Body of Knowledge® (PMBOK® Guide),* 5th edition, Newton Square, PA, 2013.

Project Management Institute (PMI): *Organizational Project Management Maturity Model (OPM3®),* 3rd edition, PMI, Newton Square, PA, 2013.

Project Management Institute (PMI): *The Standard for Program Management,* 3rd edition, PMI, Newton Square, PA, 2013.

Radatz, S.: *Das Ende allen Projektmanagements. Erfolg in hybriden Zeiten—mit der projektfreien Relationalen Organisation, Relationales Management,* Vienna, 2013.

Ravasi, D., Schultz, M.: Responding to Organizational Identity Threats: Exploring the Role of Organizational Culture, *Academy of Management Journal,* 49(3), 433–458, 2006.

Reibnitz, U. v.: *Szenario-Technik: Instrumente für die unternehmerische und persönliche Erfolgsplanung,* Gabler, Wiesbaden, 1992.

Reschke, H.: Formen der Aufbauorganisation in Projekten, in: Reschke, H., Schelle, H., Schnopp, R. (Eds.), *Handbuch Projektmanagement,* Volume 2, TÜV Rheinland, Cologne, 1989.

Robertson, B. J.: *Holacracy: Ein revolutionäres Management-System für eine volatile Welt,* Franz Vahlen, Munich, 2016.

Robertson, B. J.: Holocracy: A Complete System for Agile Organizational Governance and Steering, *Agile Project Management Executive Report,* 7(7), Cutter Consortium, 2006.

Robinson, J.: Squaring the Circle? Some Thoughts on the Idea of Sustainable Development, *Ecological Economics,* 48(4), 369–384, 2004.

Rowley, T. J.: Moving Beyond Dyadic Ties: A Network Theory of Stakeholder Influences, *Academy of Management Review,* 22(4), 887–910, 1997.

Sachs-Hombach, K.: Selbstbild und Selbstverständnis, in: Newen, A., Vogeley, K. (Eds.), *Selbst und Gehirn. Human Self-Awareness and Its Neurobiological Foundations,* p. 189–200, Mentis, Paderborn, 2000.

Schein, E.: *Career Dynamics: Matching Individual and Organizational Needs,* Addison-Wesley, Reading, MA, 1978.

Schulte-Zurhhausen, M.: *Organisation,* 6th edition, Vahlen, Munich, 2013.

Seaver, M. (ed.): *Gower Handbook of Quality Management,* Gower, Aldershot, 2003.

Senge, P.: *The Fifth Discipline: Art & Practice of the Learning Organization,* Doubleday, New York, NY, 2006.

Senge, P.: *The Fifth Discipline Fieldbook: Strategies and Tools for Building a Learning Organization,* Doubleday, New York, NY, 1994.

Serrador, P., Pinto, J.K.: Does Agile Work? A Quantitative Analysis of Agile Project Success, *International Journal of Project Management,* 33(5), 1040–1051, 2015.

Silvius, G., Schipper, R., Planko, J., van den Brink, J., Köhler, A.: *Sustainability in Project Management,* Gower, Surrey, Burlington, VT, 2012.

Society for Project Management: Projektqualität: Begriffliche und konzeptionelle Grundlagen des Qualitätsmanagements in Projekten, *projektManagement aktuell* (3), 2005.

Sowden, R., Wolf, M., Ingram, G.: *Managing Successful Programmes (MSP),* 4th edition, The Stationary Office (TSO), Norwich, 2011.Steinle, H., Bruch, H., Lawa, D. (Eds.): *Projektmanagement: Instrument moderner Dienstleistung,* edition Blickbuch Wirtschaft, Frankfurt am Main, 1995.

Steyrer, J.: Theorie der Führung, in: Kasper, H., Mayrhofer, W. (Eds.), *Personalmanagement, Führung, Organisation,* p. 25–94, Linde, Wien, 2002.

Takeuchi, H., Nonaka, I.: The New New Product Development Game, *Harvard Business Review,* 64 (1-2), 1986.

Tuckman, B. W., Jensen, M. A. C.: Stages of Small-Group Development Revisited, *Group & Organization Studies,* 2(4), 417–427, 1977.

Verburg, R. M., Bosch-Sijtsema, P., Vartiainen, M.: Getting It Done: Critical Success Factors for Project Managers in Virtual Work Settings, *International Journal of Project Management,* 31(1), 68–79, 2013.

Watzlawick, P.: *Wie wirklich ist die Wirklichkeit—Wahn, Täuschung, Verstehen,* Piper, München 1976.

Weibler J.: *Personalführung,* Franz Vahlen, München, 2001.

Weinert, A. B.: Führung und soziale Steuerung, in: Roth, E. (ed.), *Organisationspsychologie (Enzyklopädie der Psychologie,* Vol. 3), 552–577, Hogrefe, Göttingen, 1989.

Womack, J. P., Jones D. T., Roos, D.: *The Machine That Changed the World,* Simon and Schuster, New York, NY, 1990.

World Commission on Environment and Development (WCED): *Our Common Future,* Oxford, 19.

Index

(continues on next page)

D

P

Q

R

S